Theoretical and Mathematical Physics

The series founded in 1975 and formerly (until 2005) entitled *Texts and Monographs in Physics* (TMP) publishes high-level monographs in theoretical and mathematical physics. The change of title to *Theoretical and Mathematical Physics* (TMP) signals that the series is a suitable publication platform for both the mathematical and the theoretical physicist. The wider scope of the series is reflected by the composition of the editorial board, comprising both physicists and mathematicians.

The books, written in a didactic style and containing a certain amount of elementary background material, bridge the gap between advanced textbooks and research monographs. They can thus serve as basis for advanced studies, not only for lectures and seminars at graduate level, but also for scientists entering a field of research.

Editorial Board

W. Beiglböck, Institute of Applied Mathematics, University of Heidelberg, Germany
J.-P. Eckmann, Department of Theoretical Physics, University of Geneva, Switzerland
H. Grosse, Institute of Theoretical Physics, University of Vienna, Austria
M. Loss, School of Mathematics, Georgia Institute of Technology, Atlanta, GA, USA
S. Smirnov, Mathematics Section, University of Geneva, Switzerland
L. Takhtajan, Department of Mathematics, Stony Brook University, NY, USA
J. Yngvason, Institute of Theoretical Physics, University of Vienna, Austria

Pierre Collet   Jean-Pierre Eckmann

Concepts and Results in Chaotic Dynamics: A Short Course

With 67 Figures

Pierre Collet
Center of Theoretical Physics
École Polytechnique
91128 Palaiseau, France
E-mail: *Collet@cpht.polytechnique.fr*

Jean-Pierre Eckmann
Department of Theoretical Physics
and Mathematics Section
University of Geneva
1211 Geneva 4, Switzerland
E-mail: *Jean-Pierre.Eckmann@physics.unige.ch*

Math
QA
614.8
.C635
2006

Mathematics Subject Classification (2000): 37-01, 37A50

Library of Congress Control Number: 2006931058

ISBN-10 3-540-34705-4 Springer Berlin Heidelberg New York
ISBN-13 978-3-540-34705-7 Springer Berlin Heidelberg New York

This work is subject to copyright. All rights are reserved, whether the whole or part of the material is concerned, specifically the rights of translation, reprinting, reuse of illustrations, recitation, broadcasting, reproduction on microfilm or in any other way, and storage in data banks. Duplication of this publication or parts thereof is permitted only under the provisions of the German Copyright Law of September 9, 1965, in its current version, and permission for use must always be obtained from Springer. Violations are liable for prosecution under the German Copyright Law.

Springer is a part of Springer Science+Business Media
springer.com
© Springer-Verlag Berlin Heidelberg 2006

The use of general descriptive names, registered names, trademarks, etc. in this publication does not imply, even in the absence of a specific statement, that such names are exempt from the relevant protective laws and regulations and therefore free for general use.

Typesetting: by the authors and techbooks using a Springer LaTeX macro package
Cover design: eStudio Calamar, Girona, Spain

Printed on acid-free paper SPIN: 11767503 41/techbooks 5 4 3 2 1 0

To Sonia and Doris

Preface

This book is devoted to the subject commonly called Chaotic Dynamics, namely the study of complicated behavior in time of maps and flows, called dynamical systems.

The theory of chaotic dynamics has a deep impact on our understanding of Nature, and we sketch here our view on this question. The strength of this theory comes from its generality, in that it is not limited to a particular equation or scientific domain. It should be viewed as a conceptual framework with which one can capture properties of systems with complicated behavior. Obviously, such a general framework cannot describe a system down to its most intricate details, but it is a useful and important guideline on how a certain kind of complex systems may be understood and analyzed.

The theory is based on a description of idealized systems, such as "hyperbolic" systems. The systems to which the theory applies should be similar to these idealized systems. They should correspond to a *fixed* evolution equation, which, however, need to be neither modeled nor explicitly known in detail. Experimentally, this means that the conditions under which the experiment is performed should be as constant as possible. The same condition applies to analysis of data, which, say, come from the evolution of glaciations: One cannot apply "chaos theory" to systems under varying external conditions, but only to systems which have some self-generated chaos under fixed external conditions.

So, what *does* the theory allow us to do? We can measure indicators of chaos, and study their dependence on those fixed external conditions. Is the system's behavior regular or chaotic? This can be, for example, inferred by measuring Lyapunov exponents. In general, the theory tells us that complex systems should be analyzed statistically, and not, as was mostly done before the 1960s, by all sorts of Fourier-mode- and linearized, analysis. We hope that the present book and in particular Sect. 9 shows what the useful and robust indicators are.

The material of this book is based on courses we have given. Our aim is to give the reader an overview of results which seem important to us, and which are here to stay. This book is not a mathematical treatise, but a course, which tries to combine two slightly contradicting aims: On one hand to present the main ideas in a simple way and to support them with many examples; on the other to be mathematically sufficiently precise, without undue detail. Thus, we do not aim to present the most general results on a given subject, but rather explain its ideas with a simple statement and many examples. A typical instance of this restriction is that we tacitly assume enough regularity to allow for a simpler exposition.

The proofs of the main results are often only sketched, because we believe that it is more important to understand how the concepts fit together in leading to the results than to present the full details. Thus, we usually spend more space on explaining the ideas than for the proofs themselves. This point of view should enable the reader to grasp the essence of a large body of ideas, without getting lost in technicalities. For the same reason, the examples are carefully chosen so that the general ideas can be understood in a nutshell.

The level of the book is aimed at graduate students in theoretical physics and in mathematics. Our presentation requires a certain familiarity with the language of mathematics but should be otherwise mostly self-contained.

The reader who looks for a mathematical treatise which is both detailed and quite complete, may look at (de Melo and van Strien 1993; Katok and Hasselblatt 1995). For the reader who looks for more details on the physics aspects of the subject a large body of literature is available, with different degrees of mathematical rigor: (Eckmann 1981) and (Eckmann and Ruelle 1985a) deal with experiments of the early 1980s; (Manneville 1990; 2004) deals with many experimental setups; (Abarbanel 1996) is a short course for physicists; (Peinke, Parisi, Rössler, and Stoop 1992) concentrates on semiconductor experiments; (Golubitsky and Stewart 2002) has a good mix of mathematical and experimental examples. Finally, (Kantz and Schreiber 2004) deal with nonlinear time series analysis.

The references in the text cover many (but obviously not all) original papers, as well as work which goes much beyond what we explain. In this way, the reader may use the references as a guide for further study. The reader interested in more of an overview will also find references to textbooks and monographs which shed light on our subject either from different angle, or in the way of more complete treatises.

Like any such project, to remain of reasonable size, we have omitted several subjects which might have been of interest; in particular, bifurcation theory (Arnold 1978; Ruelle 1989b; Guckenheimer and Holmes 1990), topological dynamics, complex dynamics, the Kolmogorov–Arnold–Moser (KAM) theorem and many others. In particular, we mostly avoid repeating material from our earlier book (Collet and Eckmann 1980).

After a few introductory chapters on dynamics, we concentrate on two main subjects, namely hyperbolicity and its consequences, and statistical properties of interest to measurements in real systems.

To make the book easier to read, several definitions and some examples are repeated, so that the reader is not obliged to go back and forth too often.

We hope that the many illustrations, simple examples, and exercises help the reader to penetrate to the core of a beautiful and varied subject, which brings together ideas and results developed by mathematicians and physicists.

This book has profited from the questions, suggestions and reactions of the numerous students in our courses: We warmly thank them all. Furthermore, our work was financially supported by the Fonds National Suisse, the European Science Foundation, and of course our host institutions. We are grateful for this support.

Paris and Geneva,
May 2006

Pierre Collet
Jean-Pierre Eckmann

Contents

1 A Basic Problem 1

2 Dynamical Systems 5
- 2.1 Basics of Mechanical Systems 5
- 2.2 Formal Definitions 10
- 2.3 Maps 11
- 2.4 Basic Examples of Maps 12
- 2.5 More Advanced Examples 17
- 2.6 Examples of Flows 23

3 Topological Properties 27
- 3.1 Coding, Kneading 27
- 3.2 Topological Entropy 30
  - 3.2.1 Topological, Measure, and Metric Spaces 30
  - 3.2.2 Some Examples 30
  - 3.2.3 General Theory of Topological Entropy 32
  - 3.2.4 A Metric Version of Topological Entropy 34
- 3.3 Attractors 38

4 Hyperbolicity 45
- 4.1 Hyperbolic Fixed Points 46
  - 4.1.1 Stable and Unstable Manifolds 49
  - 4.1.2 Conjugation 53
  - 4.1.3 Resonances 58
- 4.2 Invariant Manifolds 60
- 4.3 Nonwandering Points and Axiom A Systems 63
- 4.4 Shadowing and Its Consequences 65
  - 4.4.1 Sensitive Dependence on Initial Conditions 69
  - 4.4.2 Complicated Orbits Occur in Hyperbolic Systems 69
  - 4.4.3 Change of Map 70
- 4.5 Construction of Markov Partitions 72

5 Invariant Measures . . . 79
5.1 Overview . . . 79
5.2 Details . . . 85
5.3 The Perron–Frobenius Operator . . . 89
5.4 The Ergodic Theorem . . . 93
5.5 Convergence Rates in the Ergodic Theorem . . . 103
5.6 Mixing and Decay of Correlations . . . 105
5.7 Physical Measures . . . 114
5.8 Lyapunov Exponents . . . 116

6 Entropy . . . 123
6.1 The Shannon–McMillan–Breiman Theorem . . . 125
6.2 Sinai–Bowen–Ruelle Measures . . . 130
6.3 Dimensions . . . 132

7 Statistics and Statistical Mechanics . . . 141
7.1 The Central Limit Theorem . . . 141
7.2 Large Deviations . . . 147
7.3 Exponential Estimates . . . 149
7.3.1 Concentration . . . 151
7.4 The Formalism of Statistical Mechanics . . . 153
7.5 Multifractal Measures . . . 156

8 Other Probabilistic Results . . . 163
8.1 Entrance and Recurrence Times . . . 163
8.2 Number of Visits to a Set . . . 172
8.3 Extremes . . . 173
8.4 Quasi-Invariant Measures . . . 175
8.5 Stochastic Perturbations . . . 178

9 Experimental Aspects . . . 187
9.1 Correlation Functions and Power Spectrum . . . 188
9.2 Resonances . . . 191
9.3 Lyapunov Exponents . . . 196
9.4 Reconstruction . . . 201
9.5 Measuring the Lyapunov Exponents . . . 205
9.6 Measuring Dimensions . . . 206
9.7 Measuring Entropy . . . 211
9.8 Estimating the Invariant Measure . . . 211

References . . . 215

Index . . . 227

1

A Basic Problem

Before we start with the subject proper, it is perhaps useful to look at a concrete physical example, which can be easily built in the laboratory. It is a pendulum with a magnet at the end, which oscillates above three symmetrically arranged fixed magnets, which attract the oscillating magnet, as shown in Fig. 1.1. When one holds the magnet slightly eccentrically and let it go, it will dance around the three magnets, and finally settle at one of the three, when friction has slowed it down enough.

The interesting question is whether one can predict where it will land. That this is a difficult issue is visible to anyone who does the experiment, because the pendulum will hover above one of the magnets, "hesitate" and cross over to another one, and this will happen many times until the movement changes to a small oscillation around one of the magnets and ends the uncertainty of where it will go. Let us call the three magnets "red," "yellow," "blue"; one can ask for every initial position from which the magnet is started (with 0 speed) *where* it will eventually land. The result of the numerical simulation, to some resolution, is shown in Fig. 1.2. The incredible richness of this figure gives an inkling of the complexity of this problem, although we only deal with a simple classical pendulum.

Exercise 1.1. *Program the pendulum equation and check the figure. The equations for the potential U are, for $q \in \mathbb{R}^2$,*

$$U(q) = \frac{3}{8}|q|^2 - \sum_{j=0}^{2} V(q - q_j)\,, \qquad (1.1)$$

where $q_j = \big(\cos(2\pi j/3), \sin(2\pi j/3)\big)$ and $V(q) = 1/|q|$. The equations of motion are

$$\dot{q} = p\,, \quad \dot{p} = -\gamma p - \nabla_q U(q)\,,$$

where $\dot{q}$ is a shorthand for $\mathrm{d}q(t)/\mathrm{d}t$. The friction coefficient is $\gamma = 0.13$.

The domains of same color are very complicated, and the surprising thing about them is that their boundaries actually coincide: If x is in the boundary ∂R of the red region, it is also in the boundary of yellow and blue: $\partial R = \partial Y = \partial B$. (This fact has

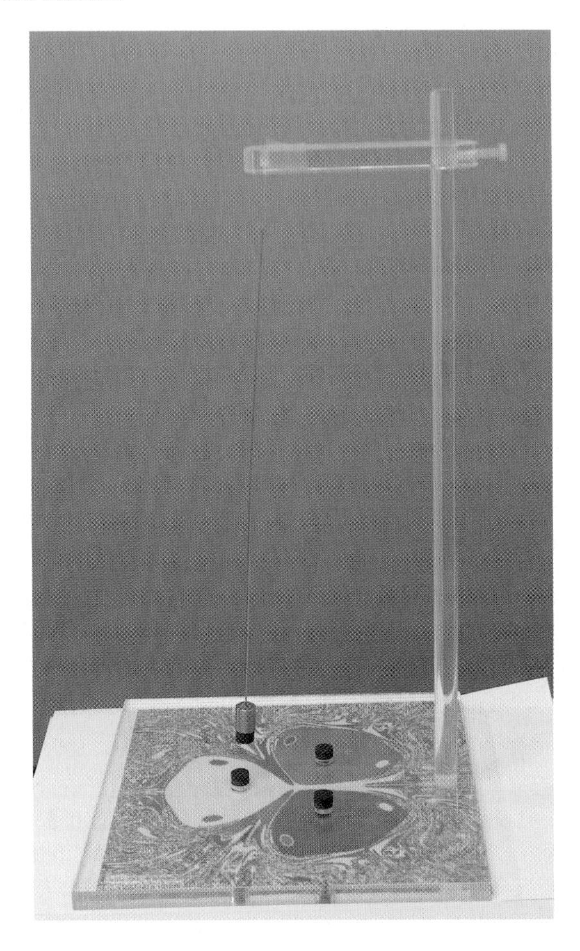

Fig. 1.1. Photograph of the pendulum with three magnets (The design is due to U. Smilansky.)

been proven for the simpler example of Fig. 3.6, and we conjecture the same result for the pendulum.)

The subject of this course is a generic understanding of such phenomena. While this example is not as clear as the one of the "crab" of Fig. 3.6, it displays a feature which will follow us throughout: **instability**. For the case at hand, there is exactly one unstable point in the problem, namely the center, and it is indicated in yellow in Fig. 1.3. Whenever the pendulum comes close to this point, it will have to "decide" on which side it will go: It may creep over the point, and the most minute change in the initial condition might change the final target color the pendulum will reach.

The aim of this book is to give an account of central concepts used to understand, describe, and analyze this kind of phenomena. We will describe the tools with which mathematicians and physicists study chaotic systems.

Fig. 1.2. The basins of attraction of the three magnets, color coded. The coordinates are the two components of the initial position: $q = (q_1, q_2)$. The three circles show the positions of the fixed magnets

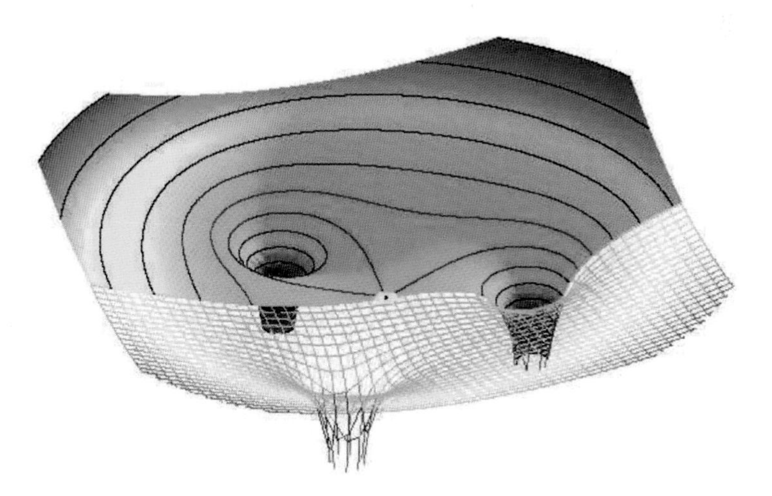

Fig. 1.3. A typical potential for the pendulum of Fig. 1.1. The equation used for the drawing is given in (1.1). The coordinates are the two components of the initial position: $q = (q_1, q_2)$, the height is the value of the potential

2

Dynamical Systems

2.1 Basics of Mechanical Systems

While we assume some familiarity with elementary mechanics, we begin here with the nonlinear pendulum in 1 dimension, to have some basis of discussion of phase space and the like. The nonlinear pendulum is a mathematical pendulum, under the influence of gravity, and with a damping $\gamma \geq 0$. Normalizing all the coefficients to 1, we consider the coordinates of this problem: momentum p and the angle φ. The pendulum can go over the top, and thus φ varies in $[0, 2\pi)$ while the momentum varies in $(-\infty, \infty)$. (We use here the notation $[a, b)$ to denote the half-open interval $a \leq x < b$.) Thus, the phase space Ω is actually a cylinder (since we identify $\varphi = 2\pi$ and $\varphi = 0$). The equations of motion for this problem are

$$\begin{aligned} \dot{\varphi} &= p\,, \\ \dot{p} &= -\sin(\varphi) - \gamma p\,, \end{aligned} \tag{2.1}$$

where both φ and p are functions of the time t, and $\dot{\varphi} = \frac{\mathrm{d}}{\mathrm{d}t}\varphi(t)$, $\dot{p} = \frac{\mathrm{d}}{\mathrm{d}t}p(t)$.

This problem has 2 fixed points, where the pendulum does not move, $p = 0$, with $\varphi = 0$ or π. The fixed point $\varphi = \pi$ is unstable, and the pendulum will fall down under any small perturbation (of its initial position, when the initial velocity is 0) while $\varphi = 0$ is stable. The nature of stability near $\varphi = 0$ changes with γ, as can be seen by linearizing (2.1) around $\varphi = p = 0$.

i) For $\gamma = 0$ the system is Hamiltonian, and the flow is shown in Fig. 2.1. Such pictures are called *phase portraits*, because they describe the phase space, to be defined more precisely in Sect. 2.2. The vertical axis is p, the circumference of the cylinder is the variable φ. The lower equilibrium point is in the front (left) of the cylinder and the upper equilibrium point is in the back, at the crossing of the red curves. These curves are called *homoclinic orbits*, they correspond to the pendulum leaving the top at zero (infinitesimally small) speed and returning to it after an infinite time. One can see two such curves, one for each direction of rotation.

ii) For $\gamma \in (0, \gamma_{\rm crit})$, with $\gamma_{\rm crit} = 2$ (*critical damping*), the orbits are as shown in Fig. 2.2. The red line is the stable manifold of the unstable fixed point: A pendulum with initial conditions on the red line will make a number of full turns and stop finally (after infinite time) at the unstable equilibrium. All other initial conditions lead to orbits spiraling into the stable fixed point. The blue line is called the unstable manifold of the unstable fixed point and it also spirals into the stable fixed point. We treat this in detail in Sect. 4.1.1.

iii) For $\gamma = \gamma_{\rm crit}$ the phase portrait is shown in Fig. 2.3. This value of γ is the smallest for which the pendulum, when started at the top ($\varphi = \pi, p = \varepsilon, \varepsilon \to 0$), will move to the stable fixed point $\varphi = 0$ without oscillating around it.

iv) For $\gamma > \gamma_{\rm crit}$, which is usually called the *supercritical* case, the phase portrait is shown in Fig. 2.4. In this figure, the red and blue lines are again the stable and unstable manifolds of the unstable fixed point, while the green line is the *strongly stable manifold* of the stable fixed point. (The tangent flow of the stable fixed point has 2 stable (eigen-)directions and the strongly stable one is the direction which attracts faster.)

v) When $\gamma = 0$, there is no friction and the flow for the pendulum is area preserving. We illustrate this in Fig. 2.5. Note that the Poincaré recurrence theorem (see also Theorem 8.2) tells us that any open set contained in a compact invariant set must return infinitely often to itself. This is clearly seen for both the red and the blue region which eventually intersect infinitely often the original ellipse from which the motion started. Note that Poincaré's theorem does *not* say that a given point must return close to itself, just that the regions must intersect. Note furthermore how the regions come close to, but avoid the unstable fixed point (since the original region did not contain that fixed point).

Some final remarks on general flows in $\mathbb{R}^d$ are in order. They are all described by differential equations of the form

$$\frac{\mathrm{d}}{\mathrm{d}t}\mathbf{x}(t) = \mathbf{F}(\mathbf{x}(t)) \, , \tag{2.2}$$

with $\mathbf{x} : \mathbb{R} \to \mathbb{R}^d$ and $\mathbf{F} : \mathbb{R}^d \to \mathbb{R}^d$ is called a *vector field*.

Theorem 2.1. *If* $\mathbf{F}$ *is Lipschitz continuous, the solution of (2.2) either exists for all times or diverges at some finite time.*

Definition 2.2. *If* $\mathbf{F}(\mathbf{x}_*) = 0$ *one calls* $\mathbf{x}_*$ *a fixed point.*

Theorem 2.3. *Outside a fixed point (that is in any small open set not containing a fixed point) any smooth flow is locally trivial, in the sense that there exists a coordinate change for which (2.2) takes the form* $\dot{\mathbf{y}} = \mathbf{A}$, *where* $\mathbf{A}$ *is the constant column vector with* d *components:* $\mathbf{A} = (1, 0, \ldots, 0)$. *(Furthermore, if (2.2) is a Hamiltonian equation, the coordinate change can be chosen to be canonical (Darboux' theorem).)*

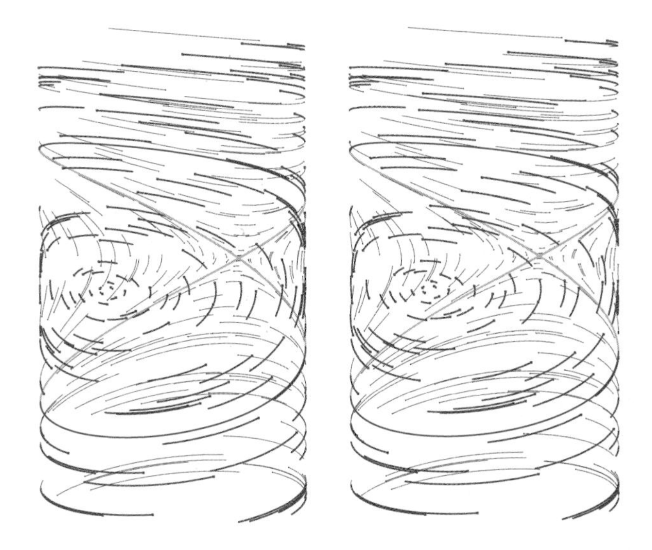

Fig. 2.1. Stereo picture for the flow of (2.1) for $\gamma = 0$. These pictures give a 3-dimensional effect if one "stares" at them to bring images of the two cylinders to convergence. The stable equilibrium point is in the front (to the left) and the unstable one is in the back of the cylinder. The black lines show short pieces of orbit. These pieces start at the dot and extend along the short lines

Fig. 2.2. The same view for a subcritical γ. The value is $\gamma = 0.25$

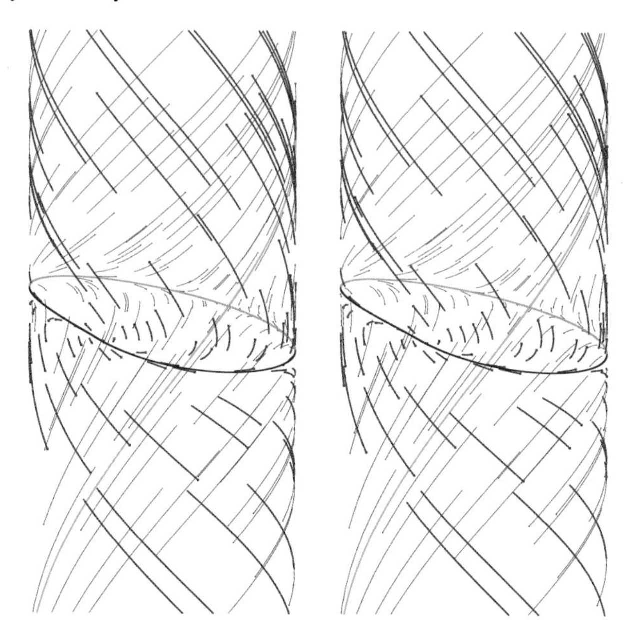

Fig. 2.3. Critical damping. The value is $\gamma = 2$

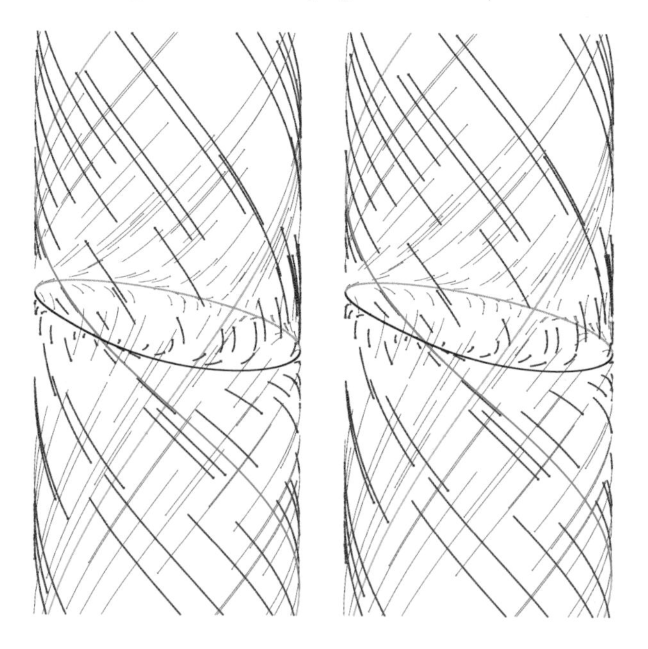

Fig. 2.4. Supercritical damping. The value is $\gamma = 2.2$

Fig. 2.5. Illustration of Poincaré's theorem for the pendulum (2.1) with $\gamma = 0$. The initial elliptical region evolves in time and its volume is preserved. The order is top left, right, bottom left, right. The corresponding times are 0, 3, 11, 52 in arbitrary units

Discussion. At a fixed point, things are more complicated than what is described in Theorem 2.3 and one must study the tangent matrix $\mathrm{D}_{\mathbf{x}_*}\mathbf{F}$, which is defined by the expansion

$$\mathbf{F}(\mathbf{x}_* + \mathbf{h}) = 0 + \mathrm{D}_{\mathbf{x}_*}\mathbf{F}\,\mathbf{h} + \mathcal{O}(|\mathbf{h}|^2)\,,$$

where $\mathbf{h} \in \mathbb{R}^d$. The local picture of the flow near the fixed point will depend on the eigenvalues of the matrix $\mathrm{D}_{\mathbf{x}_*}\mathbf{F}$. If they all have negative real part, the fixed point is *stable*, if no eigenvalue is purely imaginary (or 0) it is *hyperbolic*, and if all are imaginary, it is called *elliptic*.

A very readable exposition of mechanics is (Arnold 1978).

2.2 Formal Definitions

A (classical) physical system is described by the set Ω of all its possible *states*. It is usually called the *phase space* of the system. At any given time, all the properties of the system can be recovered from the knowledge of the instantaneous state $x \in \Omega$. The system is observed using the so-called *observables* which are real valued functions on the phase space Ω (for example, the coordinates of position or momentum). Most often the space of states Ω is a metric space (so we can speak of nearby states). In many physical situations there are additional structures on Ω: $\mathbb{R}^d$, Riemannian manifolds, Banach or Hilbert spaces.

As a simple example one can consider a mechanical system with 1 degree of freedom. The state of the system at a given time is determined by two real numbers: the position q and momentum p. The state space is therefore $\mathbb{R}^2$. A continuous material (solid, fluid) is characterized by the field of local velocities, pressure, density, temperature, and others. In that case one often uses phase spaces that are Banach spaces.

As time goes on, the instantaneous state changes (unless the system is in a situation of equilibrium). The *time evolution* is a rule describing the change of the state with time. It comes in several flavors and descriptions summarized below.

i) Discrete time evolution. This is a map f from the state space Ω into itself producing the new state from the old one after one unit of time. If x_0 is the state of the system at time zero, the state at time 1 is $x_1 = f(x_0)$ and more generally the state at time n is given recursively by the *iteration* $x_n = f(x_{n-1})$. This is often written $x_n = f^n(x_0)$ with $f^n = f \circ f \circ \cdots \circ f$ (n-times), where $\circ$ is the symbol of functional composition: $(f \circ g)(x) \equiv f\big(g(x)\big)$. The sequence $\{f^n(x_0)\}_{n \in \mathbb{Z}^+}$ is called the *trajectory* or the *orbit* of the *initial condition* x_0. (Throughout the book we use $\mathbb{Z}$ to denote the integers, $\mathbb{N}$ to denote the integers $\{1, 2, \ldots\}$ and $\mathbb{Z}^+$ to denote $\{0, 1, 2, \ldots\}$.)
If the inverse of f exists, then the orbit is defined also for negative n.

ii) Continuous time *semiflow*. This is a family $(\varphi_t)_{t \in \mathbb{R}^+}$ of maps of Ω satisfying

$$\varphi_0 = I \,, \qquad \varphi_s \circ \varphi_t = \varphi_{s+t} \,.$$

The set $\big(\varphi_t(x_0)\big)_{t \in \mathbb{R}^+}$ is called the trajectory (orbit) of the initial condition x_0. Note that if we fix a time step $\tau > 0$, and observe the state only at times $n\tau$ ($n \in \mathbb{Z}^+$), then we obtain a discrete time dynamical system given by the map $f = \varphi_\tau$. (But not every map is the discrete time version of a flow.)

iii) A differential equation on a manifold associated with a vector field $\mathbf{F}$

$$\frac{\mathrm{d}\mathbf{x}}{\mathrm{d}t} = \mathbf{F}(\mathbf{x}) \,.$$

This is for example the case of a mechanical system in the Hamiltonian formalism. Under regularity conditions on $\mathbf{F}$, the integration of this equation leads to a semiflow (and even a flow).

iv) Other, more complicated situations occur, like nonautonomous systems (in particular stochastically forced systems), systems with memory, systems with delay, etc., but we will not consider them very often in this book.

A *dynamical system* is a set of states Ω equipped with a time evolution. If there is more structure on Ω, one can put more structure on the time evolution itself. For example, for the case of discrete time, the map f may be measurable, continuous, differentiable, or have some other form of regularity.

Needless to say, dynamical systems abound in all domains of science. We already mentioned mechanical systems with a finite number of degrees of freedom and continuum mechanics. Mechanical systems with many degrees of freedom are at the root of thermodynamics (and a lot of work has been devoted to the study of billiards which are related to simple models of gases), one can also mention chemical reactions, biological and ecological systems, even many random number generators turn out to be dynamical systems.

A useful general notion is that of equivalence (*conjugation*) of dynamical systems.

Definition 2.4. *Assume two dynamical systems (Ω_1, f_1) and (Ω_2, f_2) are given. We say that they are conjugated if there is a bijection Φ from Ω_1 to Ω_2 such that*

$$\Phi \circ f_1 = f_2 \circ \Phi \ .$$

It follows immediately that $\Phi \circ f_1^n = f_2^n \circ \Phi$, and a similar definition holds for (semi-)flows. Of course the map Φ may have supplementary properties such as being measurable or continuous.

Exercise 2.5. *Show that the flows associated to two vector fields $\mathbf{X}$ and $\mathbf{Y}$ are conjugated by a diffeomorphism (1-1 invertible differentiable map with differentiable inverse) Φ if and only if for any point x*

$$\mathrm{D}_x\Phi \cdot \mathbf{X}(x) = \mathbf{Y}(\Phi(x)) \ .$$

2.3 Maps

One can reduce a continuous time dynamical system to a discrete time map by the technique of using a *Poincaré section* (see Fig. 2.6 for the construction) or by looking at the time one map. The Poincaré section is obtained by noting each intersection of the orbit with a hyperplane, and to consider the discrete (but probably irregularly spaced) times of intersection to define a map from the hyperplane to itself (provided such intersections really occur).

Similarly one can construct a continuous time system, using the technique of *suspension flow* of a discrete time system. One considers a map $f : \Omega \to \Omega$ and a *ceiling function* $h : \Omega \to \mathbb{R}^t$. A flow $t \to \Phi^t$ is then defined on $\cup_{x\in\Omega}\big(x, [0, h(x)]\big)$ with $\Phi^t(x, y) = (x, y+t)$, as long as $y+t < h(x)$ and when $y+t$ reaches $h(x)$ at time $t_0 = h(x) - y$ one defines $\Phi^{t_0}(x, y) = \big(f(x), 0\big)$. We refer the reader to the literature for more details on these constructions (see, for example, (Robinson 2004)).

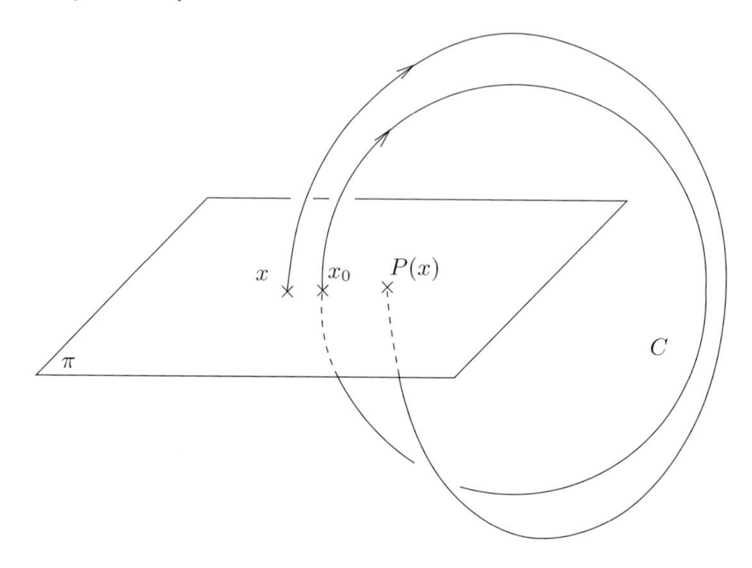

Fig. 2.6. Construction of the Poincaré map around the cycle C on the surface π, that starts at x_0 and ends at x_0. The Poincaré map P maps x to $P(x)$ and defines thus a discrete time map from a flow

2.4 Basic Examples of Maps

The aim of this section is to introduce several simple examples of maps. These examples are chosen in such a way that each typical behavior we discuss in the book will usually correspond to one of these examples. Therefore, the examples will be referred to throughout the book. As we present the examples, we already note some of their properties. However, these properties will, in general, be described again in more detail later.

Most of the examples are maps of an interval to itself, and we usually take the interval to be $[0, 1)$. We use the notation $\pmod 1$ to indicate that the result of a function is mapped back to $[0, 1)$; that is, $x \pmod 1$ is the fractional part, in $[0, 1)$, of the number $x \in \mathbb{R}$.

Example 2.6. **The Interval Map** $x \mapsto 2x \pmod 1$
By this we mean the map

$$f(x) = \begin{cases} 2x\,, & \text{if } 0 \le x < \frac{1}{2} \\ 2x - 1\,, & \text{if } \frac{1}{2} \le x < 1 \end{cases} .$$

Note that every point y has 2 preimages and therefore the map f is not invertible. On the other hand, this example will be the simplest chaotic map with which many ideas can be explained. We will see that it is indeed a very chaotic map. Its graph is shown in Fig. 2.7. This map can also be interpreted as doubling the angle on the circle.

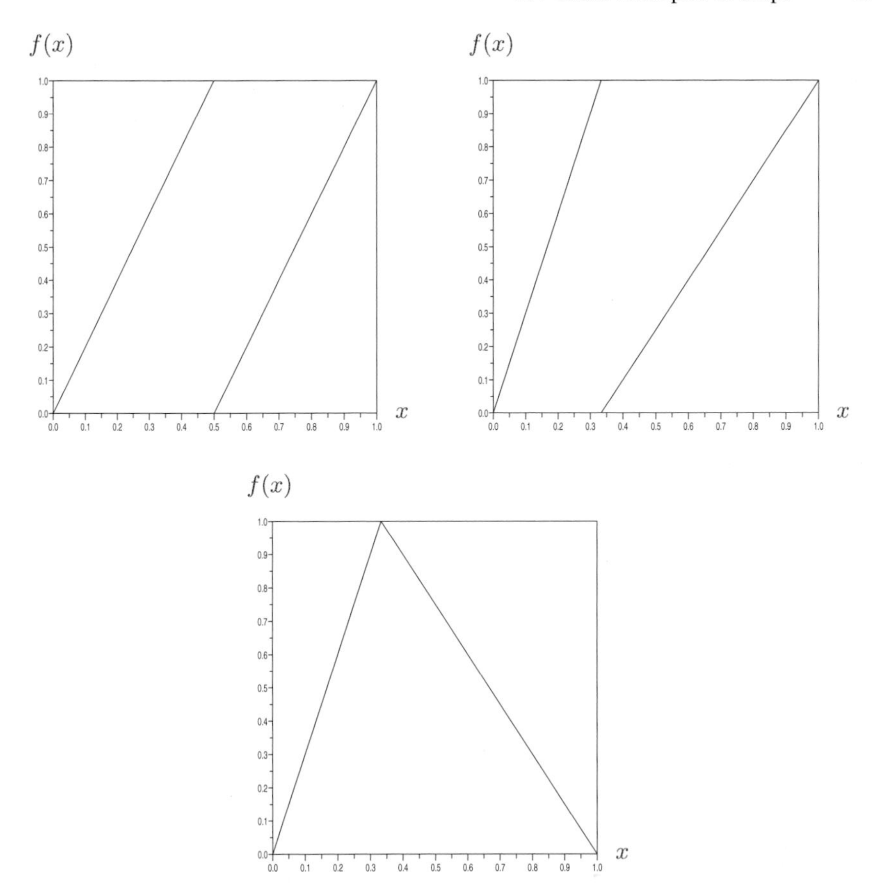

Fig. 2.7. Graphs of the maps f given in Examples 2.6–2.8. Top left: $x \mapsto 2x \pmod 1$. Top right: the skew map Eq. (2.3). Bottom: The continuous skew map Eq. (2.4)

Example 2.7. **The Skew Map**
This map is very similar to Example 2.6 with the difference that the derivative of f is not constant:

$$f(x) = \begin{cases} 3x\,, & \text{if } x \in [0, \frac{1}{3}) \\ \frac{3}{2}x - \frac{1}{2}\,, & \text{if } x \in [\frac{1}{3}, 1) \end{cases}\,. \tag{2.3}$$

Its graph is also shown in Fig. 2.7.

Example 2.8. **The Continuous Skew Map**
This example is very similar to Example 2.7, but it is a continuous function.

$$f(x) = \begin{cases} 3x\,, & \text{if } 0 \leq x < 1/3 \\ \frac{3}{2} - \frac{3x}{2}\,, & \text{if } 1/3 \leq x \leq 1 \end{cases}\,. \tag{2.4}$$

Its graph is shown in Fig. 2.7.

Example 2.9. **The Baker's Transformation**
This is a map on the domain $\Omega = [0,1) \times [0,1) \subset \mathbb{R}^2$. It is defined by

$$\begin{pmatrix} x \\ y \end{pmatrix} \mapsto \begin{pmatrix} 2x \pmod 1 \\ y' \end{pmatrix} ,$$

with

$$y' = \begin{cases} \frac{1}{2}y\,, & \text{if } 0 \leq x < \frac{1}{2} \\ \frac{1}{2}y + \frac{1}{2}\,, & \text{if } \frac{1}{2} \leq x < 1 \end{cases} .$$

The baker map is area-preserving and invertible. Its name derives from the way dough is mixed by bakers. For a drawing, see Fig. 2.12, which illustrates a dissipative version of this map (Example 2.27).

Example 2.10. **The Cat Map**
This celebrated example (following the destiny of a subset of $[0,1) \times [0,1)$ with the shape of a cat) was presented in (Arnold and Avez 1967). It maps $\Omega = [0,1)^2$ to itself. The map is given by

$$\begin{pmatrix} x \\ y \end{pmatrix} \mapsto \begin{pmatrix} 2x + y \pmod 1 \\ x + y \pmod 1 \end{pmatrix} .$$

This map is area preserving, its inverse is

$$\begin{pmatrix} x \\ y \end{pmatrix} \mapsto \begin{pmatrix} x - y \pmod 1 \\ -x + 2y \pmod 1 \end{pmatrix} ,$$

as can be seen from the matrix form

$$\begin{pmatrix} 2 & 1 \\ 1 & 1 \end{pmatrix} .$$

This map is a 2-dimensional version of Example 2.6, and is again very chaotic. All rational points in $[0,1)^2$ are periodic points. The destiny of the poor cat is shown in Fig. 2.8.

Example 2.11. **The Circle Rotation**
This map is *not* chaotic. It maps $\Omega = [0,1)$ to itself by

$$x \mapsto x + \omega \pmod 1 .$$

If ω is irrational, no point is periodic. If ω is rational, all points are periodic.

Example 2.12. **The Torus Rotation**
This is a 2-dimensional version of Example 2.11.

$$\begin{pmatrix} x \\ y \end{pmatrix} \mapsto \begin{pmatrix} x + \omega \pmod 1 \\ y + \omega' \pmod 1 \end{pmatrix} .$$

If ω, ω', and ω/ω' are irrational, all orbits are dense. (In the cat map not all orbits are dense, but the cat map is more chaotic.)

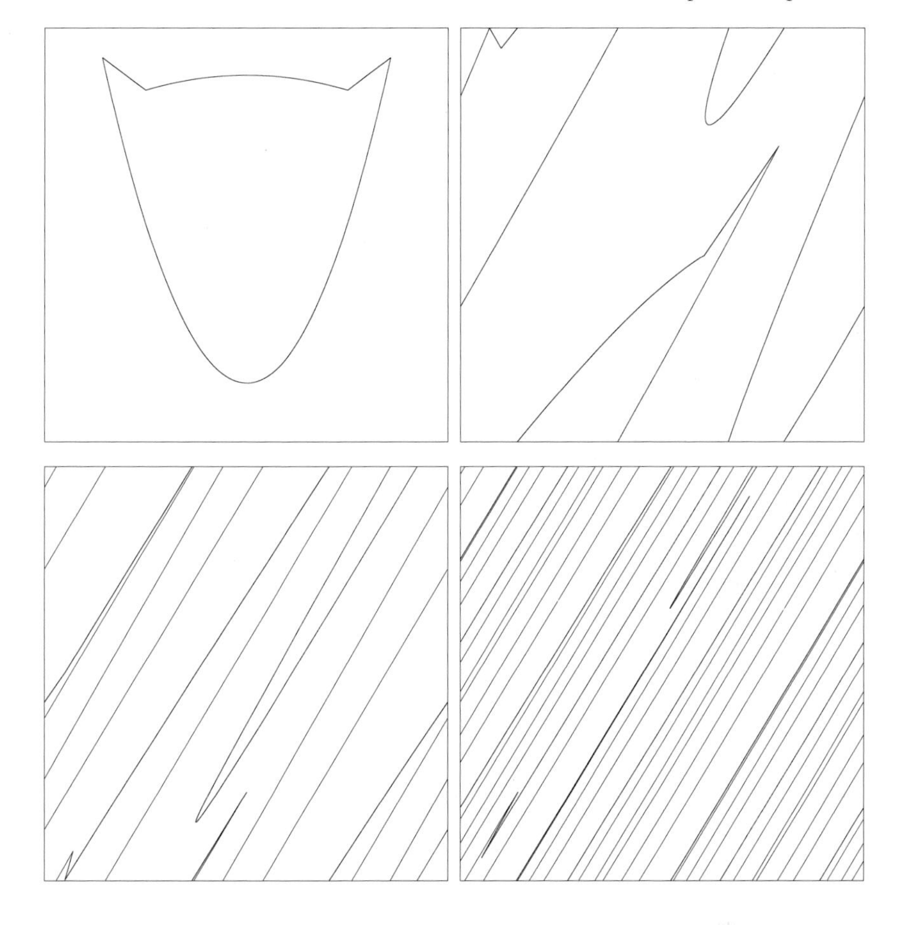

Fig. 2.8. The cat and 3 of its iterates under the cat map. Top left, right, bottom left, right. Try to find the left and right ears in the third iteration

Example 2.13. **The Solenoid**
This is a chaotic map from $\mathbb{R}^3$ to itself. It maps the torus into itself by winding twice around (see Figs. 2.9 and 2.10). In this sense, it is close to Example 2.6. In Cartesian coordinates (u, v, w) it is given by

$$\begin{pmatrix} u \\ v \\ w \end{pmatrix} \mapsto \begin{pmatrix} \cos(2\varphi)\,(1 + 0.4\cos(\varphi) + r) \\ \sin(2\varphi)\,(1 + 0.4\cos(\varphi) + r) \\ 0.25w + 0.4\sin(\varphi) \end{pmatrix} , \tag{2.5}$$

with $r = 0.25(\sqrt{u^2 + v^2} - 1)$ and $\varphi = \arctan(u/v)$. In polar coordinates $\varrho = \sqrt{u^2 + v^2}$, $\varphi = \arctan(u/v)$, w:

$$\begin{pmatrix} \varrho \\ \varphi \\ w \end{pmatrix} \mapsto \begin{pmatrix} \varrho + 0.4\cos(\varphi) \\ 2\varphi \\ 0.25w + 0.4\sin(\varphi) \end{pmatrix} .$$

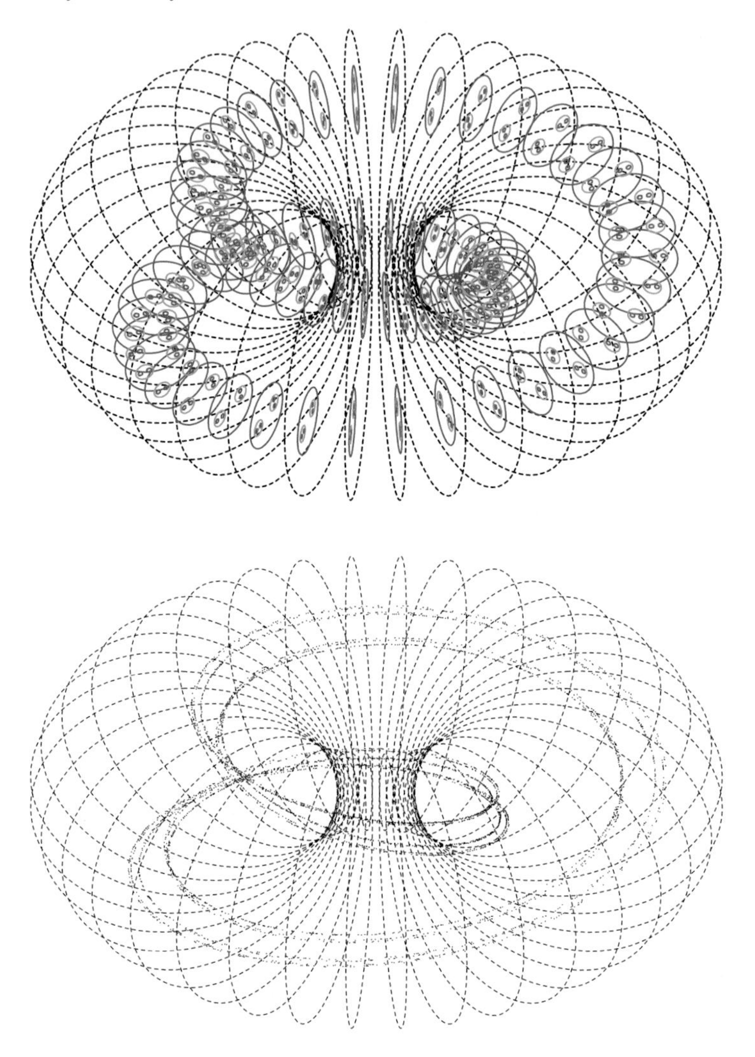

Fig. 2.9. Two views of the solenoid map (2.5). Top: some cuts of the solenoid and their first 3 iterates, orange, light blue, violet. Bottom: 5000 iterates of an initial point

Exercise 2.14. *Show that the solenoid transformation maps the interior of the torus*

$$\{u = \cos(\varphi)(1 + 0.8\sin(\vartheta)), v = \sin(\varphi)(1 + 0.8\sin(\vartheta)), w = 0.8\cos(\vartheta)\} ,$$

$\vartheta \in [0, 2\pi)$, $\varphi \in [0, 2\pi)$ *injectively into itself (see Fig. 2.10). Show that the injection is strict; that is, the image is less than the original solenoid.*

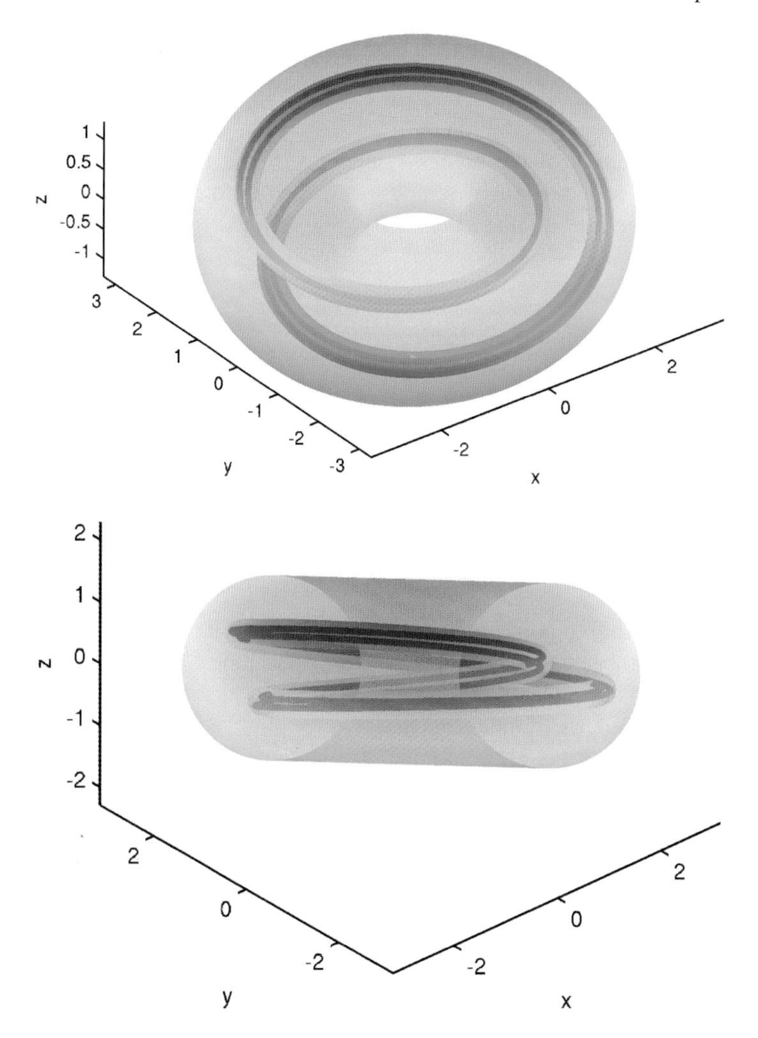

Fig. 2.10. Two views of two iterations a torus under the solenoid map of Example 2.13

2.5 More Advanced Examples

We extend the previous list of examples by slightly more complicated ones. In each of them we first specify the phase space and then the time evolution. These examples appear often in the literature in equivalent (conjugated) forms.

Example 2.15. **Piecewise Expanding Maps**
A generalization of the skew map of Example 2.7 is the class of *piecewise expanding maps of the interval*. The phase space is again the interval $\Omega = [0, 1)$. We fix a finite sequence $0 = a_0 < a_1 < \ldots < a_k = 1$ and for each interval (a_j, a_{j+1}) $(0 \le j \le k-1)$ a monotone map f_j from (a_j, a_{j+1}) to $[0, 1]$, which extends to a C^2

map on $[a_j, a_{j+1}]$. The map f on $[0, 1]$ is then given by

$$f(x) = f_j(x) \quad \text{if} \quad x \in (a_j, a_{j+1}) \ .$$

This leaves the images of the points a_j undefined, and one can choose them for example as limits from the left. We impose next the (uniform) *expanding property*: *There are an integer $m > 0$ and a constant $c > 1$ such that*

$$\left| f^{m\prime} \right| \geq c$$

at any point where f^m is differentiable.

The map $2x \pmod 1$ is of course an example of a piecewise expanding map of the interval. A piecewise expanding map of the interval is said to have the *Markov property* if the closure of the image of any defining interval (a_j, a_{j+1}) is a union of closures of such intervals. This is of course a topological notion but we will see later that it has some connection with the notion of Markov chains in probability theory.

Exercise 2.16. *Show that a finite iterate of a piecewise expanding map of the interval is also a piecewise expanding map of the interval.*

Example 2.17. **Random Number Generators**
Most random number generators turn out to be dynamical systems given by simple expressions. They generate a sequence of numbers that is meant to be a typical realization of a sequence of independent drawings of a random variable. Since computers act on a finite set of rational numbers, the theory of random number generators is mostly devoted to the study of the arithmetic properties of the map associated with the dynamical system (for example, large period, uniform distribution). Some popular examples are maps of the interval $\left[0, 2^{31} - 1\right]$ given by $f(x) = 16{,}807\, x \mod \left(2^{31} - 1\right)$ or $f(x) = 48{,}271\, x \mod \left(2^{31} - 1\right)$ which are among the best generators for 32 bits machines. We refer to (Knuth 1981; L'Ecuyer 1994; 2004), and (Marsaglia 1992) for more on the subject and for quality tests.

Example 2.18. **The Logistic Map**
The previous dynamical systems have discontinuities, and many examples have been studied which are more regular, but often more difficult to analyze. One of the most well-known examples is provided by the one-parameter family of quadratic maps. The phase space is the interval $\Omega = [-1, 1]$ and for a value of the parameter $\nu \in [0, 2]$, the time evolution is given by the map f_ν

$$f_\nu(x) = 1 - \nu x^2 \ .$$

When $\nu = 2$, the corresponding map is called the logistic map. There is a vast literature on these maps and on the more general case of *unimodal maps*. These are maps like $1 - \nu x^2$ whose graph has one maximum.

Exercise 2.19. *Prove that the invertible map $\Phi(x) = \sin(\pi x/2)$ on $[-1, 1]$ conjugates the map f_2 and the map $g(x) = 1 - 2|x|$.*

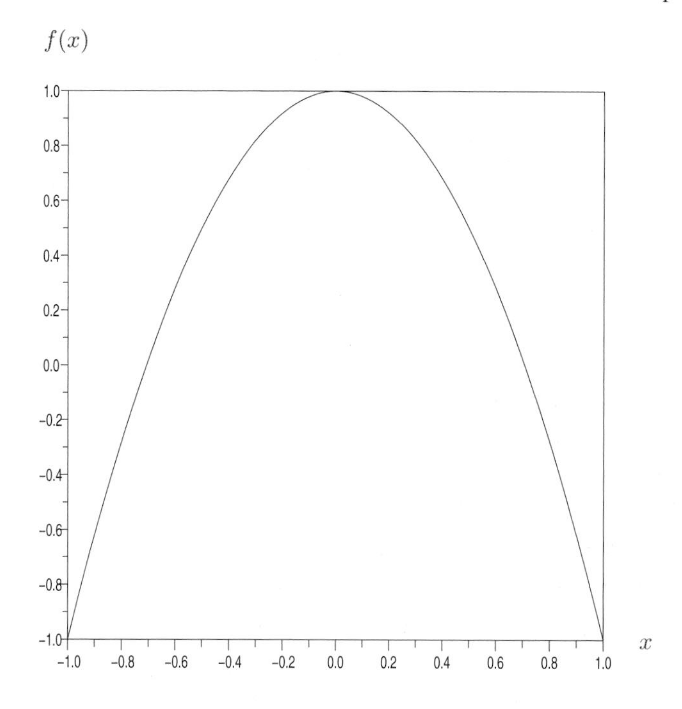

Fig. 2.11. The graph of the logistic map $x \mapsto 1 - 2x^2$

Example 2.20. **Subshifts**
We now describe a more abstract example which is in some sense the general model for uniformly chaotic systems. Let $\mathcal{A}$ be a finite set (often called a finite *alphabet*). Let M be a matrix of zeros and ones of size $|\mathcal{A}| \times |\mathcal{A}|$ ($|\mathcal{A}|$ denotes the cardinality of $\mathcal{A}$). The matrix M is often called the *incidence matrix*. Recall that $\mathcal{A}^{\mathbb{Z}}$ is the set of all bi-infinite sequences $\mathbf{x} = \{x_p\}_{p \in \mathbb{Z}}$ of elements (letters) of the alphabet $\mathcal{A}$. The phase space Ω is defined by

$$\Omega = \left\{ \mathbf{x} \in \mathcal{A}^{\mathbb{Z}} \mid M_{x_j, x_{j+1}} = 1 \, \forall j \in \mathbb{Z} \right\} .$$

The time evolution is the shift map $\mathcal{S}$ given by

$$\mathcal{S}(\mathbf{x})_j = x_{j+1} \, .$$

Exercise 2.21. *Show that $\mathcal{S}$ is an invertible map from Ω to itself (in particular describe the inverse map).*

This dynamical system $(\Omega, \mathcal{S})$ is called a *subshift of finite type*. When all the entries of the matrix M are equal to 1, the dynamical system is called the *full shift*. If one uses $\mathbb{N}$ instead of $\mathbb{Z}$ in the definition of Ω, one speaks of a *unilateral shift* (which is not invertible).

Consider again the simple case of the map $2x \pmod 1$. A point of the interval $[0, 1)$ can be coded by its dyadic decomposition, namely

$$x = \sum_{j=1}^{\infty} \varepsilon_j 2^{-j}$$

with $\varepsilon_j = 0$ or 1. The decomposition is unique except for a countable set of points.

Exercise 2.22. *Describe the countable set of points for which the decomposition is not unique. See also Example 3.1.*

It is easy to verify that the map $2x \pmod 1$ acts by shifting the dyadic sequence $\{\varepsilon_j\}$. Therefore, if we consider the alphabet with two symbols $\mathcal{A} = \{0, 1\}$, there is a conjugation between the dynamical system given by the map $x \mapsto 2x \pmod 1$ on the unit interval and the unilateral full shift over two symbols except for a countable set.

Exercise 2.23. *Explain why on most computers, for any initial condition in the interval $[0, 1)$ the orbit under the map $2x \pmod 1$ terminates (after 32 or 64 iterations) on the fixed point $x = 0$. Do the experiment.*

The preceding dyadic coding can easily be generalized to the case of piecewise expanding Markov map of the interval.

Exercise 2.24. *Consider a piecewise expanding Markov map f of the interval with monotonicity and regularity intervals (a_j, a_{j+1}), $0 \le j < k$ (see Example 2.15). To each point x of the interval we associate the following code $\boldsymbol{\sigma} \in \mathcal{A}^{\mathbb{Z}^+}$ on the alphabet of k symbols $\mathcal{A} = \{0, \ldots, k-1\}$ defined for $n = 0, 1, \ldots$ by*

$$\sigma_n = j \quad \text{if} \quad f^n(x) \in [a_j, a_{j+1}) \, .$$

Show that this gives a well-defined and invertible coding except for a countable number of exceptional points (hint: use the expansivity property). Prove that this coding defines a conjugacy between the piecewise expanding Markov map f and a unilateral subshift of finite type (start by constructing the transition matrix M of zeros and ones).

Up to now, all the more advanced examples were 1-dimensional. We now give examples in two dimensions.

Example 2.25. **Skew-product Systems**
We consider here some generalizations of the *baker's map* of Example 2.9. Recall that the phase space is the square $[0, 1)^2$ and the baker's map f is defined as follows

$$f(x, y) = \begin{cases} (2x, y/2) & \text{if } x < 1/2 \\ (2x - 1, (1 + y)/2) & \text{if } x \ge 1/2 \end{cases} \, .$$

This construction is a particular case of a *skew-product* system, of which many variants exist. A dynamical system is called a skew-product if the phase space is a product space $\Omega = \Omega_1 \times \Omega_2$, and the time evolution f has the special form

$$f(x, y) = \big(f_1(x), f_2(x, y)\big) .$$

The dynamical system (Ω_1, f_1) is often called the base of the skew product, and the map f_2 is called the action in the fibers. A *fiber* is a subset of the form $\{x\} \times \Omega_2$. In the case of the suspension flow of Sect. 2.3 the fiber was $\{x\} \times [0, h(x))$, so that the height of each fiber depends on x.

Exercise 2.26. *See also Example 3.2 for details. Show that the baker's map is invertible. To each point (x, y) of the phase space $[0, 1)^2$, we can associate a bi-infinite sequence $\boldsymbol{\sigma} \in \{0, 1\}^{\mathbb{Z}}$ by*

$$\sigma_n = \begin{cases} 0\,, & \textit{if } 0 \le f^n(x, y)_1 < 1/2 \\ 1\,, & \textit{otherwise} \end{cases} .$$

Here $f^n(x, y)_1$ denotes the first coordinate of the point $f^n(x, y) \in [0, 1)^2$. Using this coding, show that the baker's map is conjugated (except for a "small" set of points) to the full (bilateral) shift over two symbols.

Example 2.27. **The Dissipative Baker's Map**
Another example of a skew-product is the *dissipative baker's map*. The phase space is again the unit square $[0, 1) \times [0, 1)$. The map is for example given by

$$f(x, y) = \begin{cases} (3x, y/4)\,, & \text{if } 0 \le x < 1/3 \\ ((3x - 1/2), (2 + y)/3)\,, & \text{if } 1/3 \le x \le 1 \end{cases} . \tag{2.6}$$

Many variants of this example exist; see Fig. 2.12.

The above example is discontinuous, and we now present some differentiable examples.

Exercise 2.28. *Prove that the cat map of Example 2.10 is indeed a C^∞ map of the torus with a C^∞ inverse.*

Example 2.29. **The Standard Map of the Torus**
A generalization of the cat map is given by the *standard map* of the 2-torus $\Omega = \{x \in [0, 1), y \in [0, 1)\}$

$$f(x, y) = \begin{pmatrix} 2x - y + a\sin(2\pi x) \pmod 1 \\ x \end{pmatrix} .$$

Exercise 2.30. *Prove that the standard map of the torus is C^∞ with a C^∞ inverse.*

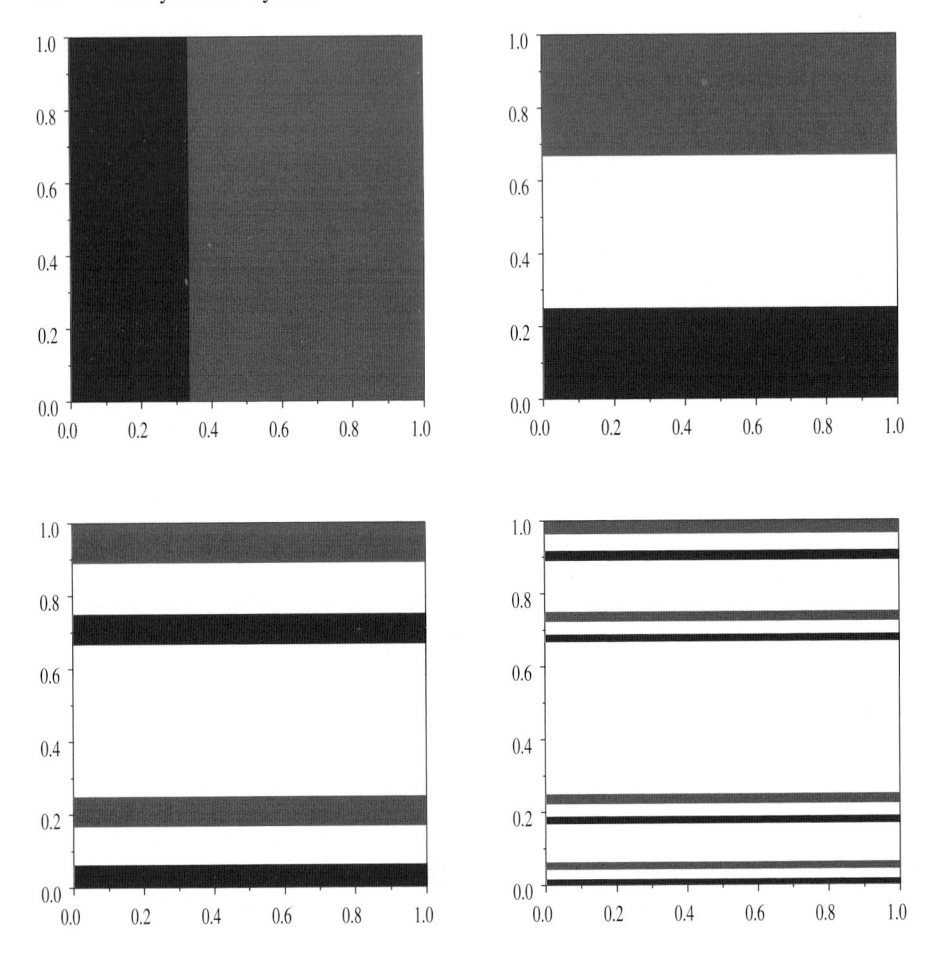

Fig. 2.12. Several iterations of the square under the dissipative baker's map (2.6)

Example 2.31. **The Hénon Map**
Another historically important example is the *Hénon map*. The phase space is $\Omega = \mathbb{R}^2$ and the two-parameter (a and b) family of maps is given by

$$f(x, y) = \begin{pmatrix} 1 - ax^2 + y \\ bx \end{pmatrix} .$$

The "historical" values of the parameters are $a = 1.4$, $b = 0.3$ (see (Hénon 1976)). The map is illustrated in Fig. 3.2.

Exercise 2.32. *Prove that the Hénon map is invertible with a regular inverse if* $b \neq 0$ *(determine the formula for the inverse).*

2.6 Examples of Flows

For the case of continuous time dynamical systems, given by vector fields, the most important example is the case of mechanical systems.

Example 2.33. **Hamiltonian Flows**
We recall that the phase space is $\Omega = \mathbb{R}^{2d}$ (or some even dimensional manifold). The first d coordinates $q_1, \ldots, q_d$ are position coordinates, and the remaining d coordinates $p_1, \ldots, p_d$ are the conjugated momentum coordinates. A real valued function $H(\mathbf{q}, \mathbf{p})$ is given on the phase space and is called the Hamiltonian of the mechanical system. The time evolution is given by Hamilton's system of equations

$$\frac{\mathrm{d}q_j}{\mathrm{d}t} = \partial_{p_j} H \,, \qquad \frac{\mathrm{d}p_j}{\mathrm{d}t} = -\partial_{q_j} H \qquad (j = 1, \ldots, n) \,.$$

One can also add friction terms. Note that systems with time-dependent forcing (nonautonomous systems) do not fall in our present description, unless one can give a dynamical description of the forcing (this often leads to a skew product system).

Example 2.34. **The Lorenz Flow**
An historically important example in dimension 3 is the *Lorenz system*. The phase space is $\Omega = \mathbb{R}^3$ and the vector field (depending on three parameters σ, r, and b) defining the time evolution is given by

$$\mathbf{X}(x, y, z) = \begin{pmatrix} \sigma(y - x) \\ -xy + rx - y \\ xy - bz \end{pmatrix} \,. \tag{2.7}$$

So, for example, $\frac{\mathrm{d}x}{\mathrm{d}t} = \sigma(y - x)$. The historical values of the parameters are $\sigma = 10$, $r = 28$, $b = 8/3$, see (Lorenz 1963). The flow is illustrated in Fig. 3.4. This equation is obtained as a truncation to three Fourier modes of an equation describing convection in hydrodynamics.

Example 2.35. **Chemical Kinetics**
The time evolution of chemical reactions is often described by kinetic equations that are systems of differential equations for the concentration of the chemical species involved in the reaction. A much-studied example is the so-called *Oregonator* which models the celebrated *Belusov–Zhabotinsky reaction*. The phase space is $\mathbb{R}^3_+$ (concentrations are nonnegative) and the (normalized) flow is given by

$$\begin{aligned} \frac{\mathrm{d}x}{\mathrm{d}t} &= s(x + y - xy - qx^2) \,, \\ \frac{\mathrm{d}y}{\mathrm{d}t} &= -\,(y + xy - fz)/s \,, \\ \frac{\mathrm{d}z}{\mathrm{d}t} &= w(x - z) \,, \end{aligned}$$

where s, q, f, and w are positive parameters. We refer to (Vidal and Lemarchand 1997) for more examples, references, and applications of dynamical systems to the study of chemical reactions.

Example 2.36. **Population Dynamics**
Dynamical systems also abound in ecology and biology. One of the most studied examples is the so-called *Lotka–Volterra* system, which describes the time evolution of the concentration of a predator–prey system. The phase space is $\mathbb{R}^2_+$ and the flow of the simplest (normalized) model is given by

$$\frac{\mathrm{d}x}{\mathrm{d}t} = x(1-y) ,$$
$$\frac{\mathrm{d}y}{\mathrm{d}t} = y(x-1) .$$

This model shows the famous oscillation for the concentration of predators (y) and preys (x). We refer to (Volterra 1990; Murray 2002; 2003) for more references and examples.

Example 2.37. **Partial Differential Equations**
The time evolution of many physical systems can be described by a nonlinear partial differential equation (PDE). As a short list we can mention hydrodynamics, nonlinear elasticity, and nonlinear optics. Chemical, biological, and ecological systems also lead to nonlinear PDEs when the spatial extension of the system is taken into account (for example by including diffusion). To give an explicit example, let us recall the Navier–Stokes equation, describing the motion of a viscous incompressible fluid in a finite container V.

The state of the system at time t is completely described by the velocity of the fluid in V. In other words, to each point $x \in V$ is attached a vector $\mathbf{v}(x,t)$ which is the velocity of the fluid particle which is at x at time t. The incompressibility assumption is equivalent to the condition $\operatorname{div} \mathbf{v} = 0$, and the presence of viscosity implies the boundary condition $\mathbf{v}\big|_{\partial V} = 0$ (∂V denotes the boundary of V). Thus, for viscous fluids, the normal component of $\mathbf{v}$ vanishes because the fluid cannot cross the wall, and the tangential component vanishes because the viscous friction would be infinite for nonzero tangential component. This is usually called the no-slip condition.

The time evolution of the velocity field $\mathbf{v}(t,x)$ is given by the Navier–Stokes equation

$$\partial_t \mathbf{v} = \mu \Delta \mathbf{v} - (\mathbf{v}\cdot\nabla)\mathbf{v} - \frac{1}{\varrho}\nabla p + \frac{\mathbf{f}}{\varrho} ,$$

where $\mu > 0$ is the viscosity, ϱ the (constant) density, $\mathbf{f}(t,x)$ an external force (gravity for example) acting in each point of the domain V, and $p(t,x)$ the local pressure. We recall that the pressure term ensures the incompressibility condition in the dynamics, and one should solve this equation in the space of divergence free vector fields. This leads to an equation for the pressure given by

$$\Delta p = -\varrho \operatorname{div}\big((\mathbf{v}\cdot\nabla)\mathbf{v}\big) ,$$

and hence to a closed equation for the velocity field $\mathbf{v}$, formally given by

$$\partial_t \mathbf{v} = \mu \Delta \mathbf{v} - (\mathbf{v}\cdot\nabla)\mathbf{v} + \nabla\Delta^{-1}\operatorname{div}\big((\mathbf{v}\cdot\nabla)\mathbf{v}\big) + \frac{\mathbf{f}}{\varrho} . \tag{2.8}$$

Note that the phase space Ω for this problem is infinite dimensional: It is the space of vector fields inside V with zero divergence and vanishing at the boundary. Even if we impose some regularity assumption (first and second derivative bounded for example), the dimension of this space is still infinite.

Exercise 2.38. *Assume V is a torus (or in other words one uses periodic boundary conditions). Use the decomposition in Fourier series to write (2.8) as a system of infinitely many coupled nonlinear differential equations.*

This infinite dimensional phase space creates difficulties in the analysis of the Navier–Stokes equation. We recall that in space dimension 3, it is still unknown if for any positive viscosity, Eq. (2.8) with bounded initial conditions leads to regular (bounded) solutions. This is known for large viscosity or for small enough initial conditions. Although these cases have a few interesting applications, these regimes are less interesting than the regime of small viscosity and large initial conditions where the phenomenon of *turbulence* develops.

The situation is somewhat better in dimension 2 where, for regular domains, regular forcings, and bounded initial conditions, the solutions are known to exist, to be unique and regular. We refer to (Constantin and Foias 1988) for more details. We will briefly comment on the results for these somewhat more complicated, although important, systems when appropriate.

3

Topological Properties

3.1 Coding, Kneading

In any experiment, one has to use a measuring apparatus which naturally has only a finite precision. In this section, we take the first steps to formalize this. One formalization will be based on partitioning the phase space into pieces and considering that the experiment will only tell us in which piece of the partition (called an atom) a point currently is, but not where inside this piece it happens to be. This is analogous to classical coarse-graining in statistical mechanics. (Later, we will also encounter partitions with different weights on the pieces.) Thus, we know only a fuzzy approximation of the true orbit of the system. As we shall see, one of the miracles appearing in hyperbolic systems is that this information alone, when accumulated over long enough time, will in fact tell us many details about the orbit. In physical applications, one often can observe only one orbit and the information one obtains is considered to be typical of the whole system. We discuss this in more detail in Chap. 9.

One can formalize this set of ideas in several ways, and two of them are particularly useful.

i) The phase space Ω is a metric space with metric d. For a given precision $\varepsilon > 0$, two points at distance less than ε are not distinguishable.
ii) One gives oneself a (measurable) *partition* $\mathcal{P}$ of the phase space,

$$\mathcal{P} = \{A_1, \ldots, A_k\} \text{ (k finite or infinite)} ,$$

with $A_j \cap A_\ell = \emptyset$ for $j \neq \ell$ and

$$\Omega = \bigcup_{j=1}^{k} A_j .$$

The elements of the partition are called *atoms*. Two points in the same atom of the partition $\mathcal{P}$ are considered indistinguishable. If a measure μ is given on the phase space, it is often convenient to use partitions modulo sets of μ-measure zero.

The notion of partition leads naturally to a *coding* of the dynamical system defined by the map f on the phase space Ω. This is a map Φ from Ω to $\{1,\ldots,k\}^{\mathbb{Z}^+}$ given by

$$\Phi_n(x) = \ell \qquad \text{if} \qquad f^n(x) \in A_\ell \ .$$

If the map is invertible, one can also use a bilateral coding. If $\mathcal{S}$ denotes the shift on sequences, it is easy to verify that $\Phi \circ f = \mathcal{S} \circ \Phi$. In general $\Phi(\Omega)$ is a complicated subset of $\{1,\ldots,k\}^{\mathbb{Z}^+}$; i.e., it is difficult to say which codes are admissible. There are however some examples of very nice codings like for Axiom A attractors (see (Bowen 1975; Keane 1991; Ruelle 2004)).

Example 3.1. This provides details to Exercise 2.22. For the map $f : x \mapsto 2x \pmod 1$, the phase space is $\Omega = [0,1)$ and one can take the partition $\mathcal{P} = I_0 \cup I_1$ with $I_0 = [0,1/2)$, $I_1 = [1/2,1)$. In this case, the coding of x is the index of the set I_j in which x lies. This is simply the first digit of the binary representation of the real number x.

Note that boundary points of the pieces I_0 and I_1 have ambiguous representations (and the same is true for their preimages under f) since, for example, the point $x = 1/2$ has the representations

$$x = 0.0111111\ldots = 0.1000\ldots \ .$$

Take now a point $x \in [0,1)$, then, apart from the ambiguity mentioned above, the representation of x in the number system of base 2 is $x = 0.i_1i_2i_3\ldots$ with $i_\ell \in \{0,1\}$. The nice thing is that f maps x to x' as

$$x = 0.i_1i_2i_3\ldots \mapsto f(x) = x' = 0.i_2i_3i_4\ldots \ .$$

Thus, the coding of f is a *shift* of the digits (to the left). Note furthermore the obvious relation

$$x = \sum_{\ell=1}^{\infty} i_\ell 2^{-\ell} \ .$$

Note that the point $f^{n-1}(x)$ is in $I_0 = [0,\frac{1}{2})$ if $i_n = 0$ and in $I_1 = [\frac{1}{2},1)$ if $i_n = 1$, for $n = 1,2,\ldots$.

Example 3.2. This provides details for Exercise 2.26 For the baker's transformation, the coding is as follows: One chooses $I_0 = \{(x,y) \mid x \in [0,\frac{1}{2})\}$, and $I_1 = \{(x,y) \mid x \in [\frac{1}{2},1)\}$. If

$$x = 0.i_1i_2\ldots \text{ and } y = 0.j_1j_2\ldots$$

then write the formal real number (in binary notation)

$$z = \ldots j_3j_2j_1.i_1i_2i_3\ldots$$

and its image z' is then just multiplication by 2; i.e., it is again a shift, but this time on the bi-infinite sequences:

$$z' = \ldots j_3 j_2 j_1 i_1 . i_2 i_3 \ldots .$$

For the coordinates, this leads to

$$x' = 0.i_2 i_3 i_4 \ldots \text{ and } y' = 0.i_1 j_1 j_2 \ldots .$$

Note that the periodic points are dense: Given x, y as above and a small ε, which we assume of the form $\varepsilon = 2^{-n+1}$, we will find a periodic point within ε of x, y. To do this, define

$$\hat{x} = 0.i_1 \ldots i_n j_n \cdots j_2 j_1 \underbrace{i_1 \cdots i_n j_n \cdots j_2 j_1}_{\text{indefinitely repeated}} ,$$

$$\hat{y} = 0.j_1 \ldots j_n i_n \cdots i_2 i_1 \underbrace{j_1 \cdots j_n i_n \cdots i_2 i_1}_{\text{indefinitely repeated}} .$$

Clearly, $|x - \hat{x}| \leq 2^{-n+1}$, $|y - \hat{y}| \leq 2^{-n+1}$, and, by construction, $(\hat{x}, \hat{y})$ is periodic of period $2n$ (just shift). Note that the periodic points have rational coordinates.

Example 3.3. The cat map f has a finite coding with 4 pieces that is not quite so simple, which we will discuss in detail later. But the curious reader can already peek at Fig. 4.11 which shows a useful partition for that case. In analogy to the previous examples, we mention the recurring fact that in such (hyperbolic) systems the periodic points are dense: For the cat map, it is again exactly the set of points with rational coordinates. To see it, consider any point (the letters a, b, p mean integers here)

$$(x, y) = \left(\frac{a}{p}, \frac{b}{p}\right) \mapsto \left(\frac{2a+b}{p}, \frac{a+b}{p}\right) \pmod 1 .$$

Note that the denominator p does not change under the mapping, and therefore there are at most p^2 different points. Thus, there must exist integers q, s, a', and b' such that

$$f^s\left(\frac{a}{p}, \frac{b}{p}\right) = \left(\frac{a'}{p}, \frac{b'}{p}\right) ,$$

and

$$f^q\left(\frac{a'}{p}, \frac{b'}{p}\right) = \left(\frac{a'}{p}, \frac{b'}{p}\right) .$$

Since f is invertible, f acts as a permutation of the p^2 different points, this shows that $(a/p, b/p)$ is a periodic point (of a period which divides p^2). Since p, a, and b are arbitrary, the assertion follows.

3.2 Topological Entropy

When describing complexity one has two choices: Topology or metric spaces. In general, we will work with metric spaces, but some concepts need only topology. Furthermore, we will need sometimes spaces with a measure, and in such a case, measure spaces are adequate.

We recall the basic axioms of these points of view.

3.2.1 Topological, Measure, and Metric Spaces

For the convenience of the reader, we recall the main differences between topological and measure spaces (see Table 3.1). While the former are more general, the latter occur more frequently in applications. A *metric space* M is a space with a distance

Table 3.1. Comparison of topological axioms and the axioms of Borel sigma-algebras

	Topological space	Measure space
Usage	Continuity	Integration
Basic ingredients	Open sets $\mathcal{O}$, closed sets	Borel sets B
		(Unions, intersections of intervals)
Unions	$\cup_i \mathcal{O}_i = \mathcal{O}$	$\cup_{\text{countable}} B_i = B$
Intersections	$\cap_{\text{finite}} \mathcal{O}_i = \mathcal{O}$	$\cap_{\text{countable}} B_i = B$
Complement	$\mathcal{O}^{\text{c}}$ = closed	B^{c} also a Borel set
Property	f is continuous if $f^{-1}(\mathcal{O}) = \mathcal{O}'$	f is measurable if $f^{-1}(B) = B'$
Measures		$\mu(\cup_{\text{disjoint}} B_i) = \sum_i \mu(B_i)$

function $d : M \times M \to \mathbb{R}^+$, namely a non-negative symmetric function satisfying the *triangle inequality* $d(x, y) + d(y, z) \geq d(x, z)$, and such that $d(x, y) = 0$ if and only if $x = y$. The open balls in a metric space obviously define a topology.

Example 3.4. A class of metrics we use quite often on sequences of symbols is given as follows: For two sequences $\mathbf{x} = \{x_n\}$ and $\mathbf{y} = \{y_n\}$ we denote by $\delta(\mathbf{x}, \mathbf{y})$ the smallest index (in absolute value) where these two sequences differ, namely

$$\delta(\mathbf{x}, \mathbf{y}) = \min \left\{ |q| \,\middle|\, x_q \neq y_q \right\} . \tag{3.1}$$

For a given number $0 < \zeta < 1$ we define a distance d_ζ (denoted simply by d when there is no ambiguity in the choice of ζ) by

$$d_\zeta(\mathbf{x}, \mathbf{y}) = \zeta^{\delta(\mathbf{x}, \mathbf{y})} .$$

3.2.2 Some Examples

As we will see several times, a recurrent theme in the study of dynamical systems is counting orbits. The idea is that a system with complex behavior has many different orbits. Here, we start with a first approach to counting pieces of a partition, which will lead to a precise version of the previous idea. This procedure is at the basis of the idea of topological entropy.

Example 3.5. **The interval map** $f : x \mapsto 2x \pmod 1$
Define

$$Q(i_0, \ldots, i_{n-1}) = \{x \mid f^j(x) \in I_{i_j} \text{ for } j = 0, \ldots, n-1\} .$$

Here, as before, $I_0 = [0, \frac{1}{2})$ and $I_1 = [\frac{1}{2}, 1)$. For example,

$$Q(101) = \{x \in \mathbb{R} \mid x = 0.101\ldots\} = [5/8, 6/8] .$$

Nonempty Q's describe which intervals the orbits visit, and in which order. In particular, for our map and for any choice of $i_0, \ldots, i_{n-1}$, one has $Q(i_0, \ldots, i_{n-1}) \neq \emptyset$, (in fact $Q(i_0, \ldots, i_{n-1})$ is an interval whose Lebesgue measure is $|Q(i_0, \ldots, i_{n-1})| = 2^{-n}$) and in particular

$$N_n \equiv \operatorname{card}\{\{i_0, \ldots, i_{n-1}\} \mid Q(i_0, \ldots, i_{n-1}) \neq \emptyset\} \tag{3.2}$$

satisfies

$$N_n = 2^n ,$$

where $\operatorname{card} S$ denotes the *cardinality* of the set S. Thus, N_n counts the number of different patterns of length n of visits to I_0 and I_1 which the orbits of f can produce. The topological entropy is then defined by

$$h_{\text{top}} = \lim_{n \to \infty} \frac{1}{n} \log N_n ,$$

which in the present case leads to $h_{\text{top}} = \log 2$. This means that there are $N_n \approx \mathrm{e}^{n h_{\text{top}}} = \mathrm{e}^{n \log 2}$ different types of orbits.

Example 3.6. **The circle rotation:** $f : x \mapsto x + \omega$
We consider the case of irrational ω. If we define N_n as in (3.2), and I_0, I_1 as before, but now for the rotation map, we find that $N_n = 2(n + 1)$, which is much smaller than in the preceding example. We call this the "*Swiss Chocolate Theorem*": Swiss chocolate has 24 pieces, and to get them all you must break the tablet 23 times, since each break produces exactly one new piece, no matter in which order you break. For the case of the map f, since the preimages are unique, and the boundaries of the Q_n's are those points which land on either $0 (= 1)$ or $\frac{1}{2}$, we see that there are exactly $2(n + 1)$ of them (as each preimage "breaks" exactly one of the intervals which already exist, and we start with 2 pieces). Thus

$$h_{\text{top}} = \lim_{n \to \infty} \frac{1}{n} \log N_n = 0$$

and therefore the number of possible types of orbits grows slower than any exponential.

Remark 3.7. One can also see this as the phenomenon of "omitted digits." By the nature of the map, the sequences of consecutive zeros are bounded by some number, m. This is indeed a very strong restriction on the set of possible sequences. An amusing example is as follows: The sum $\sum_{n=1}^{\infty} \frac{1}{n}$ is well known to diverge. However, if we restrict the sum to those n that do not contain the digit 9 in their decimal expansion, then it is bounded by 27.2.[1]

We refer to (Fogg 2002) and references therein for other examples of topological and symbolic dynamics.

3.2.3 General Theory of Topological Entropy

This subsection is mostly after (Denker, Grillenberger, and Sigmund 1976).

Open covers. Let

$$\mathcal{U} = \{U_\alpha\}$$

be a covering of a compact space X by open sets U_α, where α is in some countable index set, that is $\cup_\alpha U_\alpha = X$. This is called an *open cover*. Let

$$\mathcal{U}' = \{U'_\beta\}$$

be another such cover. Then one says that $\mathcal{U}'$ is *finer* than $\mathcal{U}$, written as

$$\mathcal{U}' \geq \mathcal{U} ,$$

if for each α there is a β for which $U'_\beta \subset U_\alpha$. One calls a cover $\mathcal{U}'$ a *subcover* of $\mathcal{U}$, written as $\mathcal{U}' \subset \mathcal{U}$, if $U'_\beta \in \mathcal{U}'$ implies that also $U'_\beta \in \mathcal{U}$. In other words, U'_β is one of the U_α. Finally,

$$\mathcal{U} \vee \mathcal{U}'$$

is the common *refinement* of $\mathcal{U}$ and $\mathcal{U}'$ and is the cover whose elements are the $U_\alpha \cap U'_\beta$, empty sets being omitted. If f is a continuous surjective map, then $f^{-1}(\mathcal{U}) = \{f^{-1}(U_\alpha)\}$.

Example 3.8. Consider again the map $f : x \mapsto 2x \pmod 1$ and $\mathcal{U} = \{I_0, I_1\}$, with I_0, I_1 as before. This is not an open cover, just two open sets which almost cover $[0, 1]$, but for the example, this is good enough. Clearly, in binary notation,

$$\begin{aligned} I_0 &= \{x \mid x = 0.0\ldots\} , \\ I_1 &= \{x \mid x = 0.1\ldots\} , \end{aligned}$$

where for the dots one can choose any sequence of 0's or 1's (except all 1's). Then

$$\begin{aligned} f^{-1}(I_0) &= \{x \mid x = 0.00\ldots \text{ or } x = 0.10\ldots\} , \\ f^{-1}(I_1) &= \{x \mid x = 0.01\ldots \text{ or } x = 0.11\ldots\} . \end{aligned}$$

[1] Better bounds are possible, this one is $(1 - \frac{9}{10})^{-1} \sum_{n=1}^{8} \frac{1}{n}$.

We next define $N(\mathcal{U})$ as the *minimal* number of pieces of all subcovers of $\mathcal{U}$. That is, a subset of the set $\{U_\alpha\}$ with a minimal number of pieces which covers X. Finally, one defines

$$H(\mathcal{U}) = \log N(\mathcal{U}) .$$

The following results are easy to check:

Proposition 3.9.

$$H(\mathcal{U}) \ge H(f^{-1}(\mathcal{U})) .$$

Proposition 3.10.

$$H(\mathcal{U} \vee \mathcal{U}') \le H(\mathcal{U}) + H(\mathcal{U}') .$$

One next defines

$$(\mathcal{U})_0^n \equiv \mathcal{U} \vee f^{-1}(\mathcal{U}) \vee \cdots \vee f^{-n}(\mathcal{U}) ,$$

and then $N\big((\mathcal{U})_0^{n-1}\big)$ is just the smallest cardinality of a family of n-tuples of elements of $\mathcal{U}$ such that for any $x \in X$ there exists an n-tuple $(U_{\alpha_0}, \dots, U_{\alpha_{n-1}})$ so that

$$f^k(x) \in U_{\alpha_k} \text{ for } k = 0, \dots, n-1 .$$

Lemma 3.11. *For any open cover $\mathcal{U}$ the limit*

$$H(\mathcal{U}, f) \equiv \lim_{n\to\infty} \frac{1}{n} \log N((\mathcal{U})_0^{n-1}) \tag{3.3}$$

exists.

$H(\mathcal{U}, f)$ is called the *topological entropy* of the map f for the open cover $\mathcal{U}$.

Definition 3.12. *The expression*

$$h_{\text{top}}(f) = \sup\{H(\mathcal{U}, f) \mid \mathcal{U} \textit{ an open cover of } X\} ,$$

is called the **topological entropy of** f.

Remark 3.13. For the Example 3.8 one has $H(\mathcal{U}, f) = \log 2$ as well as $h_{\text{top}} = \log 2$.

Proof of Lemma 3.11. One has, by Props. 3.9 and 3.10,

$$H((\mathcal{U})_0^1) = H(\mathcal{U} \vee f^{-1}(\mathcal{U})) \le H(\mathcal{U}) + H(f^{-1}(\mathcal{U})) \le 2H(\mathcal{U}) ,$$

so that $\frac{1}{n} H((\mathcal{U})_0^{n-1}) \le H(\mathcal{U})$ by induction. If we call $h_n = H((\mathcal{U})_0^{n-1})$, then we have

$$h_{m+n} \le h_m + h_n , \tag{3.4}$$

which is called the *subadditivity property*. It implies

$$\lim_{n\to\infty} \frac{h_n}{n} = \inf_m \frac{h_m}{m} . \tag{3.5}$$

Proof of Eq. (3.5). Fix $m \in \mathbb{N}$. Every $n \in \mathbb{N}$ can be uniquely decomposed as $n = km + \ell$ with k and $\ell < m$ nonnegative integers. Then, by applying repeatedly (3.4), one gets

$$h_n \le kh_m + \ell h_1 \,,$$
$$\frac{h_n}{n} \le \frac{h_m}{m} + \frac{\ell}{n} h_1 \,,$$

so that

$$\limsup_{n\to\infty} \frac{h_n}{n} \le \frac{h_m}{m} \,.$$

Since this is true for every m and $\inf \le \liminf$, we get

$$\limsup_{n\to\infty} \frac{h_n}{n} \le \inf_m \frac{h_m}{m} \le \liminf_{n\to\infty} \frac{h_n}{n} \,,$$

and this means that the limit in (3.5) exists and is equal to the infimum.

Translating all this back to $h_n = H((\mathcal{U})_0^{n-1}) = \log N((\mathcal{U})_0^{n-1})$, we see that (3.3) holds.

3.2.4 A Metric Version of Topological Entropy

This is due to Bowen and Dinaburg (Bowen 1971; Dinaburg 1971). We now assume the space X has a distance function $d(x, y)$. All we need is the triangle inequality. In addition, we fix a map f.

Definition 3.14. *The finite set $S_{n,\varepsilon}$ of points is called* (n, ε)-separated *(also called* ε-different before time n*) if for all $x \neq y \in S_{n,\varepsilon}$ there is a k, $0 \le k < n$, for which*

$$d\big(f^k(x), f^k(y)\big) > \varepsilon \,.$$

We let $s(n, \varepsilon) = \max |S_{n,\varepsilon}|$, where the maximum is over all possible choices of $S_{n,\varepsilon}$.

The idea is to place as many points as possible into X but to make sure that the orbits of any 2 of them are at least once separated by ε during the first $n - 1$ iterations.

Example 3.15. If $X = [0, 1)$, then obviously, $s(0, \varepsilon) \approx 1/\varepsilon$, since we can just place that many points at distance ε, and we are considering 0 iterations.

Example 3.16. For the rotation $x \mapsto x + \omega \pmod 1$, one finds that

$$s(n, \varepsilon) \approx \frac{1}{\varepsilon} \,,$$

for all n, because for $n = 0$ we are in the case of Example 3.15 and because the rotation map preserves distances between points.

Example 3.17. For the map $x \mapsto 2x \pmod 1$, one has, e.g.,

$$S_{3,\frac{1}{4}} = \left\{0, \frac{1}{8}, \frac{2}{8}, \ldots, \frac{7}{8}\right\} ,$$

namely the points $\{0.000, 0.001, \ldots, 0.111\}$. (The images are at least once at distance $\frac{1}{2}$.) Generalizing, we find

$$s\left(n, \frac{1}{4}\right) \sim 2^n .$$

Definition 3.18. *One defines*

$$s(\varepsilon) = \limsup_{n\to\infty} \frac{1}{n} \log s(n, \varepsilon) ,$$

and then

$$h_{\mathrm{d}} = \lim_{\varepsilon \searrow 0} s(\varepsilon) .$$

The limit exists because $s(\varepsilon)$ is a nonnegative decreasing function of ε.

Proposition 3.19. *If X is a compact space then $h_{\mathrm{top}} = h_{\mathrm{d}}$.*

We first need a second, "dual" definition to $s(n, \varepsilon)$.

Definition 3.20. *The finite set $G_{n,\varepsilon}$ of points is called (n, ε)-*spanning *(we use G for generating) if for every $x \in X$, there is a $y \in G_{n,\varepsilon}$ such that for all k, $0 \le k < n$,*

$$d\big(f^k(x), f^k(y)\big) \le \varepsilon . \tag{3.6}$$

We let $g(n, \varepsilon) = \inf |G_{n,\varepsilon}|$, where the minimum is over all possible $G_{n,\varepsilon}$.

The idea here is that the points in $G_{n,\varepsilon}$ are sufficiently dense so that the first n points of the orbit of *any* $x \in X$ are not more than ε away from the orbit of some point in $G_{n,\varepsilon}$.

Although the two definitions seem quite different, they are closely related: One has the inequality

Lemma 3.21. $g(n, \varepsilon) \le s(n, \varepsilon) \le g(n, \varepsilon/2)$.

Proof of Lemma 3.21.

i) Every $S_{n,\varepsilon}$ which maximizes $s(n, \varepsilon)$ is a $G_{n,\varepsilon}$. Indeed, suppose the contrary. Since $S_{n,\varepsilon}$ cannot serve as a $G_{n,\varepsilon}$, there is at least one x such that (3.6) fails for all $y \in S_{n,\varepsilon}$, that is for some k, $0 \le k < n$, one has $d(f^k(x), f^k(y)) > \varepsilon$. But this means that x can be added to $S_{n,\varepsilon}$, which contradicts the assumption that it was maximal. So, the first inequality of Lemma 3.21 is proved.

ii) Here, we use the triangle inequality: Let $G_{n,\varepsilon/2}$ be a set which realizes the minimum $g(n,\varepsilon/2)$. Then, for all $x \in X$ there is a $z(x) \in G_{n,\varepsilon/2}$ such that (3.6) holds:

$$d(f^k(x), f^k(z(x))) \leq \varepsilon/2 \text{ for all } k = 0, \ldots, n-1 \,.$$

If $x_1 \neq x_2 \in S_{n,\varepsilon}$, then it follows that $z(x_1) \neq z(x_2)$, since otherwise

$$d(f^k(x_1), f^k(x_2)) \leq d(f^k(x_1), f^k(z(x_1))) + d(f^k(z(x_2)), f^k(x_2)) \leq \varepsilon \,,$$

for all $k = 0, \ldots, n$ which contradicts the definition of $S_{n,\varepsilon}$. Thus $|S_{n,\varepsilon}| \leq |G_{n,\varepsilon/2}|$, and the second inequality of Lemma 3.21 is shown as well. □

Proof of Prop. 3.19.

i) We first show $h_{\mathrm{d}} \leq h_{\mathrm{top}}$. Given an $\varepsilon > 0$, let S be an (n,ε)-separated subset of X, and let $\mathcal{U}$ be an open cover of X by sets U_α of diameter less than ε. By construction of S, two points $x_1 \neq x_2$ cannot lie in the same n-tuple $(U_{\alpha_0}, \ldots, U_{\alpha_{n-1}})$ and therefore $s(n,\varepsilon) \leq N((\mathcal{U})_0^{n-1})$ from which we conclude $s(\varepsilon) \leq H(\mathcal{U}, f)$.

ii) The inequality $h_{\mathrm{top}} \leq h_{\mathrm{d}}$. Let $\mathcal{U}$ be an open cover of X. There is an $\varepsilon > 0$ such that for any $x \in X$ the ball $B_\varepsilon(x)$ of radius ε centered at x is contained in *one* $U_{\alpha(x)}$. We now use a G which is a minimal (n,ε)-spanning set with $|G_{n,\varepsilon}| = g(n,\varepsilon)$. Fix $z \in G$ and choose for each $k = 0, \ldots, n-1$ an element $U_{\alpha_k(z)}$ of $\mathcal{U}$ which contains the ball $B_\varepsilon(f^k(z))$. By the definition of G, there is for every $x \in X$ a $z \in G$ so that $f^k(x) \in B_\varepsilon(f^k(z))$, for $k = 0, \ldots, n-1$. Thus, $f^k(x) \in U_{\alpha_k(z)}$ and the family

$$\{U_{\alpha_0(z)} \cap \cdots \cap f^{-(n-1)}(U_{\alpha_{n-1}(z)}) \mid z \in G\}$$

is a subcover of $(\mathcal{U})_0^{n-1}$. Therefore, $N((\mathcal{U})_0^{n-1}) \leq |G| = g(n,\varepsilon)$ and the assertion follows. □

Useful properties. We enumerate here a few useful consequences (without proof) of the above definitions and results.

i) For two different (but equivalent) metrics d, d' on X one has

$$h_d = h_{d'}$$

which shows that the complexity measured by topological entropy does not depend on how the distances are measured (and we can omit the index d).

ii)

$$h(f^k) = kh(f) \,.$$

iii) If f is a Lipschitz map on $\mathbb{R}^p$ with $|f(x) - f(y)| < C \cdot |x - y|$, then

$$h(f) \leq \max(0, p \log C) .$$

(In particular, for the map $f : x \mapsto 2x \pmod 1$ we get the bound $h(f) \leq \log 2$, which is saturated as we have seen before.)

iv) For *unimodal maps*, such as $f : x \mapsto 1 - \nu x^2$ on $[-1, 1]$ (with $\nu \in [0, 2]$), we let $M_n(f)$ be the number of extrema of the graph of f^n. Then

$$h_{\text{top}}(f) = \lim_{n \to \infty} \frac{1}{n} \log M_n(f) .$$

v) The topological entropy is a topological invariant:

$$h(f) = h(g^{-1} \circ f \circ g) ,$$

where g is any continuous coordinate change. The importance of this for physics is that clearly, h does not depend on any deformation of the signal by a measuring apparatus.

vi) One would like to know whether a system is chaotic, so one would like to have *lower* bounds on h. Those are difficult to get. The most spectacular such result is the following

Theorem 3.22. *If f is unimodal, and has a periodic orbit whose period is of the form $p \cdot 2^n$, with p odd, $p \geq 3$ and n an integer, then $h(f) > 2^{-n} \log(\lambda_p)$ where λ_p is the largest positive solution of $x^p - 2x^{p-2} - 1 = 0$. Thus $h > 0$ which means that one has topological chaos.*

This theorem is a generalization of a celebrated result of Li and Yorke (Li and Yorke 1975): "Period 3 implies chaos."

Up to now, we have only *counted* how often certain regions are reached, but the topological entropy has no notion of how long one stays in a given region. In this respect, it corresponds to the microcanonical ensemble, which gives equal weight to all points. Starting from Chap. 5, we will consider these weights sense.

We come back to the Definition 3.18 as

$$h_{\text{top}} = \lim_{\varepsilon \searrow 0} \limsup_{n \to \infty} \frac{1}{n} \log N_n(\varepsilon) ,$$

where $N_n(\varepsilon) = s(n, \varepsilon)$ of that definition. We show in Fig. 3.1 the topological entropy as a function of the parameter for the one-parameter family f_ν of quadratic maps of Example 2.18.

A positive topological entropy means that there are a large number of qualitatively different trajectories. It is an indication of chaos. A transverse homoclinic crossing already implies a positive topological entropy; we introduce this concept in Sect. 4.4.2. However, this chaos may occur in a very small, irrelevant, part of the

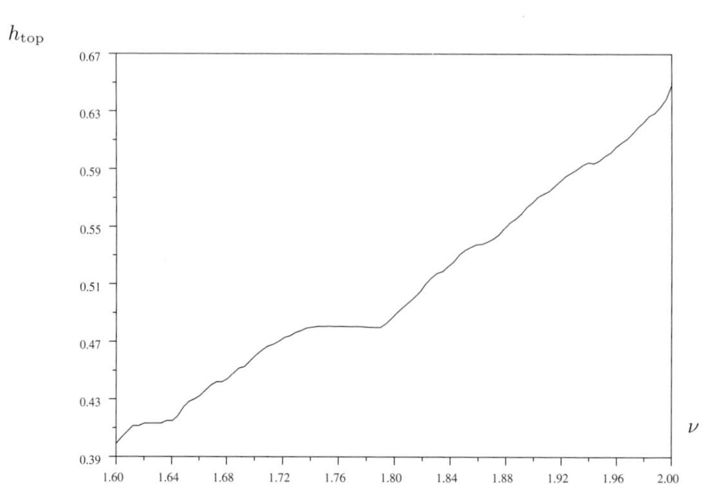

Fig. 3.1. Topological entropy as a function of the parameter for quadratic family of Example 2.18. The entropy is known to be monotonically increasing, with flat pieces corresponding to the "windows" one can see in Fig. 3.5

phase space. For example, one can have a stable periodic orbit whose *basin of attraction* (the set of points that converge to the periodic orbit) is almost all the phase space and a fractal repeller of small dimension, zero volume, supporting all the positive topological entropy. This is what happens, for example, for (unimodal) maps of the interval with a stable periodic orbit of period 3, and has led to the famous statement that period 3 implies chaos (see (Li and Yorke 1975) and, e.g., (Collet and Eckmann 1980) for its generalization). As we have already mentioned, in general one does not observe all the trajectories but only the typical trajectories with respect to a measure (a Physical measure for example).

3.3 Attractors

In the study of dynamical systems there is an interplay between the geometric and ergodic approach. One of the first examples comes from the notion of attractor in dissipative systems. Several possible definitions for an attractor exist. We adopt the following one, formulated for discrete time evolution. There is an analogous notion for continuous time evolution. We say that a subset $\mathscr{A}$ of the phase space is invariant under the map f if $f(\mathscr{A}) \subset \mathscr{A}$.

Definition 3.23. *A (compact) invariant subset $\mathscr{A}$ of the phase space is an* attracting set *if there is a neighborhood V of $\mathscr{A}$ such that for any neighborhood U of $\mathscr{A}$, there is an integer n_U such that for any $n > n_U$, $f^n(V) \subset U$.*

In particular, all orbits with initial condition in V accumulate on $\mathscr{A}$.

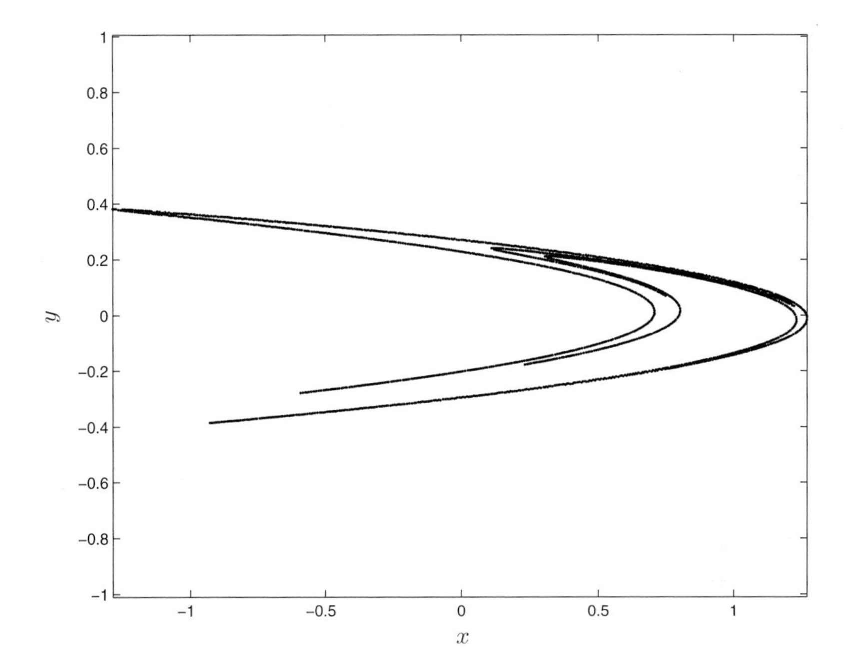

Fig. 3.2. Attractor of the Hénon map, $a = 1.4$, $b = 0.3$. Starting from the point $(x, y) = (0.1, 0.3)$ one iterates first 200 times without plotting to "reach" the attractor. Shown is a representative subset of the next 100,000 iterations

Definition 3.24. *We will say that a (compact) invariant subset $\mathscr{A}$ of the phase space is an* attractor *if it is an attracting set containing a dense orbit.*

We refer to (Ruelle 1989b) and (Guckenheimer and Holmes 1990) for more on these notions. The *basin of attraction* of an attracting set $\mathscr{A}$ is the set of initial conditions whose orbit accumulates on $\mathscr{A}$. Attractors may have complicated geometry in which case they are called *strange attractors* (see also (Eckmann and Ruelle 1985a; Milnor 1985a;b) for further discussions of how to define attractors). We now present some pictures of attractors.

Figure 3.2 is a plot of the "attractor" for the Hénon map of Example 2.31. This picture was obtained by starting with an initial condition at $(0.1, 0.3)$, iterating a large number of times to be as near as possible to the attractor, and then plotting a certain number of the following iterations supposed to visit densely the attractor.

A subset V for checking that we have an attracting set and its first two iterates are shown in Fig. 3.3, again for the Hénon map. The attractor of the Lorenz system (2.7) is shown in Fig. 3.4.

Remark 3.25.

i) The simplest attractors are the stable fixed points (stationary solutions) and the stable periodic orbits (stable invariant cycles).

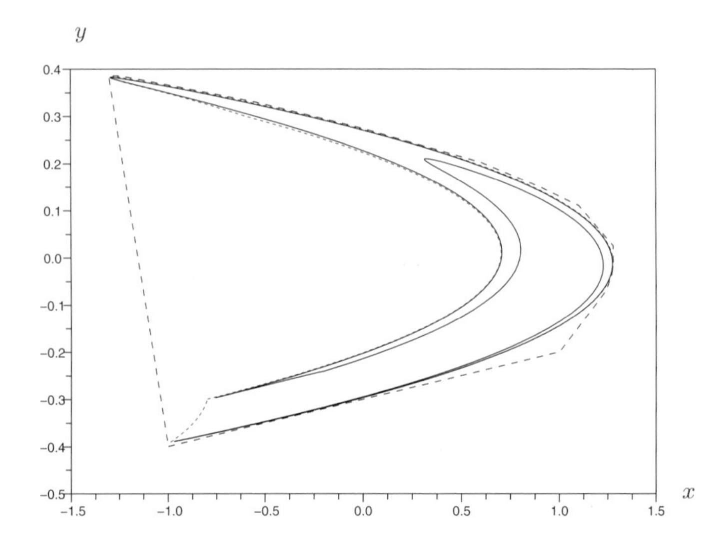

Fig. 3.3. The neighborhood V (dashed blue) and two of its iterates (dashed red and black) for the Hénon attractor, $a = 1.4$, $b = 0.3$

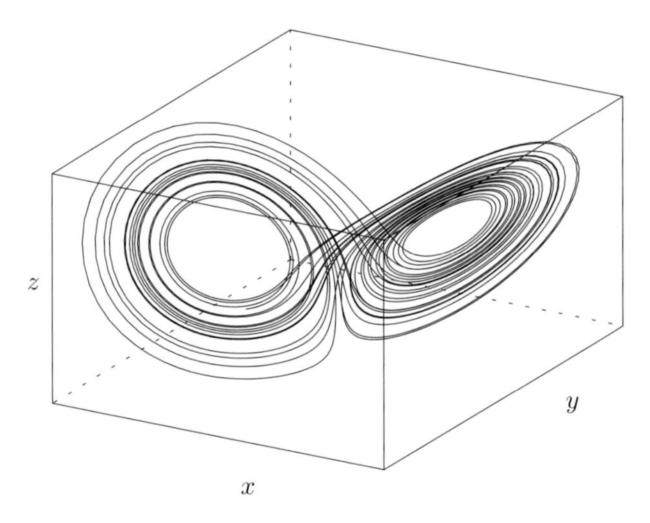

Fig. 3.4. Attractor of the Lorenz system, for the parameter values $\sigma = 10$, $r = 28$, and $b = 8/3$

ii) If the dynamical system depends on a parameter, the attractor will also in general change with the parameter, not only quantitatively but also qualitatively as occurs for example in bifurcations. In Fig. 3.5 we draw the attractor of the quadratic family of Example 2.18 as a function of the parameter. One sees in particular the famous sequence of period doubling bifurcations.

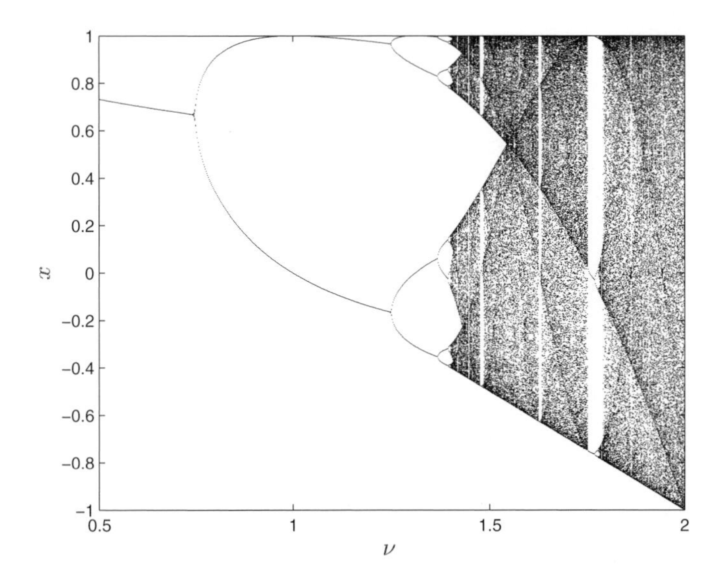

Fig. 3.5. The attractors of the quadratic family as a function of the parameter ν. We chose 600 equally spaced values of ν. For each of them, we start at $x_0 = 0.3$ and iterate the map $f_\nu : x \mapsto 1 - \nu x^2$ for 600 times, leading to $x_{600} = f_\nu^{600}(x_0)$. We then plot the next 300 points x_{601} to x_{900} on the vertical axis

iii) A dynamical system may have several attractors in the same phase space, as for example the pendulum of Fig. 1.1.

iv) Each attractor has its own basin: this is the set of initial conditions that are attracted to it.

v) The boundaries of the basins are in general complicated invariant sets, repelling transversally (toward the attractors). A well-known example is provided by the Newton method applied to the equation $z^3 = 1$. The map, in the (extended) complex plane, is given by $z_{n+1} = f(z_n) = z_n - (z_n^3 - 1)/(3z_n^2) = (2z_n + 1/z_n^2)/3$. It has the three stable fixed points (attractors): $\mathrm{e}^{2\pi \mathrm{i} n/3}$, $n = 0, 1, 2$ (which are the solutions of $z^3 = 1$). Figure 3.6 is a drawing of the three basins of attraction of these three fixed points. In other words, just as with the pendulum example of Chap. 1, the three basins of attraction are open sets whose boundaries coincide: Each boundary point of two countries is also a boundary point of the third one. The boundary points are found as the successive preimages of $z = 0$; see also (Eckmann 1983).

vi) The concept of attractor is only suitable for the study of dissipative systems. Volume preserving systems do not have attractors, in particular mechanical systems without friction do not have attractors.

Exercise 3.26. *Show that the attractor of the dissipative baker's map (2.6) is the product of a* Cantor set *by a segment (see Fig. 2.12). The reader who is not familiar*

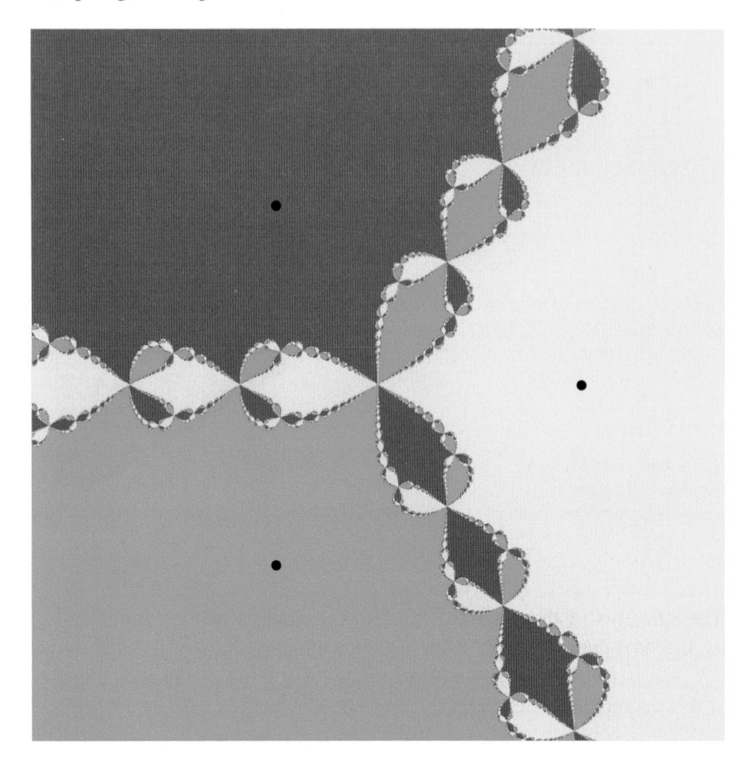

Fig. 3.6. The three attracting regions for the map of the Newton algorithm which finds the roots of $z^3 = 1$. The picture shows the domain $|\mathrm{Re}\, z|$, $|\mathrm{Im}\, z| < 2$. The three fixed points $\exp(2\pi i n)$, $n = 0, 1, 2$, are indicated in black. Note that the boundary points of the three regions coincide

with Cantor sets can peek at Exercise 5.19 and Fig. 5.5 to get a preliminary idea of the concept.

We now make some comments about the attractors for partial differential equations (see Example 2.37). As was mentioned before, in this case the phase space is infinite dimensional, and compactness is a much stronger requirement than being bounded. It turns out that in many situations, it has been rigorously shown that there exists a compact attractor. This holds in particular for dissipative systems. The reason is that the time evolution often regularizes the solution. Consider for example the Navier–Stokes equation (2.8) in space dimension 2 with periodic boundary conditions (in other words on the torus) with a regular forcing. One can show that if one starts with a bounded but not so regular initial condition, then in a finite time the solution becomes very regular, even analytic if the forcing is analytic. This regularizing effect is a source of compactness often used in the proofs of existence of compact attractors. We refer to (Temam 1997; Ruelle 1984) for more details.

Note however that a compact set in an infinite dimensional phase space can be somewhat weird. It can look somewhat like a sea urchin throwing pins in infinitely many orthogonal directions. A better result would be to know that an attractor is finite dimensional (we discuss more precisely in Sect. 6.3 the notion of dimension of a set). Unfortunately, a set of finite dimension in an infinite dimensional space can also be pretty weird, although we will see later on (see Mañé's Theorem in Sect. 9.4), that it can be faithfully represented in a finite dimensional space.

A better notion has been investigated which is called an *inertial manifold*. This is a compact piece of (finite dimensional) manifold (hypersurface), which is an attracting set (and hence invariant) (see (Foiaş and Temam 1979)). In particular, it attracts transversally all the orbits. For some partial differential equations, the existence of such inertial manifolds has been proven. Unfortunately, this is still an open question for the 2-dimensional Navier–Stokes equation. Another drawback is that often one is unable to show that these inertial manifolds are regular enough. The picture is however quite nice: there is a finite dimensional manifold that attracts all the trajectories (at least locally) and which therefore contains the asymptotic dynamics. It is therefore enough to restrict the study of the system to these finite dimensional objects. In that sense we have finally gotten rid of the infinite dimensional phase space. We refer to (Temam 2001; Novo, Titi, and Wynne 2001) for results and applications to numerical simulations.

From a physical point of view, partial differential equations with their infinite dimensional phase space have infinitely many degrees of freedom. If there exists a finite dimensional inertial manifold, the picture is much nicer. The coordinates along the manifold describe the relevant degrees of freedom, while the transverse directions (which are contracted at least on average) describe the infinitely many degrees of freedom which are slaved to the relevant ones.

4

Hyperbolicity

This chapter deals with the sources of chaotic behavior. The theory relies on two fundamental concepts. The first, called hyperbolicity, deals with the issue of instability. It generalizes the notion of unstable (or stable) fixed point, to points which are neither fixed nor periodic. The second concept is called Axiom A (of Smale) and its definition asks for an abundance of periodic points (in a hyperbolic set). Important consequences of this definition are a certain stability of the results with respect to (small) changes of the dynamics, and this is of course very important for physical applications when the exact equations are not known. As a second consequence, one can obtain for Axiom A systems a coding of the type we have seen in Example 3.8 for the special case of the map $f : x \mapsto 2x \pmod{1}$.

One of the earliest examples of a dynamical system (flow) with hyperbolic behavior is the *geodesic flow* on the tangent space of compact surfaces with constant negative curvature. The norm of the velocity is conserved (the analog of the conservation of the energy), and one can look at the motion on the unit tangent bundle. In this time evolution, each point moves along a geodesic on the manifold as time progresses, and the velocity direction changes accordingly. The negative curvature makes the orbits of nearby points separate exponentially. Many ideas and concepts of dynamical systems were first developed for the geodesic flow and then extended to general hyperbolic systems. We refer to (Arnold and Avez 1967; Anosov 1969) and (Katok and Hasselblatt 1995) for more references, the history, and the detailed study of these systems.

We begin by looking at the tangent map for a smooth dynamical system. We consider a compact manifold M (of dimension d). The *tangent bundle* (the collection of tangent spaces) is defined as

$$\mathrm{T}M = \bigcup_{x\in M} \mathrm{T}_x M ,$$

where $\mathrm{T}_x M$ is the tangent space to M at x. On M there acts a differentiable map f. This means that $f : M \to M$ and (in case $M = \mathbb{R}^d$), Taylor's formula can be written as

$$f(x+\xi) = f(x) + \mathrm{D}_x f \xi + \mathcal{O}(\xi^2) , \qquad (4.1)$$

where $\mathrm{D}_x f$ is the differential of f evaluated at x (the $d \times d$ matrix of partial derivatives). If $\xi \in \mathrm{T}_x M$ then $\mathrm{D}_x f \xi$ is, by (4.1), in $\mathrm{T}_{f(x)} M$ and thus, $\mathrm{D}_x f$ maps $\mathrm{T}_x M$ to $\mathrm{T}_{f(x)} M$. In other words, it is a map from the tangent space at x to the tangent space at $f(x)$, sometimes called the *tangent map*. Note that if x is a fixed point of f, then $\mathrm{T}_x M = \mathrm{T}_{f(x)} M$.

Definition 4.1. *In the case of* $M = \mathbb{R}^d$, *if* x_0 *is a fixed point of a map* f, *the* affine map $x \mapsto x_0 + \mathrm{D}_{x_0} f(x - x_0)$ *is called the* linearization *(or* linearized map *of* f *at* x_0*).*

The chain rule of differentiation takes the form

$$f^2(x+\xi) = f(f(x+\xi)) = f(f(x)) + \mathrm{D}_{f(x)} f \cdot \mathrm{D}_x f \cdot \xi + \mathcal{O}(\xi^2) ,$$

so that the tangent map of f^2 is obtained by matrix multiplication, more generally:

$$\mathrm{D}_x(f^n) = \mathrm{D}_{f^{n-1}(x)} f \cdots \mathrm{D}_{f(x)} f \cdot \mathrm{D}_x f .$$

In any multidimensional situation, the order of multiplication of the matrices is important.

We will usually assume that f is a *diffeomorphism*, that is, a 1–1 invertible differentiable map with differentiable inverse, so that the construction above can also be done for the inverse map f^{-1}.

Exercise 4.2. *Check that* $\mathrm{D}_{f(x)}(f^{-1}) \cdot \mathrm{D}_x f$ *is the identity matrix.*

4.1 Hyperbolic Fixed Points

In this section, we study what happens near a (hyperbolic) fixed point of the map f. We begin by the analysis of linear maps and show then how the results extend to the nonlinear case. We then show how one changes to normal coordinates in Sect. 4.1.2. In Sect. 4.1.3 we study what kinds of obstacles one finds to smooth conjugation.

Let x_0 be a fixed point for a transformation f acting on $\mathbb{R}^d$ (namely, $f(x_0) = x_0$). We have seen in (4.1) that locally the dynamics is governed by the differential $\mathrm{D}_{x_0} f$ of the map at the fixed point. The $n\mathrm{th}$ iterate of the tangent map is given by the matrix $(\mathrm{D}_{x_0} f)^n$.

In the case where the tangent map $\mathrm{D}_{x_0} f$ is hyperbolic (namely it has no eigenvalue of modulus one), some vectors are contracted exponentially fast by the iterates of the linearized map (we say they belong to the stable subspace $E^{\mathrm{s}}_{x_0}$) while others are exponentially expanded (we say they belong to the unstable subspace $E^{\mathrm{u}}_{x_0}$).

Here, the linear space of stable directions $E^{\mathrm{s}}_{x_0}$ is formed by the (spectral) subspace corresponding to the eigenvalues of $\mathrm{D}_{x_0} f$ of modulus less than 1, while the unstable subspace $E^{\mathrm{u}}_{x_0}$ corresponds to those of modulus greater than 1.

Definition 4.3. *The point* x_0 *is a* hyperbolic fixed point *of the map* f *if* $f(x_0) = x_0$ *and if* $\mathrm{D}_{x_0} f$ *has no eigenvalues of modulus equal to 1. We call then* $D_{x_0} f$ *a* hyperbolic matrix.

Remark 4.4. From the definition of hyperbolicity, it follows that $\mathrm{T}_{x_0} M = E^{\mathrm{s}}_{x_0} \oplus E^{\mathrm{u}}_{x_0}$. We also emphasize that the spectral theory is done in the reals. The matrix $\mathrm{D}_{x_0} f$ has real entries but may have complex eigenvalues. The corresponding eigenvectors may have nonreal components and are not defined in the real setting (i.e., in the real vector space $\mathrm{T}_{x_0} M$). On the other hand, the spectral subspaces $E^{\mathrm{s}}_{x_0}$ and $E^{\mathrm{u}}_{x_0}$ *are* well defined in the real vector space $\mathrm{T}_{x_0} M$ (roughly speaking, for a nonreal eigenvector one should consider the 2-dimensional real subspace generated by its real and imaginary part).

We now consider in more detail the dynamics near the point x_0.

We start with the simple case where f has x_0 as a fixed point and is *affine* on $\mathbb{R}^d$. We then change the coordinate system so that the fixed point x_0 coincides with the origin. Then we can write our transformation as

$$f(x) = A\,x,$$

where $A = \mathrm{D}_0 f$ defines a linear map; that is, A is a $d \times d$ matrix.

If all the eigenvalues of A are of modulus less than 1, then the iterate of any initial point under f converges to the origin. To understand the case where some eigenvalues have modulus smaller than 1 and the others have modulus larger than 1, we consider first the even simpler case of dimension $d = 2$, with A diagonal, namely

$$A = \begin{pmatrix} a & 0 \\ 0 & b \end{pmatrix} ,$$

with $a > 1 > b > 0$. We illustrate the dynamics in Fig. 4.1. For the point $x = \begin{pmatrix} x^{(1)} \\ x^{(2)} \end{pmatrix} \in \mathbb{R}^2$ we find

$$f \begin{pmatrix} x^{(1)} \\ x^{(2)} \end{pmatrix} = \begin{pmatrix} a x^{(1)} \\ b x^{(2)} \end{pmatrix} . \tag{4.2}$$

After n iterates we get for the image of the point $x = \begin{pmatrix} x^{(1)} \\ x^{(2)} \end{pmatrix}$ the value

$$f^n(x) = A^n x = \begin{pmatrix} a^n x^{(1)} \\ b^n x^{(2)} \end{pmatrix} .$$

From this expression, we first conclude that if a point belongs to the $x^{(2)}$ axis ($x^{(1)} = 0$), then its entire trajectory follows this axis and converges to the origin, since $|b| < 1$. Moreover, any point of $\mathbb{R}^2$ whose first component $x^{(1)}$ is different from 0 has a trajectory that diverges from the origin. In other words, if we consider a ball B of radius $r > 0$ around the origin, the only points whose trajectory stays forever in B

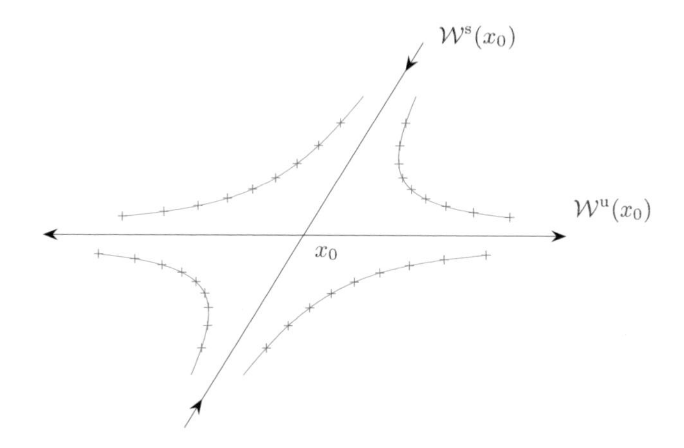

Fig. 4.1. The hyperbolic fixed point x_0 for a linear map f, with its stable and unstable manifolds $\mathcal{W}^{\mathrm{s}}(x_0)$ and $\mathcal{W}^{\mathrm{u}}(x_0)$ (which are linear) and a few points of four orbits of f

are those on the $x^{(2)}$ axis. In addition, these are also the only points whose orbit converges to the origin. Therefore, for our example, the $x^{(2)}$ axis is called the *stable manifold* of the fixed point.

Exercise 4.5. *Let A be a 2×2 matrix with real coefficients, determinant 1 and trace larger than 2. Show that the two eigenvalues are real and positive, one larger than 1, the other smaller than 1. Consider the 2-dimensional linear map $x \mapsto Ax$. Show that each orbit belongs to a hyperbola (the pair of eigendirections corresponding to a degenerate situation).*

Considering the inverse transformation f^{-1} which is given by the matrix

$$A^{-1} = \begin{pmatrix} 1/a & 0 \\ 0 & 1/b \end{pmatrix} ,$$

namely

$$f^{-1}(x) = \begin{pmatrix} x^{(1)}/a \\ x^{(2)}/b \end{pmatrix} ,$$

one can define in the same way the *unstable manifold* of the fixed point as the stable manifold of the inverse map. This is the set of points in the ball B whose orbit under f^{-1} stays forever in the ball B, and moreover this is the set of points whose orbit under f^{-1} converges to the fixed point (the origin). In this simple situation, the unstable manifold is the $x^{(1)}$ axis ($x^{(2)} = 0$).

Remark 4.6. It follows from the above discussion that a hyperbolic fixed point is automatically isolated: No other fixed points can be in its neighborhood.

4.1.1 Stable and Unstable Manifolds

Generalizing the discussion of the linear case, we now define important geometrical objects for the study of dynamical systems: the *invariant manifolds*. We start with the easy case of fixed points for maps. This immediately carries over to the case of periodic points. We then give the general definition in Sect. 4.2.

A first example of stable and unstable manifolds was provided by the example of the pendulum in Sect. 2.1. There, we saw, in Figs. 2.1–2.4, the stable and unstable manifolds of the two fixed points of the pendulum: The stable equilibrium when the pendulum is at rest at the bottom, and the unstable equilibrium when it is at the top. The invariant manifolds were then special orbits that approach or leave these equilibrium points according to special directions. The present subsection presents this construction for more general dynamical systems.

We begin by considering the general case of a regular nonlinear map f with a hyperbolic fixed point x_0. Near x_0 one can approximate f by an affine map, and the construction for the nonlinear case can be viewed as a perturbation of the affine case, at least locally. However, in general, the straight (invariant) lines of Fig. 4.1 will be replaced by bent curves.

Definition 4.7. *A manifold $\mathcal{W}^{\mathrm{s}}_{\mathrm{loc}}(x_0)$ is called a* local stable manifold *of the map f in the neighborhood $B_{x_0}(r)$ of a fixed point x_0 if*

i) $f\big(\mathcal{W}^{\mathrm{s}}_{\mathrm{loc}}(x_0)\big) \subset \mathcal{W}^{\mathrm{s}}_{\mathrm{loc}}(x_0)$.
ii) The orbit of any point in $\mathcal{W}^{\mathrm{s}}_{\mathrm{loc}}(x_0)$ converges to x_0.
iii) Every point in $B_{x_0}(r)$ whose orbit never leaves $B_{x_0}(r)$ is in $\mathcal{W}^{\mathrm{s}}_{\mathrm{loc}}(x_0)$.

A similar definition, with f^{-1}, is used to define *local unstable manifolds.*

Remark 4.8. The definition takes into account that, as we will see below, for the *global* invariant manifold, orbits can leave the local neighborhood and then come back to converge to the fixed point. In Fig. 4.3 one can see how the global manifold meanders and accumulates at the fixed point. In this case, we are not considering a fixed point but in fact a periodic point of period 6 (which is a fixed point of f^6).

In the purely linear Example 4.2, the local stable and unstable manifolds are segments of the $x^{(2)}$ (respectively $x^{(1)}$) axis centered at the origin. For this purely linear example, we see that the radius r of the ball B is not important. Later on, when we discuss the nonlinear case, this radius will have to be chosen small enough so that the nonlinear effects will be controllable.

The following theorem describes the local (un)stable manifolds.

Theorem 4.9. *Let f be a regular invertible map with a fixed point x_0. Assume that the spectrum of the differential $\mathrm{D}_{x_0} f$ is the (disjoint) union of two (finite) subsets σ^{u} and σ^{s} of $\mathbb{C}$, with σ^{u} contained in the complement of the closed unit disk, and σ^{s} contained in the open unit disk. Then $\mathcal{W}^{\mathrm{s}}_{\mathrm{loc}}(x_0)$ exists on a sufficiency small ball $B_{x_0}(r)$. Denote by E^{u} and E^{s} the spectral subspaces corresponding to σ^{u} and σ^{s}*

respectively. The manifold $\mathcal{W}^{\mathrm{s}}_{\mathrm{loc}}(x_0)$ is tangent to E^{s} at x_0 and has the same dimension. Similarly, there is a local unstable manifold $\mathcal{W}^{\mathrm{u}}_{\mathrm{loc}}(x_0)$ with similar properties with respect to the inverse map. These two local manifolds depend continuously on the map.

Figure 4.2 shows the local stable and unstable manifolds around a 2-dimensional hyperbolic fixed point.

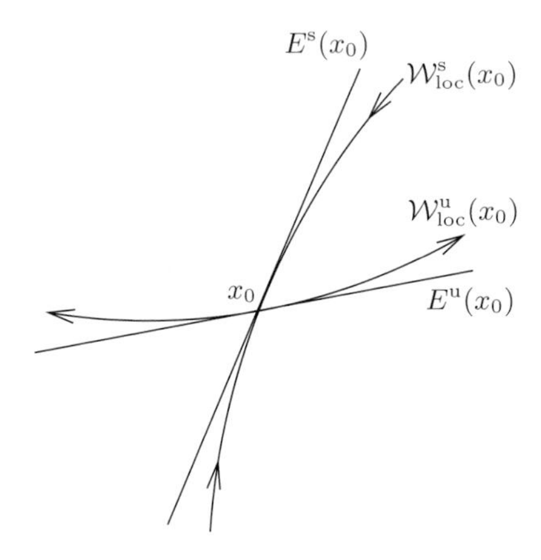

Fig. 4.2. Local stable and unstable manifolds around a 2-dimensional hyperbolic fixed point, x_0

Proof (sketch). We do not prove Theorem 4.9, and refer to (Hirsch, Pugh, and Shub 1977) for a complete proof and extensions. The basic idea of the proof goes as follows: The local stable manifold is constructed as a piece of graph of a map from the stable space to the unstable space. More precisely, using near the fixed point (assumed to be at the origin) coordinates given by the decomposition $E^{\mathrm{s}} \oplus E^{\mathrm{u}}$, one looks for a map g from E^{s} to E^{u} whose graph is invariant, namely $f\left(\eta, g(\eta)\right) = \left(\eta', g(\eta')\right)$ for any η in a neighborhood of 0 in E^{s}, and η' some point in E^{s} (depending on η). With obvious notations, we get the system of equations

$$\begin{aligned} \eta' &= f^{\mathrm{s}}\left(\eta, g(\eta)\right) , \\ g(\eta') &= f^{\mathrm{u}}\left(\eta, g(\eta)\right) . \end{aligned}$$

In order to obtain a closed equation for g, one would like to extract η as a function of η' in the first equation. Note that $f^{\mathrm{s}}\left(\eta, g(\eta)\right) = \mathrm{D}_0 f^{\mathrm{s}} \eta + \mathcal{O}(1)\left(\|\eta\|^2 + \|g(\eta)\|^2\right)$.

Since $\mathrm{D}_0 f^{\mathrm{s}}$ is invertible, and the other terms are small, one can show, using the inverse function theorem that near the origin η can be expressed as a function of η' (depending on g). Using a fixed point theorem one can show by this method the existence and uniqueness of g. □

Exercise 4.10. *Assuming f is C^∞, use the above argument to compute the first few terms of Taylor series of g at the origin. Observe in particular that g starts with quadratic terms.*

We can now define the global invariant manifolds by extending the local ones. One then obtains a picture that looks like that of Fig. 4.3.

Definition 4.11. *The (global)* stable manifold *of a fixed point is the set of points in the phase space whose orbit converges to this fixed point.*

If an orbit converges to a fixed point, it will by definition penetrate any ball around this fixed point, whatever its radius. Therefore, if the point x belongs to the stable manifold of a fixed point x_0, there is an integer n such that $f^n(x)$ belongs to the local stable manifold of x_0. Therefore, if $\mathcal{W}^{\mathrm{s}}(x_0)$ denotes the (global) stable manifold of the fixed point x_0, we have

$$\mathcal{W}^{\mathrm{s}}(x_0) = \bigcup_n f^{-n}(\mathcal{W}^{\mathrm{s}}_{\mathrm{loc}}(x_0)) \ .$$

Of course in the case of the affine map (4.2), we deduce immediately that the (local) stable manifold is the vertical axis. For a linear map, the global stable and unstable manifolds are just the linear extensions of the local ones (see Fig. 4.1).

One can make the analogous construction for the (global) *unstable manifold* $\mathcal{W}^{\mathrm{u}}(x_0)$, using the iterates of the map, and in particular one defines

$$\mathcal{W}^{\mathrm{u}}(x_0) = \bigcup_n f^{n}(\mathcal{W}^{\mathrm{u}}_{\mathrm{loc}}(x_0)) \ .$$

In these definitions, the global manifolds are obtained by gluing together pieces of manifolds which join in a smooth way. However, there are infinitely many such pieces which may get more and more contorted as can be seen in the example of the Hénon–Heiles map on $\mathbb{R}^2$, as shown in Fig. 4.3. The map is given by $(x', y') = f(x, y)$, where

$$\begin{aligned} x' &= x + 1.6y(1 - y^2) \ , \\ y' &= y - 1.6x'(1 - (x')^2) \ . \end{aligned} \tag{4.3}$$

(see (Bartlett 1976)). This map was obtained by Hénon and Heiles as an approximation to a Hamiltonian problem (Hénon and Heiles 1964). The approximation (4.3) is an area-preserving map.

Exercise 4.12. *Check that (4.3) is area-preserving, and compute its inverse. Find the fixed points, and discuss their hyperbolicity.*

The nonlinearities of the map f can fold the global invariant manifolds in a complicated way. They may, for example, come back very near to the fixed point as is

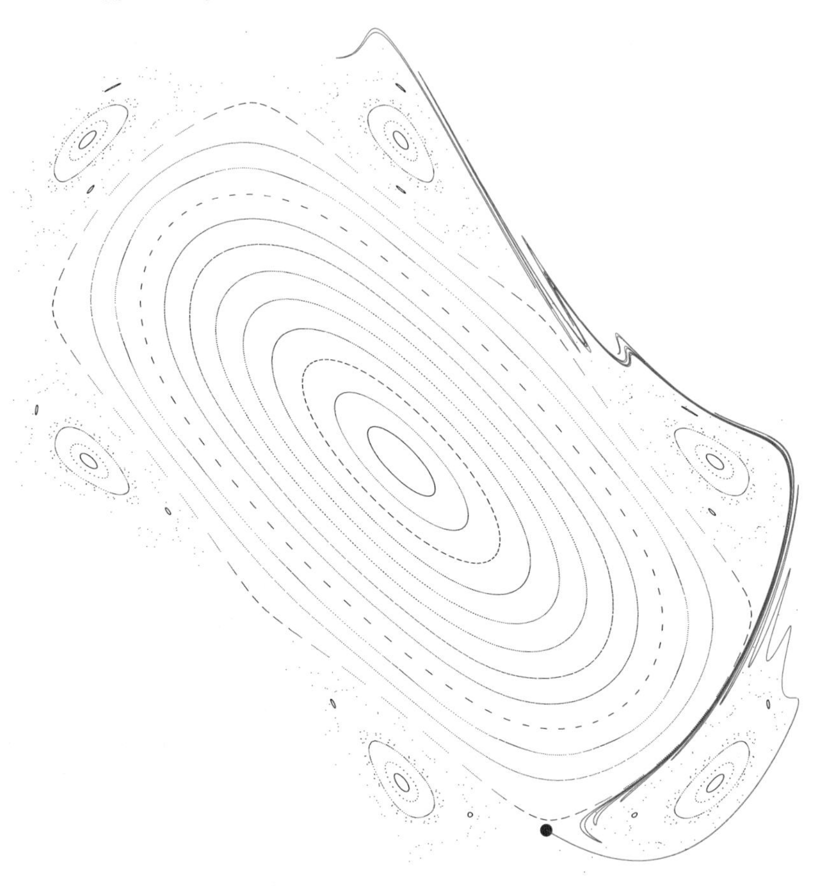

Fig. 4.3. A finite, but large, part of one half of the unstable manifold (red) of a periodic point (blue) of period 6 ($x \approx 0.2845$, $y \approx -0.7029$) for the Hénon–Heiles map (4.3). Showing more of the unstable manifold would eventually surround all six islands and return (infinitely often) close to the blue fixed point, without actually hitting it. The black points are several orbits of the map. Each elliptical set of points near the center is a distinct orbit, as are the sixfold smaller elliptical sets which one can see further away from the center. The "dust" corresponds to a further orbit. Such phenomena are covered by the Kolmogorov–Arnold–Moser theory (see (de la Llave 2001))

the case in homoclinic crossings. A neighborhood of the fixed point may then contain many pieces of the global unstable manifold. However, only one of these pieces contains the fixed point, it is the local unstable manifold.

In Theorem 4.9 we have carefully avoided the case where the spectrum of $\mathrm{D}_{x_0} f$ contains eigenvalues of modulus 1. In that case there is a notion of center manifold, and we refer the reader to (Hirsch, Pugh, and Shub 1977) for details. There one finds also a general approach to the unstable manifold even for the case where the map is not invertible. Note that for the case of flows (except at fixed points) there is always one eigenvalue of the tangent flow equal to 1. This is related to the observation that if two orbits start, say, 1 second apart from the same point, they will not separate (or approach) exponentially fast, as they will stay forever 1 second apart. This eigenvalue 1 does not spoil things, as the stable and unstable directions are in some sense perpendicular to the time direction.

4.1.2 Conjugation

We mentioned earlier that near a hyperbolic fixed point a map behaves like its differential. We now state this more precisely.

Theorem 4.13 (Hartman–Grobman theorem). *Let f be a C^2 map from $E = \mathbb{R}^d$ to itself, with a hyperbolic fixed point x_0. Assume that $\mathrm{D}_{x_0} f$ is invertible. Then there exist a homeomorphism Φ of E fixing x_0 and a neighborhood V of x_0 such that on V we have*

$$f = \Phi^{-1} \circ \mathrm{D}_{x_0} f \circ \Phi \, .$$

In other words, we can find a change of coordinates (see Definition 2.4) near x_0 such that in the new coordinate system the map is linear.

Remark 4.14. The theorem says that the change of coordinate is continuous, but the proof will establish that it is slightly more regular (Hölder continuous). Resonances, which we discuss in the next subsection, are often restrictions to higher differentiability. For a detailed proof of Theorem 4.13, see, for example, (Nitecki 1971). The proof we sketch below extends essentially unchanged to Banach spaces, provided differentiable cutoff functions with bounded support exist. They exist for all Hilbert spaces, but counterexamples are known for certain Banach spaces (for example ℓ^1). The interested reader may look up (Bonic and Frampton 1965).

Remark 4.15. The invertibility of the differential of the map at the fixed point is crucial in the Hartman–Grobman theorem. Consider, for example, the map $f : x \mapsto x^2$ on the reals. The origin is a fixed point, and the differential in this point is 0, $f'(0) = 0$. The linearized map f_0 is given by $f_0(x) = 0$ for every x. On the other hand, in any neighborhood of 0 one can find two points x_1 and x_2 for which $f(x_1) \neq f(x_2)$. Therefore, f and f_0 cannot be continuously conjugated in any neighborhood of 0.

We next define the important notions of stable and unstable vectors (directions) at the point x (not necessarily a fixed point) for diffeomorphisms f:

Definition 4.16.

i) *A vector ξ in $\mathrm{T}_x M$ is called (exponentially)* stable *if there are a $\lambda < 1$ and a $C < \infty$ such that for all $n \geq 0$*

$$\|\mathrm{D}_x f^n \xi\|_{\mathrm{T}_{f^n(x)} M} \leq C \lambda^n \|\xi\|_{\mathrm{T}_x M} \, . \tag{4.4}$$

ii) A vector ξ in $\mathrm{T}_x M$ is called (exponentially) unstable *if there are a $\lambda < 1$ and a $C < \infty$ such that for all $n \geq 0$*

$$\|\mathrm{D}_x f^{-n}\xi\|_{\mathrm{T}_{f^{-n}(x)}M} \leq C\lambda^n \|\xi\|_{\mathrm{T}_x M} \,. \tag{4.5}$$

Definition 4.17. *A fixed point x of f is called a (exponentially)* stable fixed point *if all tangent vectors are stable with the same $\lambda < 1$. It is called an* unstable fixed point *if at least one vector in $\mathrm{T}_x M$ is unstable. Note that unstable fixed points can have both stable and unstable directions.*

Remark 4.18. A vector ξ might be neither stable nor unstable.

Remark 4.19. The norms $\|\cdot\|_{\mathrm{T}_x M}$ are the norms at the tangent space $\mathrm{T}_x M$, and may depend on x. This complication is useful, since defining different norms will allow us to eliminate the constant C in the definition; see below.

Elimination of the constant C in (4.4) and (4.5). The presence of the constant C is not very convenient, because when $C > 1$ then $C\lambda^n$ is smaller than 1 only for some n which might need to be large. Therefore, one would like bounds with $C = 1$. This can be done by an *adapted norm*—also called *Lyapunov metric*—which is equivalent to the original norm. We only show this for the case of Hilbert spaces, where the construction is easier. Let $(\cdot,\cdot)_x$ be the scalar product corresponding to the norm $\|\cdot\|_{\mathrm{T}_x M}$. One defines a new scalar product $\langle\cdot,\cdot\rangle_x$ by summing. For example, if both ξ and η are stable, then

$$\langle\xi,\eta\rangle_x \equiv \sum_{i=0}^{\infty} \frac{1}{\lambda^{2i(1-\varepsilon)}} \left(\mathrm{D}_x f^i\xi, \mathrm{D}_x f^i\eta\right)_{f^i(x)} \,. \tag{4.6}$$

Clearly, the sum converges by definition (4.4). An easy calculation gives the result:

$$\begin{aligned}
\langle\mathrm{D}_x f\xi, \mathrm{D}_x f\xi\rangle_{f(x)} &= \sum_{i=0}^{\infty} \frac{1}{\lambda^{2i(1-\varepsilon)}} \left(\mathrm{D}_x f^{i+1}\xi, \mathrm{D}_x f^{i+1}\xi\right)_{f^{i+1}(x)} \\
&= \sum_{j=1}^{\infty} \frac{\lambda^{2(1-\varepsilon)}}{\lambda^{2j(1-\varepsilon)}} \left(\mathrm{D}_x f^j\xi, \mathrm{D}_x f^j\xi\right)_{f^j(x)} \\
&\leq \lambda^{2(1-\varepsilon)}\langle\xi,\xi\rangle_x \,.
\end{aligned}$$

and so we have effectively replaced λ by $\lambda^{1-\varepsilon}$ and C by 1, as asserted. The construction for the unstable directions is similar, and if ξ is stable and η unstable (or vice versa) one just defines $\langle\xi,\eta\rangle_x = 0$.

Consider now the simpler case of a hyperbolic fixed point x_0. In that case the above construction defines a new norm, and one would like to compare it with the old one. We will denote by E^{u} and E^{s} the unstable and stable spectral subspaces of $\mathrm{D}_{x_0} f$. We first control the angle between these two subspaces in the initial scalar product.

Exercise 4.20. *Let A be a linear operator in $\mathbb{R}^d$, and assume that ξ and η are unit vectors in $\mathbb{R}^d$ such that $\|A\xi\| \geq \varrho^{-1}$ and $\|A\eta\| \leq \varrho$ for some number $\varrho < 1$. Show that the angle ϑ between ξ and η satisfies*

$$1 - \cos\vartheta \geq \frac{(\varrho^{-1} - \varrho)^2}{2\|A\|^2} .$$

Hint: write $\xi = \eta + (\xi - \eta)$, and estimate the norm of $\xi - \eta$ in terms of $\cos\vartheta$. *Then apply A and use the hypothesis.*

We apply the above exercise with $A = \mathrm{D}_{x_0} f^n$ with n large enough so that $C\lambda^n < 1$. Note that if $\xi \in E^{\mathrm{u}}$, $\xi' = \mathrm{D}_{x_0} f^n \xi \in E^{\mathrm{u}}$, and $\|\mathrm{D}_{x_0} f^{-n} \xi'\| \leq C\lambda^n \|\xi'\|$ implies $\|\mathrm{D}_{x_0} f^n \xi\| \geq C^{-1}\lambda^{-n}\|\xi\|$. This allows to control the projection onto E^{u} parallel to E^{s} (and onto E^{s} parallel to E^{u}).

Exercise 4.21. *Show that there is a constant $D > 1$ such that for any vector $v \in TM_{x_0}$*

$$D^{-1}(v, v) \leq \langle v, v\rangle \leq D(v, v) .$$

Hint: write $v = \xi + \eta$ with $\xi \in E^{\mathrm{u}}$ and $\eta \in E^{\mathrm{s}}$.

Proof (of Theorem 4.13). Again, we show the proof only for $M = \mathbb{R}^d$. Without loss of generality (i.e., conjugating with a translation) we can assume that $x_0 = 0$. If we denote $L = \mathrm{D}_0 f$, we have

$$f(x) = Lx + Q(x)$$

with $\mathrm{D}_0 Q = 0$; that is, Q denotes terms of higher order (near 0). Since L is hyperbolic, we can decompose as before $E = E^{\mathrm{u}} \oplus E^{\mathrm{s}}$, and if we denote by L^{u} the restriction of L to the invariant subspace E^{u} (respectively L^{s}, the restriction of L to the invariant subspace E^{s}), there is a number $\beta > 1$ such that the spectrum of L^{u} is outside the disk of radius β in the complex plane, centered at the origin, while the spectrum of L^{s} is inside the disk of radius β^{-1}. We can now construct as we did above an equivalent norm on E such that there is a number $\varrho < 1$ satisfying

$$\|L^{\mathrm{s}}|_{E^{\mathrm{s}}}\| < \varrho \qquad \|(L^{\mathrm{u}})^{-1}|_{E^{\mathrm{u}}}\| < \varrho .$$

We next modify the map outside a neighborhood of the origin to be able to work globally later. Let $\varepsilon \in (0, 1)$ be a (small) number to be chosen adequately later. This number gives the order of magnitude of the neighborhood V. Let φ be a real valued C^1 function on E with compact support contained in the unit ball centered at the origin and equal to 1 in a neighborhood of the origin. We define the map Q_ε by

$$Q_\varepsilon(x) = \varphi(x/\varepsilon)Q(x) ,$$

and also

$$f_\varepsilon(x) = Lx + Q_\varepsilon(x) .$$

The advantage of f_ε over f is that outside a neighborhood of size of order ε of the origin, it is equal to L (and hence conjugated to L). On the other hand, near the origin, f_ε and f coincide. Therefore, if we can conjugate f_ε to L on the whole of E, we obtain a local conjugation between f and L near the origin.

The map Q_ε is differentiable and it is easy to verify that there is a constant C independent of ε such that

$$\|\mathrm{D}Q_\varepsilon\| \le C\varepsilon\ .$$

We will use also that $\mathrm{D}Q_\varepsilon(x) = 0$ if $\|x\| \ge \varepsilon$.

Exercise 4.22. *Prove this statement by observing that near the origin* $\|Q(x)\| \le \mathcal{O}(1)\|x\|^2$.

We now consider the conjugation equation and rearrange it in a suitable form. We look for a map $\varPhi$ of the form

$$\varPhi(x) = x + R(x)\ ,$$

where R is small in a suitable sense to be defined below, and satisfies $R(0) = 0$. The conjugation relation Definition 2.4 is written

$$\varPhi(Lx) = Lx + R(Lx) = f_\varepsilon(\varPhi(x)) = Lx + LR(x) + Q_\varepsilon(x + R(x))\ ,$$

or in other words

$$(\mathrm{Id} + R) \circ L = (L + Q_\varepsilon) \circ (\mathrm{Id} + R)\ .$$

It is now convenient to use the decomposition $E = E^{\mathrm{u}} \oplus E^{\mathrm{s}}$. Projecting on E^{u} and E^{s}, we obtain a system of two equations (with obvious notations)

$$\begin{aligned} &L^{\mathrm{u}}x^{\mathrm{u}} + R^{\mathrm{u}}(L^{\mathrm{u}}x^{\mathrm{u}}, L^{\mathrm{s}}x^{\mathrm{s}}) \\ &= L^{\mathrm{u}}x^{\mathrm{u}} + L^{\mathrm{u}}R^{\mathrm{u}}(x^{\mathrm{u}}, x^{\mathrm{s}}) + Q_\varepsilon^{\mathrm{u}}\big(x^{\mathrm{u}} + R^{\mathrm{u}}(x^{\mathrm{u}}, x^{\mathrm{s}}), x^{\mathrm{s}} + R^{\mathrm{s}}(x^{\mathrm{u}}, x^{\mathrm{s}})\big)\ , \\ &L^{\mathrm{s}}x^{\mathrm{s}} + R^{\mathrm{s}}(L^{\mathrm{u}}x^{\mathrm{u}}, L^{\mathrm{s}}x^{\mathrm{s}}) \\ &= L^{\mathrm{s}}x^{\mathrm{s}} + L^{\mathrm{s}}R^{\mathrm{s}}(x^{\mathrm{u}}, x^{\mathrm{s}}) + Q_\varepsilon^{\mathrm{s}}\big(x^{\mathrm{u}} + R^{\mathrm{u}}(x^{\mathrm{u}}, x^{\mathrm{s}}), x^{\mathrm{s}} + R^{\mathrm{s}}(x^{\mathrm{u}}, x^{\mathrm{s}})\big)\ . \end{aligned}$$

We now simplify and rearrange this system of equations to

$$\begin{aligned} &R^{\mathrm{u}}(x^{\mathrm{u}}, x^{\mathrm{s}}) \\ &= (L^{\mathrm{u}})^{-1}R^{\mathrm{u}}(L^{\mathrm{u}}x^{\mathrm{u}}, L^{\mathrm{s}}x^{\mathrm{s}}) \\ &\quad - (L^{\mathrm{u}})^{-1}Q_\varepsilon^{\mathrm{u}}\big(x^{\mathrm{u}} + R^{\mathrm{u}}(x^{\mathrm{u}}, x^{\mathrm{s}}), x^{\mathrm{s}} + R^{\mathrm{s}}(x^{\mathrm{u}}, x^{\mathrm{s}})\big)\ , \end{aligned} \tag{4.7}$$

$$R^{\mathrm{s}}(x^{\mathrm{u}}, x^{\mathrm{s}}) = L^{\mathrm{s}}R^{\mathrm{s}}((L^{\mathrm{u}})^{-1}x^{\mathrm{u}}, (L^{\mathrm{s}})^{-1}x^{\mathrm{s}}) + Q_\varepsilon^{\mathrm{s}}(\hat{x}^{\mathrm{u}}, \hat{x}^{\mathrm{s}}))\ ,$$

with

$$\begin{aligned} \hat{x}^{\mathrm{u}} &= (L^{\mathrm{u}})^{-1}x^{\mathrm{u}} + R^{\mathrm{u}}((L^{\mathrm{u}})^{-1}x^{\mathrm{u}}, (L^{\mathrm{s}})^{-1}x^{\mathrm{s}})\ , \\ \hat{x}^{\mathrm{s}} &= (L^{\mathrm{s}})^{-1}x^{\mathrm{s}} + R^{\mathrm{s}}((L^{\mathrm{u}})^{-1}x^{\mathrm{u}}, (L^{\mathrm{s}})^{-1}x^{\mathrm{s}})\ . \end{aligned}$$

We will not work with the C^1 topology but with *Hölder norms*. For $0 < \alpha \leq 1$, we define the Banach space $\mathscr{H}_\alpha(E)$ as the set of continuous maps g from E to itself, satisfying $g(0) = 0$ and such that

$$\|g\|_{\mathscr{H}_\alpha(E)} = \sup_{x \neq y} \frac{\|g(x) - g(y)\|}{\|x - y\|^\alpha} < \infty .$$

The word *Lipshitz norm* is reserved for Hölder norms with exponent $\alpha = 1$.

Exercise 4.23. *Prove that equipped with the above norm, $\mathscr{H}_\alpha(E)$ is a Banach space, using that $g(0) = 0$.*

We now define a map Ψ from the metric space $\mathscr{H}_\alpha(E)$ to itself by

$$\begin{aligned}
&\Psi^{\mathrm{u}}(R^{\mathrm{u}}, R^{\mathrm{s}})(x^{\mathrm{u}}, x^{\mathrm{s}}) \\
&= (L^{\mathrm{u}})^{-1} R^{\mathrm{u}}(L^{\mathrm{u}} x^{\mathrm{u}}, L^{\mathrm{s}} x^{\mathrm{s}}) \\
&\quad - (L^{\mathrm{u}})^{-1} Q^{\mathrm{u}}_\varepsilon\big(L^{\mathrm{u}} x^{\mathrm{u}} + R^{\mathrm{u}}(x^{\mathrm{u}}, x^{\mathrm{s}}), L^{\mathrm{s}} x^{\mathrm{s}} + R^{\mathrm{s}}(x^{\mathrm{u}}, x^{\mathrm{s}})\big) ,
\end{aligned}$$

and

$$\Psi^{\mathrm{s}}(R^{\mathrm{u}}, R^{\mathrm{s}})(x^{\mathrm{u}}, x^{\mathrm{s}}) = L^{\mathrm{s}} R^{\mathrm{s}}((L^{\mathrm{u}})^{-1} x^{\mathrm{u}}, (L^{\mathrm{s}})^{-1} x^{\mathrm{s}}) + Q^{\mathrm{s}}_\varepsilon(\hat{x}^{\mathrm{u}}, \hat{x}^{\mathrm{s}})) .$$

It is obvious that solving the system (4.7) is equivalent to finding a fixed point of Ψ.

Exercise 4.24. *Show that if $R \in \mathscr{H}_\alpha(E)$, then $\Psi(R)(0, 0) = 0$.*

We now choose the values of ε and α so that Ψ contracts a ball in $\mathscr{H}_\alpha(E)$ centered at the origin. Let

$$\lambda = \max\left\{\|L^{\mathrm{u}}\|, \|(L^{\mathrm{s}})^{-1}\|\right\} .$$

Note that $\lambda > 1$ is independent of ε and α.

For r and r' belonging to $\mathscr{H}_\alpha(E)$ we have

$$\|\Psi(r)\|_{\mathscr{H}_\alpha(E)} \leq \varrho\lambda^\alpha \|r\|_{\mathscr{H}_\alpha(E)} + C\varepsilon\lambda\big(1 + \|r\|_{\mathscr{H}_\alpha(E)}\big)$$

and

$$\|\Psi(r) - \Psi(r')\|_{\mathscr{H}_\alpha(E)} \leq \varrho\lambda^\alpha \|r - r'\|_{\mathscr{H}_\alpha(E)} + C\varepsilon\lambda \|r - r'\|_{\mathscr{H}_\alpha(E)} .$$

Therefore, if we choose $\alpha > 0$ small enough and $\varepsilon > 0$ small enough, we can satisfy the inequality

$$\varrho\lambda^\alpha + 2C\varepsilon\lambda < 1 .$$

This implies that the map Ψ contracts the ball of radius 1 and centered at the origin of the metric space $\mathscr{H}_\alpha(E)$. By the contraction mapping principle (see (Dieudonné 1968)), Ψ has a unique fixed point in this ball and the theorem is proven.

Exercise 4.25. *Modify the above proof so that it works for $f \in C^1$.*

4.1.3 Resonances

Having found a conjugation map in the previous subsection, one may ask how smooth that conjugation can be. If the map is not smooth, there is not much to expect, but it turns out that the smoothness of the conjugation depends also on fine details of the eigenvalues of the linearization $\mathrm{D}_x f$ of the map f. This is called the problem of resonances, which can prevent the conjugation from being regular.

Let as before f be a regular map, acting on in $\mathbb{R}^d$. We also assume that the origin is a fixed point. Let $L = \mathrm{D}_0 f$ be the differential at the origin. To simplify the discussion, we make several assumptions: First, that L can be diagonalized (in the reals) and we denote by $\lambda_1, \ldots, \lambda_d$ its eigenvalues. We further assume that they are all different from 0 and 1, but some may coincide. Therefore, Theorem 4.13 applies. Finally, we choose a coordinate system consisting of eigendirections. Since f is regular, we can write for any $\ell \in \mathbb{Z}^+$ by Taylor's formula

$$f(x) = \sum_{j=1}^{\ell} f_j(x) + \mathcal{O}(1)\|x\|^{\ell+1} ,$$

where each f_j is a vector valued homogeneous function of degree j. More precisely, we have

$$f_j(x) = \sum_{k_1,\ldots,k_j} f_{j,k_1,\ldots,k_j} \prod_{r=1}^{j} x_{k_r} ,$$

and the $f_{j,k_1,\ldots,k_j}$ are fixed vectors (the sum extends over $1 \le k_i \le d$, $i = 1, \ldots, j$). Note that $f_1(x) = Lx$. Let Φ be a local conjugation between f and the linear map $x \mapsto Lx$ near the origin, namely

$$\Phi \circ f = L\Phi . \tag{4.8}$$

If Φ is regular, we can also write

$$\Phi(x) = \sum_{j=1}^{\ell} \Phi_j(x) + \mathcal{O}(1)\|x\|^{\ell+1} ,$$

and use this expansion in Eq. (4.8). We can then consider on both sides homogeneous terms of identical degree. Identifying them will lead to equations for the unknown Φ_j. For example, there is no term of degree 0. The terms of degree 1 give

$$\Phi_1 \circ L = L\Phi_1 .$$

To see what this equation means, it is convenient to use coordinates (recall that $\Phi_1(x)$ is a vector and we use the exponent to index its coordinates). We get immediately for any $1 \le k \le d$ and $1 \le l \le d$

$$\lambda_k \Phi^l_{1,k} = \lambda_l \Phi^l_{1,k} .$$

An obvious solution to this set of equations is given by

$$\Phi_{1,k}^{l} = \delta_{k}^{l} .$$

In other words, we have chosen Φ_1 to be the identity.

Remark 4.26. If $\lambda_k = \lambda_l$ ($k = l$ or not), there are other solutions, but we will consider only this one which is the most natural one. The other solutions correspond to linear transformations commuting with L.

Let us now look at the second order. We get

$$\Phi_2 \circ L + \Phi_1 \circ f_2 = L\Phi_2 ,$$

which is more conveniently written

$$L\Phi_2 - \Phi_2 \circ L = \Phi_1 \circ f_2 .$$

This is an equation for Φ_2 since we already know Φ_1. Let us try to solve it in coordinates. We obtain for any $1 \le k_1 \le d$, $1 \le k_2 \le d$ and $1 \le l \le d$

$$\left(\lambda_l - \lambda_{k_1}\lambda_{k_2}\right)\Phi_{2,k_1,k_2}^{l} = f_{2,k_1,k_2}^{l} .$$

This equation leads to

$$\Phi_{2,k_1,k_2}^{l} = \frac{f_{2,k_1,k_2}^{l}}{\lambda_l - \lambda_{k_1}\lambda_{k_2}} .$$

Clearly, this defines Φ_{2,k_1,k_2}^{l} unless there is a triplet l, k_1, k_2 for which

$$\lambda_l = \lambda_{k_1}\lambda_{k_2} .$$

In this case one says that the triplet is a *resonance*. When there is a resonance, the equation for Φ_{2,k_1,k_2}^{l} has only a solution (in fact infinitely many) if $f_{2,k_1,k_2}^{l} = 0$. Otherwise, when $f_{2,k_1,k_2}^{l} \neq 0$, we conclude that Φ cannot be C^2. If none of the triplets l, k_1, k_2 is a resonance, Φ is C^2 and one can work on the next (third order) of homogeneity. The general order has always the same structure, namely one gets an equation

$$L\Phi_j - \Phi_j \circ L = R_j$$

where R_j involves the Φ_r with $1 \le r \le j - 1$ and is therefore known at step j. The nonresonance condition at step j is therefore that no eigenvalue can be expressed as a product of j eigenvalues (possibly with repetitions), otherwise the map is said to have a resonance at the fixed point. Note that if one eigenvalue is equal to 1, then the system is resonant. In the case of flows, the nonresonance condition is that no eigenvalue be the sum of more than one eigenvalue. We refer to (Nelson 1969; Belickiĭ 1978) for more details.

4.2 Invariant Manifolds

Up to now we discussed the stable and unstable manifolds of fixed points. Similar results hold for the stable and unstable manifolds of periodic orbits of maps by taking a large enough iterate of the map such that each point of the periodic orbit becomes a fixed point. For example, we have shown in Fig. 4.3 the case of an unstable manifold of a point of period 6.

Here, we generalize the construction to points which are neither fixed nor periodic. The idea is to have little pieces of manifolds at each point of an orbit that map nicely one into the other. In the case of the local stable manifold of a hyperbolic fixed point discussed above, there is also a control on the rate of convergence to the fixed point through the iterations. The next definition generalizes these ideas.

Definition 4.27. *An (open) submanifold $\mathcal{W}$ of the phase space Ω with a metric d is called a* local stable manifold with exponent $\gamma > 0$ *if there is a constant $C > 0$ such that for any x and y in $\mathcal{W}$ and any integer n we have*

$$d\big(f^n(x), f^n(y)\big) \le C\mathrm{e}^{-\gamma n} .$$

Remark 4.28. To keep things easy, we omitted the question of smallness of neighborhoods. In all applications, we will consider sufficiently small neighborhoods, so that these local stable manifolds are close to being linear.

Note that it follows immediately from the definition that if $\mathcal{W}$ is a local stable manifold with exponent γ, then for any integer p, $f^p(\mathcal{W})$ is also a local stable manifold with exponent γ. We also notice that a local stable manifold with exponent γ is also a local stable manifold with exponent γ' for any $0 < \gamma' < \gamma$. When f is invertible, one can define the local unstable manifolds as the local stable manifolds of the inverse. We refer to (Hirsch, Pugh, and Shub 1977) for the general case and more details. A local stable manifold containing a point x is called a *local stable manifold at x*.

Remark 4.29. A clever argument involving the stretching in the unstable direction can be used to define unstable manifolds for maps whose inverse does not exist (see (Hirsch, Pugh, and Shub 1977)).

For example, if one considers the dissipative baker map (2.6), it is easy to verify that the horizontal segments are unstable manifolds, while the vertical segments are the stable manifolds. Similarly, in the solenoid example (2.13), the local stable manifolds are pieces of vertical planes containing the z axis. Figure 4.4 shows several pieces of local stable manifolds for the Hénon map (2.31). One notices that they vary rapidly from one point to the other, and some of them have a rather large curvature. This is a manifestation of the so-called nonuniform hyperbolicity.

We next state the important

Definition 4.30. *A set $\Lambda \subset M$ is called a* uniformly hyperbolic set *for the diffeomorphism f if*

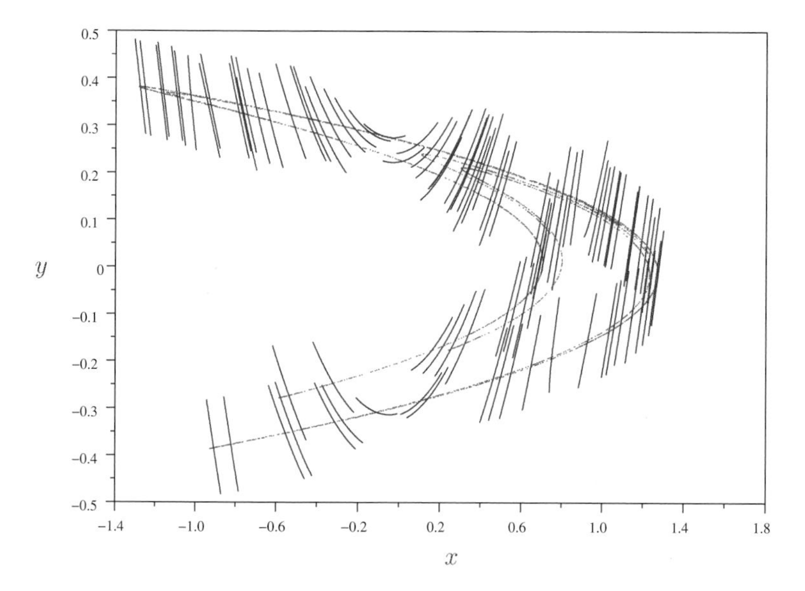

Fig. 4.4. Several local stable manifolds (in blue) for the Hénon map (2.31)

i) it is invariant: $f(\Lambda) \subset \Lambda$,
ii) for all $x \in \Lambda$*, the tangent space* $\mathrm{T}_x M$ *splits into stable and unstable subspaces* $\mathrm{T}_x M = E^{\mathrm{s}}_x \oplus E^{\mathrm{u}}_x$*, and*
iii) the vectors in E^{s}_x*, resp.* E^{u}_x *are stable, resp. unstable in the sense of Definition 4.16, with the constant* $\lambda < 1$ *independent of* x.

Definition 4.31. *The diffeomorphism* f *appearing in Definition 4.30 is called a* hyperbolic map *(on* Λ*).*

Remark 4.32. The set Λ can have many holes and be a Cantor set.

Remark 4.33. It is easy to verify that since f is a diffeomorphism, the vector spaces E^{s}_x and E^{u}_x satisfy

$$\mathrm{D}_x f E^{\mathrm{s}}_x = E^{\mathrm{s}}_{f(x)} \quad \text{and} \quad \mathrm{D}_x f E^{\mathrm{u}}_x = E^{\mathrm{u}}_{f(x)} \,.$$

Remark 4.34. By the construction of Remark 4.19 we may (and will) assume that equivalent norms have been chosen so that

$$\|\mathrm{D}_x f^n \xi\|_{E^{\mathrm{s}}_{f^n(x)}} < \lambda^n \|\xi\|_{E^{\mathrm{s}}_x} \,,$$

and

$$\|\mathrm{D}_x f^{-n} \eta\|_{E^{\mathrm{u}}_{f^{-n}(x)}} < \lambda^n \|\eta\|_{E^{\mathrm{u}}_x} \,,$$

with a slightly larger $\lambda < 1$ for all $x \in \Lambda$ and all $\xi \in E^{\mathrm{s}}_x$ and $\eta \in E^{\mathrm{u}}_x$.

Exercise 4.35. *As in the case of a hyperbolic fixed point, give a lower bound on the angle between E^{s}_x and E^{u}_x. Argue as before to prove that the new scalar product in the tangent space (which depends on x in general) defines a norm uniformly equivalent to the original one.*

Proposition 4.36. *The spaces E^{s}_x and E^{u}_x depend continuously on x.*

We omit the proof; see, for example, (Hirsch, Pugh, and Shub 1977, p. 21).

Remark 4.37. The theory developed in (Hirsch, Pugh, and Shub 1977) shows more, and we will also use these additional facts:

i) Continuity means continuity in Λ.
ii) Continuity implies that the dimensions of the spaces E^{s}_x and E^{u}_x are constant. Thus, the dimensions do not depend on $x \in \Lambda$.
iii) On compact spaces, the projections $\Pi_{E^{\mathrm{s}}_x}$ and $\Pi_{E^{\mathrm{u}}_x}$ have uniformly bounded norms. In a Hilbert space setting, one would say that the angle between E^{s}_x and E^{u}_x is uniformly (in $x \in \Lambda$) bounded away from 0.

Example 4.38. The solenoid map (see Figs. 2.9 and 2.10) has $\dim E^{\mathrm{s}}_x = 2$ and $\dim E^{\mathrm{u}}_x = 1$ for all x. The spaces E^{s}_x and E^{u}_x "point" in different directions at different points x. The stable directions are across the sausage, while the unstable are along the spaghetti. They both depend continuously on x. The set Λ is the intersection of the disklike regions $\times[0, 2\pi)$ in Fig. 2.9. Note that in this case $\Lambda \neq M$.

Example 4.39. A hyperbolic fixed point x is a special case of a hyperbolic set $\Lambda = \{x\}$: In that case, $f(x) = x$ and the eigenvalues of $\mathrm{D}_x f$ do not lie on the unit circle. In this case the rates of growth or decay of the tangent vectors are described by the eigenvalues. If the spectrum of $\mathrm{D}_x f$ is $\lambda_1, \lambda_2, \ldots, \lambda_d$ with

$$|\lambda_1| \geq |\lambda_2| \geq \cdots \geq |\lambda_j| > 1 > |\lambda_{j+1}| \geq \cdots \geq |\lambda_d| ,$$

then the bound in the unstable direction is $\lambda = |\lambda_j|^{-1}$, and in the stable direction $\lambda = |\lambda_{j+1}|$.

Example 4.40. The cat map: In this case $\Lambda = M$. Maps where $\Lambda = M$ are called *Anosov maps*. We have

$$\mathrm{D}_x f^n = \begin{pmatrix} 2 & 1 \\ 1 & 1 \end{pmatrix}^n ,$$

independently of x. The eigenvalues are $\lambda_\pm = \frac{3\pm\sqrt{5}}{2}$ and for the stable and unstable directions E^{s} and E^{u} one has $\mathrm{D}_x f^n \xi = \lambda_-^n \xi$ for $\xi \in E^{\mathrm{s}}$ and $\mathrm{D}_x f^{-n} \xi = \lambda_+^{-n} \xi$ for $\xi \in E^{\mathrm{u}}$. In this case $\lambda_- = \lambda_+^{-1}$.

Example 4.41. The case of rotations $x \mapsto x + \omega \pmod 1$ is not hyperbolic because the derivative is 1; that is, the eigenvalue of the differential is on the unit circle. The map $x \mapsto 2x \pmod 1$ is hyperbolic in the forward direction (with expansion rate 2), but in the backward direction, the map is not well-defined.

4.3 Nonwandering Points and Axiom A Systems

In this section, we introduce the generalization of periodic points alluded to before. In a way, all that is needed is that orbits of points come back close to where they started and that they are recurrent in some weak sense. This will be defined below. Finally, an Axiom A system will be a system with many hyperbolic nonwandering points.

This recurrence is characterized in the following

Definition 4.42. *A point x is called* wandering *if there exists an open set B containing x such that for some $n_0 < \infty$ and for all $n > n_0$ one has $f^n(B) \cap B = \emptyset$. (In other words, the neighborhood eventually never returns.) A point x is called* nonwandering *if it is not wandering. The definition implies that in this case, for every open set B containing x, one has $f^n(B) \cap B \neq \emptyset$ for infinitely many n.*

Lemma 4.43. *The set of wandering points is open, and that of nonwandering points is closed.*

The proof is obvious from the definition.

Example 4.44. A fixed point is nonwandering, independently of its stability (since any image of a neighborhood of the point intersects that neighborhood). Note that this also holds for the fixed point $x = 0$ of the map $x \mapsto x + x^2$, which is neither stable nor unstable.

Example 4.45. A periodic point is also nonwandering (since, if the period is p, one comes back to the point after p iterations).

Example 4.46. Nonwandering points might be neither fixed nor periodic. For example, for $x \mapsto x + \omega \pmod 1$ with irrational ω: *Every* point is nonwandering.

Example 4.47. Classical mechanics: Every point of a compact energy surface is nonwandering by Poincaré's recurrence theorem (see Theorem 8.2).

Example 4.48. Bowen–Katok flow: see Fig. 4.5. The orbit of every initial point (except P_1 and P_2) accumulates on the figure "∞," but spends longer and longer times near P_0. However, the nonwandering points are formed by the figure "∞" and the 2 unstable fixed points. (This can be seen by following the orbit of an open ball covering a segment of the figure ∞.)

Definition 4.49 (Smale 1967). *A dynamical system (C^1-diffeomorphism) is called an* Axiom A *system if*

i) The set of nonwandering points is uniformly hyperbolic.
ii) The periodic points are dense in the set $\mathcal{N}$ of nonwandering points.

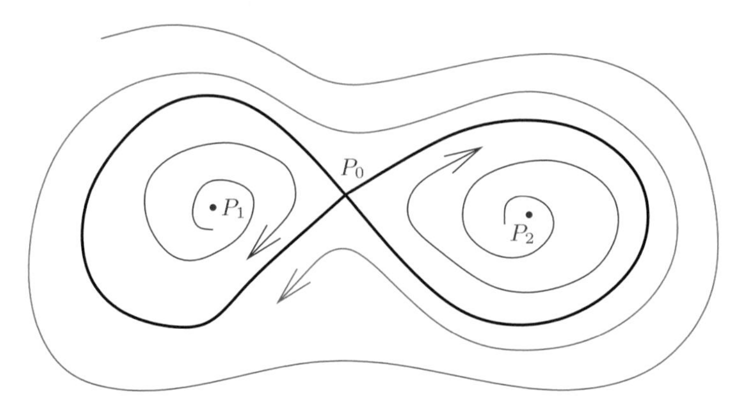

Fig. 4.5. The flow of Example 4.48. This example has more nonwandering points than just the attractor. The nonwandering points are the figure "eight" plus the three fixed points, P_0, P_1, P_2. On the other hand, any (red or blue) orbit not starting on P_1 or P_2 will approach the figure eight and spend more and more time near P_0. Therefore, the Dirac measure at P_0 is a Physical measure (see Sect. 5.7). The points P_1 and P_2 are unstable fixed points and the (blue) orbits will stay very long near P_0. Similarly, the red orbit is attracted to the figure "eight" and stays longer and longer near P_0 as time goes on

Remark 4.50. There are numerous results which hold for weaker variants of this definition, in particular, a lot of work deals with not quite uniformly hyperbolic systems. In the interest of simplicity, we do not present any of these results. They are, however, interesting from a conceptual point of view, because, for applications in Nature, one would like to know how much one really has to assume to get results similar to those obtained for Axiom A systems. From a practical point of view, one will have to assume anyway that the system behaves "as if" the strong assumptions were valid. We will discuss this again in Chap. 9. For a set of mathematical results, see, for example, (Alves, Bonatti, and Viana 2000; Bonatti and Viana 2000).

Example 4.51. The archetypical example of an Axiom A map was given by Smale in (Smale 1967). It is called the *horseshoe map*, mapping a square onto a $\supset$-shaped region which intersects the square again. See Fig. 4.6.

Example 4.52. The cat map. We have already seen that the periodic points are dense in M, and that the map is hyperbolic everywhere, so $\mathcal{N} = M$. The map is not only Axiom A but Anosov.

Example 4.53. The solenoid is Axiom A (but not Anosov), that is $\mathcal{N} \subsetneq M$.

Example 4.54. The Bowen–Katok flow of Fig. 4.5 is not Axiom A (since the figure ∞ is not densely filled with periodic points).

Example 4.55. If f is an irrational rotation of the torus, it is not Axiom A.

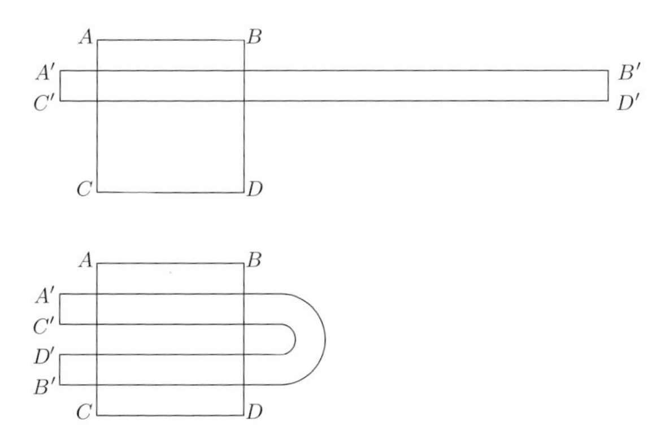

Fig. 4.6. The horseshoe map of Smale. The square $ABCD$ is first mapped onto the region $A'B'C'D'$ (top), which is then folded back across the square (bottom)

The definition of Axiom A is just the right choice, as we shall see. In particular, Axiom A systems are robust under small perturbations of the evolution equation, while this is not necessarily true for non-Axiom A systems.

We refer to (Afraimovich and Hsu 2003; Alligood, Sauer, and Yorke 1997; Devaney 2003; Guckenheimer and Holmes 1990; Ott, Sauer, and Yorke 1994) and (Ruelle 1989b) for general presentations, results, and references.

4.4 Shadowing and Its Consequences

This is a very interesting structural theory of hyperbolic systems (the Axiom A property is not needed here). It implies a certain rigidity of such systems. We begin with the somewhat counterintuitive, but very fruitful notion of *shadowing*.

Definition 4.56. *Let f be a map $M \to M$. A sequence $\{x_n\}_{n\in\mathbb{Z}}$ is called an ε-*pseudo-orbit *if $|f(x_i) - x_{i+1}| < \varepsilon$ for all $i \in \mathbb{Z}$. In other words, at every iteration, the pseudo-orbit makes an "error" of at most ε.*

Definition 4.57. *A point y_0 is said to δ-*shadow *a sequence $\{x_i\}_{i\in\mathbb{Z}}$, if*

$$|x_i - f^i(y_0)| < \delta ,$$

for all $i \in \mathbb{Z}$.

Remark 4.58. In some examples, we will consider maps which are not diffeomorphisms, such as $x \mapsto 2x \pmod 1$, and then we will consider the definitions above only for indices in $\mathbb{Z}^+$.

Theorem 4.59. *Let f be a C^2 diffeomorphism and let Λ be a compact, hyperbolic set for f. For every sufficiently small $\delta > 0$ there is an $\varepsilon > 0$ such that every ε-pseudo-orbit (lying in Λ) is δ-shadowed by a y_0. Furthermore, this y_0 is unique.*

Remark 4.60. The condition that the ε-pseudo-orbit lies in Λ is not necessary, it suffices that the pseudo-orbit lies close to Λ. Because if $\{x_n\}$ is an ε-pseudo-orbit at a distance ε' from Λ, then by the differentiability of f and the compactness of Λ one can find an ε''-pseudo-orbit lying in Λ with $\varepsilon'' < \mathrm{const.}(\varepsilon + \varepsilon')$ where the constant depends only on the derivative of f.

Exercise 4.61. *Prove this statement.*

The counterintuitive thing about Theorem 4.59 is that although the orbits of f separate exponentially with time, and the ε-pseudo-orbit is only approximate, there is such a y_0. One should note that y_0 is in general not one of the x_i above, but a very cleverly chosen other point. The following example illustrates how the apparent contradiction between expansion and shadowing is resolved.

Example 4.62. One can see the basic idea for the map $f : x \mapsto 2x \pmod 1$, when viewed on the circle rather than on the interval $[0, 1)$, although this is not a diffeomorphism. Assume that we are given ε_i with $|\varepsilon_i| < \varepsilon$ and that the points x_i satisfy $x_{i+1} - f(x_i) = \varepsilon_i$. Astonishingly, one can construct y_0 explicitly in this case:

$$y_0 = x_0 + \tfrac{1}{2} \sum_{k=0}^{\infty} \varepsilon_k 2^{-k} .$$

We verify that this works (modulo 1):

$$|y_0 - x_0| \le \tfrac{1}{2} \sum_{k=0}^{\infty} |\varepsilon_k| 2^{-k} \le \varepsilon \tfrac{1}{2} \sum_{k=0}^{\infty} 2^{-k} \le \varepsilon .$$

Similarly, modulo 1, we have $x_1 = 2x_0 + \varepsilon_0$, $x_2 = 2x_1 + \varepsilon_1 = 4x_0 + 2\varepsilon_0 + \varepsilon_1$,

$$\begin{aligned} x_n &= f^n(x_0) = 2^n x_0 + 2^{n-1}\varepsilon_0 + 2^{n-2}\varepsilon_1 + \cdots + \varepsilon_{n-1} , \\ y_n &= f^n(y_0) = 2^n x_0 + 2^{n-1}\varepsilon_0 + 2^{n-2}\varepsilon_1 + \cdots + \varepsilon_{n-1} + \tfrac{1}{2}\varepsilon_n + \tfrac{1}{4}\varepsilon_{n+1} + \cdots , \end{aligned}$$

so that

$$|y_n - x_n| \le \tfrac{1}{2} \sum_{k=0}^{\infty} 2^{-k} |\varepsilon_{n+k}| \le \varepsilon .$$

We see that "2" is the expansion rate, and the convergence of the sum is precisely guaranteed because the inverse of the expansion rate is < 1.

We reformulate Theorem 4.59 as follows: We assume that $f : M \to M$ is a hyperbolic diffeomorphism with the following bounds on the matrix norms

$$\begin{aligned} \|\mathrm{D}_x f^n|_{E_x^{\mathrm{s}}}\| &\le \lambda^n , \\ \|\mathrm{D}_x f^{-n}|_{E_x^{\mathrm{u}}}\| &\le \lambda^n , \end{aligned}$$

with $\lambda < 1$. We may assume this form by a change of metric as in (4.6).

Theorem 4.63 (Shadowing Lemma). *Let $\Lambda \subset M$ be a compact hyperbolic set. We assume $\delta > 0$ is sufficiently small. There is a $K > 1$ such that with $\varepsilon = \delta/K$, every ε-pseudo-orbit in Λ is δ-shadowed by a unique y_0.*

Proof. One reduces the proof to a fixed point theorem in a Banach space. To simplify the discussion, we will assume that the manifold M equals $\mathbb{R}^d$. Let $\mathcal{B}$ be the Banach space of all sequences $\mathbf{x} = \{x_n\}_{n\in\mathbb{Z}}$ equipped with the sup norm

$$\|\mathbf{x}\|_{\mathcal{B}} = \sup_{n\in\mathbb{Z}} |x_n| .$$

We define a map $A : \mathcal{B} \to \mathcal{B}$ by

$$(A\mathbf{x})_n = f(x_{n-1}) .$$

Note that if $A\mathbf{z} = \mathbf{z}$, this means that $f(z_{n-1}) = z_n$; that is, the fixed points of A are equivalent to being orbits of f. And if $\mathbf{x}$ is an ε-pseudo-orbit, this is expressed as

$$\|A\mathbf{x} - \mathbf{x}\|_{\mathcal{B}} \le \varepsilon . \tag{4.9}$$

The claim of the theorem can now be formulated as follows:
If $\mathbf{x}$ satisfies (4.9) then there is a $\mathbf{y}$ for which $\|A\mathbf{y} - \mathbf{y}\|_{\mathcal{B}} = 0$ and $\|\mathbf{y} - \mathbf{x}\|_{\mathcal{B}} \le K\varepsilon = \delta$. This is what we want, because

$$\sup_n |y_n - x_n| = \sup_n |f^n(y_0) - x_n| \le K\varepsilon .$$

We now make some preparatory observations. We note that the map A is twice differentiable, with a first differential given by

$$\big(\mathrm{D}_{\mathbf{x}}A(\mathbf{h})\big)_n = \big(\mathrm{D}_{x_{n-1}}f\big)h_{n-1} ,$$

and a second differential given by

$$\big(\mathrm{D}^2_{\mathbf{x}}A(\mathbf{h},\mathbf{h}')\big)_n = \big(\mathrm{D}^2_{x_{n-1}}f\big)(h_{n-1}, h'_{n-1}) .$$

In particular, we have

$$\big\|\mathrm{D}_{\mathbf{x}}A\big\|_{\mathcal{B}} \le \sup_x \big\|\mathrm{D}_x f\big\| ,$$

and

$$\big\|\mathrm{D}^2_{\mathbf{x}}A\big\|_{\mathcal{B}} \le \sup_x \big\|\mathrm{D}^2_x f\big\| . \tag{4.10}$$

The important point is that the linear map $(I - \mathrm{D}_{\mathbf{x}}A)$ is invertible. To see this, we must solve

$$(I - \mathrm{D}_{\mathbf{x}}A)\mathbf{h} = \mathbf{v}$$

for any $\mathbf{v} \in \mathcal{B}$. Here, I is the identity operator. For each $\mathbf{x} \in \mathcal{B}$, we define a linear operator $L_{\mathbf{x}}$ on $\mathcal{B}$ by the Neumann series

$$(L_{\mathbf{x}}\mathbf{v})_n = \sum_{j=0}^{\infty} \mathrm{D}_{x_{n-j}} f^j \Pi_{E^{\mathrm{s}}_{x_{n-j}}} v_{n+j} - \sum_{j=1}^{\infty} \mathrm{D}_{x_{n+j}} f^{-j} \Pi_{E^{\mathrm{u}}_{x_{n+j}}} v_{n+j} ,$$

where $\Pi_{E^s_x}$ and $\Pi_{E^u_x}$ denote the (orthogonal) projections on the stable and unstable subspaces at x.

Using the hyperbolicity assumption Definition 4.16—see below for a caveat—it follows at once that these operators are bounded and moreover

$$\sup_{\mathbf{x}\in\mathcal{B}} \left\|L_{\mathbf{x}}\right\|_{\mathcal{B}} \leq \frac{2C}{1-\lambda} . \tag{4.11}$$

Exercise 4.64. *It is left to the reader to check that for any* $\mathbf{x} \in \mathcal{B}$, $L_{\mathbf{x}}$ *is the inverse of* $I - \mathrm{D}_{\mathbf{x}}A$.

To prove the existence of a fixed point $\mathbf{y}$ for A near $\mathbf{x}$, we use Newton's method (with fixed slope) for another map Φ given by

$$\Phi(\mathbf{z}) = \mathbf{z} + w(\mathbf{z}) ,$$

where

$$w(\mathbf{z}) = -\mathbf{z} + (\mathrm{D}_{\mathbf{x}}A - I)^{-1}(A(\mathbf{z}) - \mathbf{z}) .$$

Clearly $\Phi(\mathbf{z}) = 0$ is equivalent to $A\mathbf{z} = \mathbf{z}$. From the definition of w it follows immediately that

$$\mathrm{D}_{\mathbf{z}}w\big|_{\mathbf{z}=\mathbf{x}} = 0 ,$$

and using the estimate (4.11) and (4.10) we get for any $\mathbf{z} \in \mathcal{B}$

$$\left\|\mathrm{D}_{\mathbf{z}}w\right\|_{\mathcal{B}} \leq 2C\frac{\sup_x \left\|\mathrm{D}^2_x f\right\|}{1-\lambda}\left\|\mathbf{z}-\mathbf{x}\right\|_{\mathcal{B}} .$$

Define $\beta > 0$ by

$$\beta = \frac{1-\lambda}{8C\sup_x \left\|\mathrm{D}^2_x f\right\|} ,$$

and let V be the ball of radius β around $\mathbf{x}$. We have for any $\mathbf{y}_1$ and $\mathbf{y}_2$ in V,

$$\left\|w(\mathbf{y}_1) - w(\mathbf{y}_2)\right\|_{\mathcal{B}} \leq \frac{1}{2}\left\|\mathbf{y}_1 - \mathbf{y}_2\right\|_{\mathcal{B}} .$$

We now use the fixed point theorem in Banach spaces, (see (Crandall and Rabinowitz 1971) or (Dieudonné 1968)): If

$$\left\|w(\mathbf{x})\right\|_{\mathcal{B}} < \frac{\beta}{4} ,$$

then Φ is a homeomorphism of V onto a neighborhood W of 0. This immediately implies that in V there is a unique fixed point of A satisfying the conclusions.

We cheated somewhat because we assumed that the hyperbolicity assumption Definition 4.16 could be applied to the pseudo-orbit. This is obviously true if the fields of stable and unstable directions are constant as for example in the case of the dissipative baker map Example 2.27. In the general case this can be fixed using a continuity argument leading to an inessential complication of the otherwise conceptually very simple proof. □

4.4.1 Sensitive Dependence on Initial Conditions

This is the celebrated *butterfly effect* (see Fig. 5.11). Let us assume the conditions of Theorem 4.59. Then we have

Lemma 4.65. *Consider a hyperbolic dynamical system. There is an* $\varepsilon > 0$ *such that the following is true. Let* $\{x_n\}$ *and* $\{y_n\}$ *be any two orbits. Then either* $x_k = y_k$ *for all* n *or there is at least one* k *for which* $|x_k - y_k| \geq \varepsilon$.

Definition 4.66. *The property of separation of any two orbits is called* expansivity.

In other words, in a hyperbolic system, two orbits cannot stay forever arbitrarily close to each other. Note that ε does not depend on $\{x_n\}$ or $\{y_n\}$.

Remark 4.67. Some confusion appears frequently about what one can do with the butterfly effect since the popular literature is abundant with butterflies that create tornados in (mostly) far-away places (see (Witkowski 1995) for an inventory of these horror scenarios). But, in fact, the lemma tells us only that each small perturbation risks growing, but in no way it tells us how this will happen. Thus, there is, fortunately, no way to place a butterfly cleverly in such a way as to produce a tornado in any prescribed place or to produce even any tornado at all.

Proof (of Lemma 4.65). In this proof and in the examples of Sects. 4.4.2–4.4.3, and 4.5, we will always use a clever choice of a pseudo-orbit. Here, for example, we choose

$$z_n = \begin{cases} x_n , & \text{if } n \text{ is even} \\ y_n , & \text{if } n \text{ is odd} \end{cases} .$$

Then one has $|f(z_n) - z_{n+1}| = |x_{n+1} - y_{n+1}|$ for all n. If $|x_n - y_n| < \varepsilon$ for all n then $\{z_n\}$ is an ε-pseudo-orbit and there is a *unique* orbit which $K\varepsilon$-shadows it, but since both $\{x_n\}$ and $\{y_n\}$ are candidates, we conclude that $x_n = y_n$ for all n (since both are shadowing $\{z_n\}$). And the other alternative is indeed $|x_n - y_n| \geq \varepsilon$ for at least one n. □

We come back to this concept in Sect. 5.8.

4.4.2 Complicated Orbits Occur in Hyperbolic Systems

Assume that x_* is a fixed point of a map f and that the stable and unstable manifolds of x_* intersect at some other point $\bar{x}$. This is called a *homoclinic point*. To keep things simple, we assume throughout the necessary hyperbolicity. Choose a small neighborhood $\mathcal{N}$ of x_* and a sequence of integers

$$\ldots, n_{-2}, n_{-1}, n_0, n_1, n_2, \ldots ,$$

which are sufficiently large, $n_i > N$ for all i. Then there is an orbit which "tours" the loop $x_* \to \bar{x} \to x_*$ as follows: It stays n_i times in $\mathcal{N}$, comes close to $\bar{x}$, returns to $\mathcal{N}$ and stays n_{i+1} times there, and so on.

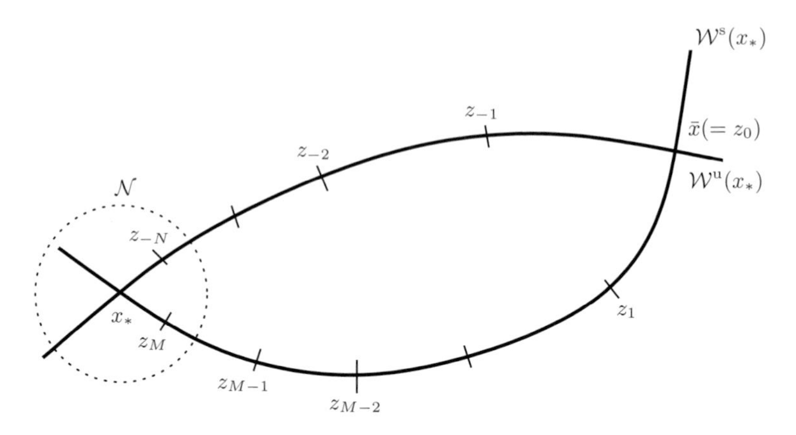

Fig. 4.7. The construction of an approximate periodic orbit visiting a homoclinic point $\bar{x}$ and the fixed point x_*

Again, such an orbit can easily be found by considering first the following pseudo-orbit (see Fig. 4.7). Let $z_i = f^i(\bar{x})$ be the images and preimages of $\bar{x}$. They will accumulate near x_* because $\bar{x}$ is a homoclinic point for x_*, and in particular on the stable manifold of x_*. Assume that z_{-N} is the first backward point in $\mathcal{N}$ and z_M the first forward point (see Fig. 4.7). Then we choose a pseudo-orbit

$$\ldots,\underbrace{x_*,\ldots,x_*}_{n_{-1}},z_{-N},z_{-N+1},\ldots,z_M,\underbrace{x_*,\ldots,x_*}_{n_0},\ldots$$

that is, we put the "right" number of x_* and then a sequence of z's, repeating indefinitely with multiplicity of the x_* being $\ldots,n_{-1},n_0,n_1,\ldots$. Note that $f(z_i)-z_{i+1}=0$, and $f(x_*)-x_*=0$. And, if the diameter of $\mathcal{N}$ is small enough, then $|f(x_*)-z_{-N}|<\varepsilon$ and $|f(z_M)-x_*|<\varepsilon$ so we have a pseudo-orbit which does the right gymnastics. And we can conclude immediately that there is a true orbit which does the same, previously prescribed gymnastics.

4.4.3 Change of Map

Using the gymnastics with shadowing, one can prove the following:

Theorem 4.68. *Let f be a map satisfying Axiom A. Let g be a map which is sufficiently close to f:*

$$\sup_{x\in M}|f(x)-g(x)|\leq\varepsilon_0\;.$$

Then the periodic points of g are in 1 to 1 correspondence with those of f and there is a continuous *change Φ of coordinates such that*

$$\Phi\circ f=g\circ\Phi\;.$$

Remark 4.69. The map Φ is defined only on nonwandering points. One should not expect Φ to be differentiable, because differentiability of Φ would mean that the derivatives at the fixed points of f and g must be the same. And the same must hold for the product of derivatives around any periodic point.

Exercise 4.70. *Check these statements.*

The idea of the proof is to observe that each orbit of g is a pseudo-orbit of f, since they are close to each other. In particular, the periodic orbits of g lead to periodic pseudo-orbits and therefore there is for each periodic point x of g a true periodic point y of f close-by. And now Φ is defined by $\Phi(y) = x$. Since the periodic points are supposed to be dense in Λ the map Φ is well-defined (and continuous). For the detailed proof of Theorem 4.68, see (Bowen 1975).

Remark 4.71. One can illustrate the Theorem 4.68 by observing what happens in a concrete case numerically. If we perturb the cat map respectively a torus rotation, the difference is very visible (see Fig. 4.8). The torus map is $x \mapsto x + 0.1234 \pmod 1$, $y \mapsto y + 0.1234 \cdot \sqrt{2} \pmod 1$. The perturbation which is added after each iteration $\pmod 1$ is

$$\varepsilon \exp\Big(-\delta\big((x-0.5)^2 + (y-0.5)^2\big)\Big) ,$$

with $\varepsilon = 0.3$ and $\delta = 0.65$. Each figure has 20,000 points. (This example is due to Ruelle.)

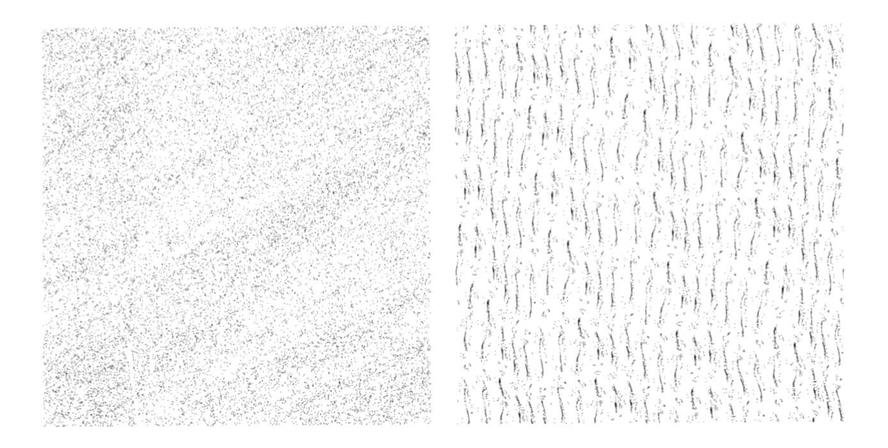

Fig. 4.8. Perturbations of the cat map (left) and of a torus rotation (right) by a (Gaussian) function with maximum at the center of each image. Note that the orbit structure of the cat map is still quite homogeneous while that of the torus map is strongly perturbed

4.5 Construction of Markov Partitions

This is probably the most spectacular result in hyperbolic dynamics. It tells us that, roughly, hyperbolic systems are not much more complicated than our friend $x \mapsto 2x \pmod 1$ or the cat map. And it will also allow a comparison of hyperbolic systems to problems in statistical mechanics, with its wealth of concepts and results.

But let us begin with a basic concept for hyperbolic systems: Given two points x, y (which are close to each other) we define

$$[x,y] = \mathcal{W}^{\mathrm{s}}_{\mathrm{loc}}(x) \cap \mathcal{W}^{\mathrm{u}}_{\mathrm{loc}}(y) .$$

This *bracket operation* is a continuous function of its arguments (see also Fig. 4.9).

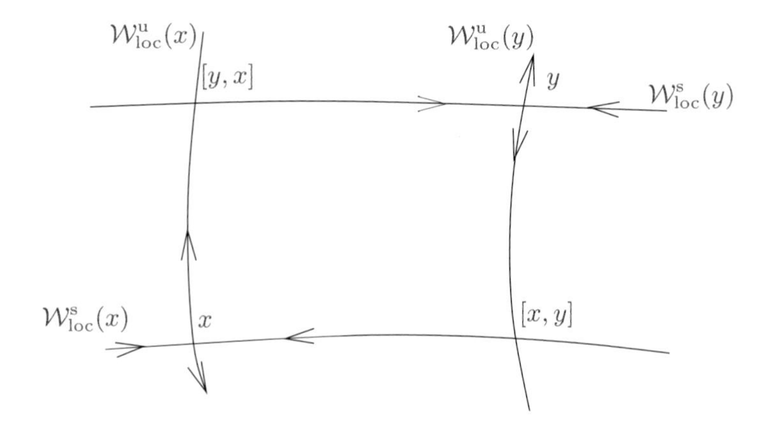

Fig. 4.9. The bracket operation. The point $[x,y]$ is the intersection of the local stable manifold of x and the local unstable manifold of y

Definition 4.72. *A* rectangle $\mathcal{R}$ *is a set which is closed under the* $[\cdot,\cdot]$ *operation.*

In other words, if $x, y \in \mathcal{R}$ then also $[x,y]$ and $[y,x]$ are in $\mathcal{R}$. Note that a conventional rectangle in $\mathbb{R}^2$ is a rectangle in the sense of Definition 4.72 (if we assume that the coordinate axes are stable and unstable manifolds). While this helps the intuition, it should be noted that the rectangles we encounter below in dynamical systems (when intersected with the attractor) can be quite worse, for example, a product of a Cantor set and a line as in the dissipative baker's map (see Fig. 2.12).

Definition 4.73. *A* Markov partition *of a hyperbolic set* Λ *is a* finite *cover of* Λ *by proper rectangles* $\mathcal{R}_i$ *with disjoint interiors, such that if* $x \in \mathrm{int}(\mathcal{R}_i)$ *and* $f(x) \in \mathrm{int}(\mathcal{R}_j)$ *then*

$$f\big(\mathcal{W}^{\mathrm{s}}_{\mathrm{loc}}(x;\mathcal{R}_i)\big) \subseteq \mathcal{W}^{\mathrm{s}}_{\mathrm{loc}}(f(x);\mathcal{R}_j) ,$$
$$f\big(\mathcal{W}^{\mathrm{u}}_{\mathrm{loc}}(x;\mathcal{R}_i)\big) \supseteq \mathcal{W}^{\mathrm{u}}_{\mathrm{loc}}(f(x);\mathcal{R}_j) .$$

Here, int is the interior of a set. This definition says that boundaries of rectangles must be mapped in a special way. It is a generalization of the idea that the end points of the partition $[0,1] = [0,1/2] \cup [1/2,1]$ (namely each of the 3 points $0, 1/2, 1$ is mapped by $x \mapsto 2x \pmod 1$ onto one of the 3 points).

Remark 4.74. Some authors require in addition that the partition be a *generating partition*; that is, that each infinite code corresponds to a *unique* point in phase space.

Example 4.75. In Figs. 4.10 and 4.11 we show a Markov partition with three pieces for the cat map (which is not generating). If we number the rectangles in Fig. 4.10 as red $= 1$, blue $= 2$, green $= 3$, we see that transitions are possible according to the *transition matrix* (with $A_{ij} = 1$ if the transition from j to i is possible and $A_{ij} = 0$, otherwise):

$$A = \begin{pmatrix} 1 & 0 & 1 \\ 1 & 1 & 1 \\ 0 & 1 & 1 \end{pmatrix} .$$

Figures 4.12–4.13 show a partition for the simpler "square root" of the cat map, given by

$$\begin{pmatrix} x \\ y \end{pmatrix} \mapsto \begin{pmatrix} x + y \pmod 1 \\ x \end{pmatrix} , \tag{4.12}$$

and this partition *is* generating. If we number the rectangles in Fig. 4.12 as red $= 1$, blue $= 2$, green $= 3$, violet $= 4$, we see that transitions are possible according to the Markov matrix

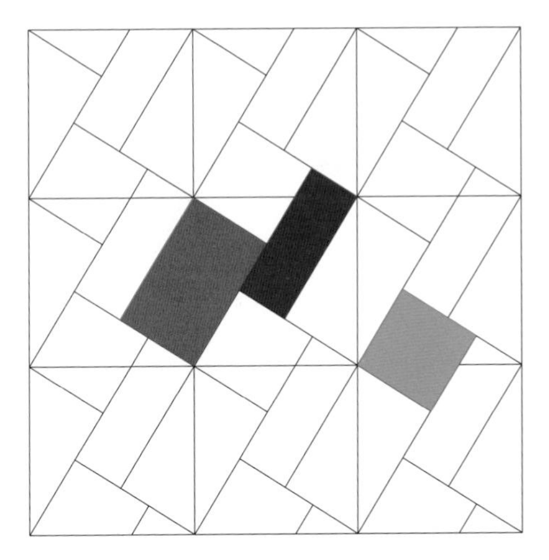

Fig. 4.10. Nine copies of the unit square for the cat map, and three rectangles forming a partition. The nine copies serve to be able to show each rectangle as a whole, but one should remember that the playground is one square with periodic boundary conditions

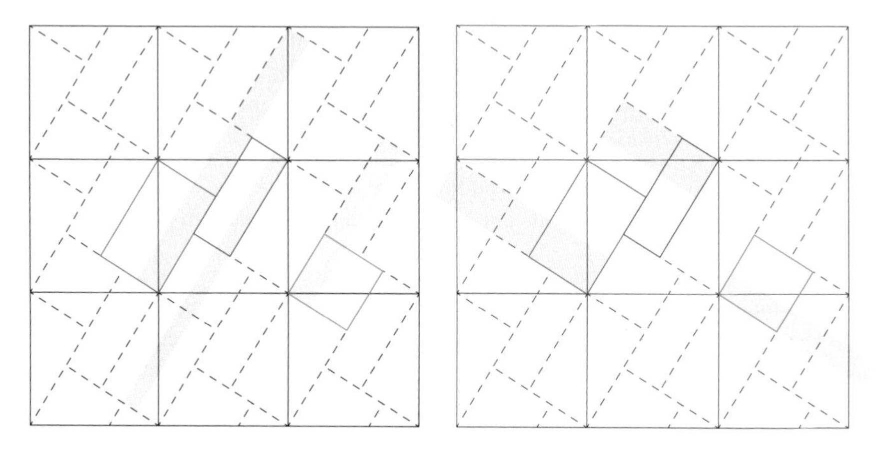

Fig. 4.11. The left part shows how the stable boundaries of each rectangle are mapped by the cat map f onto stable boundaries (those are the lines from north-east to south-west). The right part shows how f^{-1} maps the unstable boundaries onto unstable boundaries

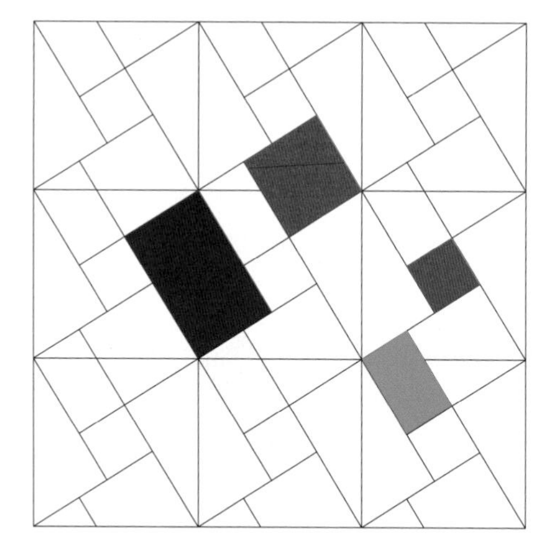

Fig. 4.12. Nine copies of the unit square for the "square root" map (4.12), and four rectangles forming a generating partition. The nine copies serve to be able to show each rectangle as a whole, but one should remember that the playground is one square with periodic boundary conditions

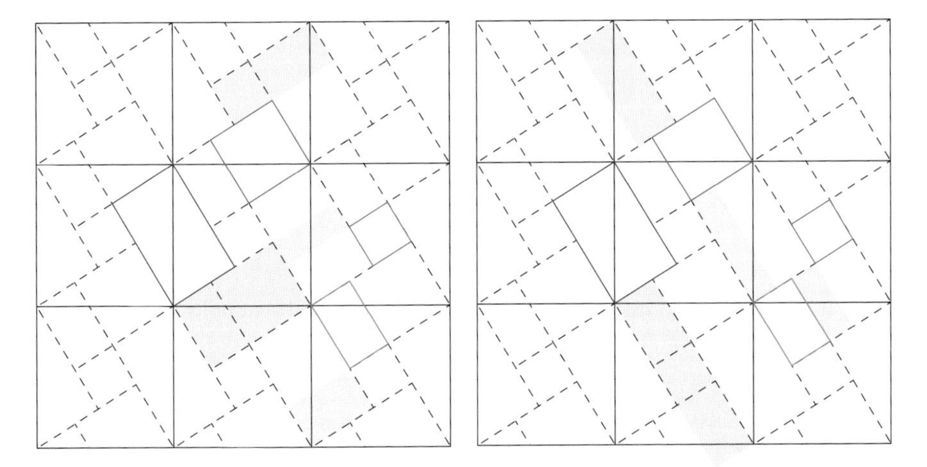

Fig. 4.13. The left part shows how the stable boundaries of each rectangle are mapped by the cat map f onto stable boundaries (those are the lines from north-east to south-west). The right part shows how f^{-1} maps the unstable boundaries onto unstable boundaries. The representations have been moved mod 1 to make the different rectangles nonoverlapping

$$A = \begin{pmatrix} 0\,1\,0\,1 \\ 1\,0\,1\,0 \\ 0\,1\,0\,0 \\ 0\,1\,0\,0 \end{pmatrix} .$$

The largest eigenvalue λ of this matrix is equal to unstable eigenvalue $(1+\sqrt{5})/2$ of the transformation (4.12). See (Baladi 2000; Collet and Eckmann 2004) for a detailed discussion of these eigenvalues.

Exercise 4.76. *Show that there is a Cantor set of points of the partition of Fig. 4.10, all of whose forward codes are the same. Hint: These are points which fall on the red rectangle or its translate in the unstable direction. Thus, the partition is not generating.*

Theorem 4.77 (Bowen). *Let Λ be a hyperbolic set for the dynamical system f and let $\varepsilon > 0$ be sufficiently small. Then there is a generating Markov partition whose (finite number of) pieces have diameter less than ε.*

The proof is complicated in details, but quite straightforward conceptually. It uses extensively the shadowing lemma to construct the rectangles.

Before we start with the proof, we describe how to characterize stable and unstable manifolds in terms of coding, by illustrating this for the baker's map.

Example 4.78. We recall the baker's transformation

$$\begin{pmatrix} x \\ y \end{pmatrix} \mapsto \begin{pmatrix} 2x \pmod 1 \\ y' \end{pmatrix} ,$$

with

$$y' = \begin{cases} \frac{1}{2}y\,, & \text{if } 0 \le x < \frac{1}{2}\,, \\ \frac{1}{2}y + \frac{1}{2}\,, & \text{if } \frac{1}{2} \le x < 1\,. \end{cases}$$

It has a Markov partition with the pieces

$$I_0 = \{(x,y) \mid x \le \tfrac{1}{2}\}\,,$$
$$I_1 = \{(x,y) \mid x \ge \tfrac{1}{2}\}\,.$$

A point (x, y) can then be coded by (see Exercise 2.28 and Example 3.2)

$$\boldsymbol{\sigma}(x,y) = \dots j_3 j_2 j_1 . i_1 i_2 i_3 \dots$$

with the $i_k, j_k \in \{0, 1\}$ defined by the binary expansion

$$x = 0.i_1 i_2 i_3 \dots\,,$$
$$y = 0.j_1 j_2 j_3 \dots\,.$$

We shall fix the convention that $\sigma_0 = i_1$, $\sigma_k = i_{k+1}$, $\sigma_{-k} = j_k$, when $k > 0$. The baker's map maps the point (x, y) to the point

$$x' = 0.i_2 i_3 \dots\,,$$
$$y' = 0.i_1 j_1 j_2 j_3 \dots\,,$$

and therefore we get $\sigma_0(x', y') = i_2$ and the map is nothing but a *shift of the symbol sequence*. Note that σ_0 indicates whether x is in I_0 or I_1. It is easy to understand now that the stable manifold of the point (x, y) is just given by all those (x', y') whose symbolic sequence $\{\sigma'_j\}$ *coincides* for all sufficiently large j with $\{\sigma_j\}$:

$$\mathcal{W}^{\mathrm{s}}(x,y) \Leftrightarrow \{\boldsymbol{\sigma} \mid \sigma_{j-1} = i_j \text{ for } j > j_0\}\,. \tag{4.13}$$

After n iterations, any point in $\mathcal{W}^{\mathrm{s}}(x, y)$ will have the property that

$$f^n(x,y) \text{ will correspond to an element of } \{\boldsymbol{\sigma} \mid \sigma_{j-1-n} = i_j \text{ for } j > j_0\}\,,$$

which means that the sequences coincide for all indices j with $|j| < n - j_0$. As $n \to \infty$ this means that the images of the points converge to each other.

Proof of Theorem 4.77.

i) Choose a δ_0 sufficiently small so that for all $x \in \Lambda$,

$$\mathcal{W}^{\mathrm{s}}(x) \cap \mathcal{W}^{\mathrm{u}}(x) \cap B_{\delta_0}(x) = \{x\}\,.$$

In other words, in this small neighborhood the stable and unstable manifolds cross only at x itself, nowhere else. Such a neighborhood exists, because, as we already saw, the angle between E^{u}_x and E^{s}_x is bounded below away from zero uniformly in x. Furthermore, one can show that the curvatures of $\mathcal{W}^{\mathrm{s}}(x)$ and $\mathcal{W}^{\mathrm{u}}(x)$ are uniformly bounded.

ii) Choose $\delta = \delta_0/10$.
iii) Using the notation of Theorem 4.63, choose $\varepsilon(= \delta/K)$ such that every ε-pseudo-orbit is δ-shadowed by a true orbit.
iv) Choose $\gamma \in (0, \varepsilon)$ such that $|x - y| < \gamma$ implies $|f(x) - f(y)| < \varepsilon/2$.
v) Choose p points $y_1, \dots, y_p$ such that the balls $B_{\gamma/2}(y_i)$, $i = 1, \dots, p$ cover Λ.
vi) Define the $p \times p$ transition matrix A by

$$A_{ij} = \begin{cases} 1\,, & \text{if } |f(y_i) - y_j| < \varepsilon \\ 0\,, & \text{otherwise} \end{cases} .$$

One defines the *admissible sequences* $\Sigma(A)$ by

$$\Sigma(A) = \{\boldsymbol{\sigma} = \{\dots, \sigma_{-1}, \sigma_0, \sigma_1, \dots\}\} ,$$

with $\sigma_i \in \{1, \dots, p\}$ and $A_{\sigma_i, \sigma_{i+1}} = 1$ for all i. In other words, the sequences of numbers such that $|f(y_{\sigma_i}) - y_{\sigma_{i+1}}| < \varepsilon$.
vii) By construction, to every $\boldsymbol{\sigma} \in \Sigma(A)$ there is associated the ε-pseudo-orbit $\dots, y_{\sigma_{-1}}, y_{\sigma_0}, y_{\sigma_1}, \dots$. Thus, by the shadowing lemma there is a *unique* true orbit, which δ-shadows it. Let x_0 be the point of this true orbit which corresponds to y_{σ_0}. We define the map Θ by

$$\Theta : \Sigma(A) \to \Lambda\,, \qquad \Theta(\boldsymbol{\sigma}) = x_0\,.$$

viii) We define the set

$$T_k = \Theta(\Sigma_k(A))\,,$$

where $\Sigma_k(A)$ are all those sequences for which $\sigma_0 = k$. That is, T_k is the image under Θ of all these sequences. *We shall see that these T_k are basically the desired rectangles, except that they overlap.*[1]
ix) From the definition of Θ one can conclude that

$$f \circ \Theta = \Theta \circ \text{shift}$$

and furthermore that the image is *onto*: $\Theta(\Sigma(A)) = \Lambda$. The first property is obvious from the definition. To prove the second, choose $x \in \Lambda$, and let i_j be given by $|f^j(x) - y_{i_j}| < \gamma$. From iv. it follows that $|f(y_{i_j}) - y_{i_{j+1}}| < \varepsilon$. Thus the sequence $\{y_{i_j}\}$ forms an ε-pseudo-orbit and therefore, the sequence $\boldsymbol{\sigma}$ formed by the $\sigma_j = i_j$ is in $\Sigma(A)$. It is δ-shadowed by the orbit of x and thus by the uniqueness of shadowing, $\Theta(\boldsymbol{\sigma}) = x$.
x) The map Θ is continuous. If $\boldsymbol{\sigma} - \boldsymbol{\sigma}' \to 0$ (in the sense of the ultrametric defined earlier (3.1)), this means that the sequences coincide on longer and longer stretches. Then the points $x = \Theta(\boldsymbol{\sigma})$ and $x' = \Theta(\boldsymbol{\sigma}')$ have the property that $|f^j(x) - f^j(x')| < \varepsilon$ on longer and longer stretches. But this means that $x \to x'$ due to hyperbolicity.

[1] But that will be fixed later.

xi) Here comes another surprising fact: Define $[\boldsymbol{\sigma}, \boldsymbol{\sigma}']$ by fixing its $j\mathrm{th}$ component as

$$[\boldsymbol{\sigma}, \boldsymbol{\sigma}']_j = \begin{cases} \sigma_j\,, & \text{if } j \geq 0 \\ \sigma'_j\,, & \text{if } j \leq 0 \end{cases} .$$

If $\boldsymbol{\sigma}$ and $\boldsymbol{\sigma}'$ are in the same set $\Sigma(A)_k$, that is, $\sigma_0 = \sigma'_0 = k$, then

$$\Theta([\boldsymbol{\sigma}, \boldsymbol{\sigma}']) = [\Theta(\boldsymbol{\sigma}), \Theta(\boldsymbol{\sigma}')] .$$

This is obvious from (4.13).

xii) So, we have finally found the required rectangles. A small problem remains, namely that the rectangles might overlap. This is repaired by cutting up the rectangles into smaller pieces, by prolonging the edges one has obtained so far. For an account of this, see (Gallavotti, Bonetto, and Gentile 2004, pp. 129ff). One obtains at most p^2 rectangles in this way, and the construction is terminated.

xiii) The partition one constructs in this way is generating since the map Θ is 1–1.

Exercise 4.79. *Check that at the end of the construction, the rectangles have the Markov property as defined in Definition 4.73.*

5

Invariant Measures

5.1 Overview

We start by giving a brief overview of some ideas and results to which we will come back later in detail.

We consider again the standard model

$$f_\nu : x \mapsto 1 - \nu x^2 , \tag{5.1}$$

which for fixed $\nu \in [0, 2]$ is a dynamical system mapping the interval $I = [-1, 1]$ into itself. Fix now a ν and choose an $x_0 \in I$. We iterate,

$$x_n = f_\nu(x_{n-1}) , \text{ for } n = 1, 2, \dots .$$

What will happen? In the first case, $\nu < 0.75$ most initial points have orbits converging to the fixed point, whose coordinates are given by $x_* = 1 - \nu x_*^2$ (with $x_* \in [-1, 1]$). This point attracts orbits because the derivative satisfies $|f'_\nu(x_*)| < 1$.

Exercise 5.1. *Show that there can be at most* one *such attracting point for a given* ν.

Remark 5.2. One can in fact show that any f_ν of (5.1) (when $\nu \in (0, 2]$) can have at most one stable periodic orbit. For a generalization to maps with negative Schwarzian derivative, see, for example, (Collet and Eckmann 1980).

However, things are much more interesting when $\nu = 2$, as can be seen in Fig. 5.1, where we histogrammed the first 50,000 iterates of the map into 400 bins (in $[0, 1]$). The curve which fits the histogram is (if area is normalized to 1)

$$h(x) = \frac{1}{\pi\sqrt{1 - x^2}} .$$

Note the important fact that one obtains the *same* distribution for almost all initial points (with respect to the Lebesgue measure). Exceptional points would be for example the fixed points $x_* = -1$ and $x_* = 1/2$ and their preimages.

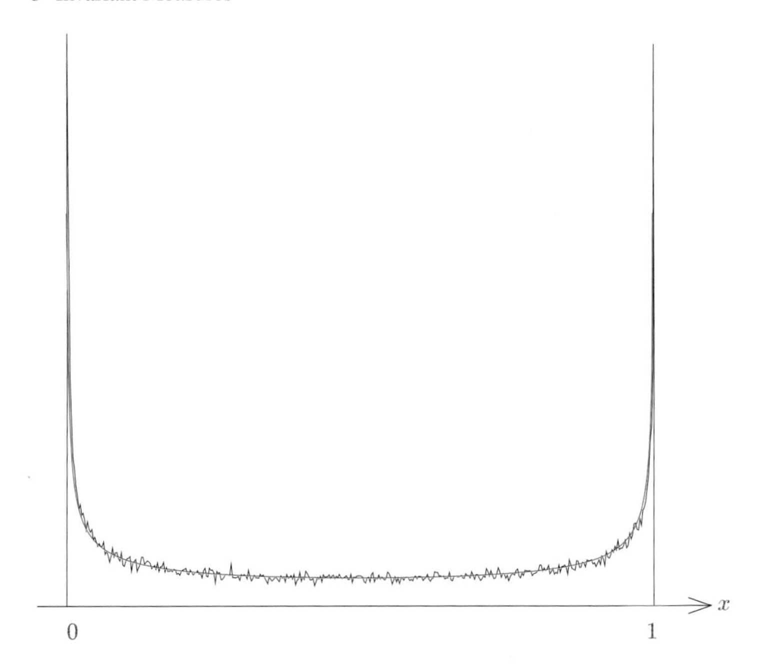

Fig. 5.1. The histogram and the invariant density $h(x)$ (in red) for the map $f(x) = 1 - 2x^2$. The histogram uses 50,000 iterates in 400 bins, $x_0 = 0.7$. The vertical axis is arbitrary

As different values of ν are chosen, the invariant density can take a large variety of forms, not at all like that of Fig. 5.1 as is shown for the Misiurewicz–Ruelle map (Ruelle 1977; Misiurewicz 1981) or for the Feigenbaum map in Fig. 5.2 for $f(x) = 1 - \nu_{\mathrm{M}} x^2$ and $f(x) = 1 - \nu_{\mathrm{F}} x^2$ with $\nu_{\mathrm{M}} \approx 1.5436$ and $\nu_{\mathrm{F}} \approx 1.401155$. Note that the histogram for the Feigenbaum map is very ragged, indeed, the support of the invariant measure is a Cantor set.

From the construction of the histogram we see that a measure μ is defined on I. Its weight on $J \subset I$ is given by

$$\mu(J) = \lim_{n\to\infty} \frac{\#\text{ of } i \in \{1,\ldots,n\} \text{ for which } f^i(x_0) \in J}{n} . \tag{5.2}$$

These measures are invariant:

Definition 5.3. *Let f be a measurable map $M \to M$. A measure μ is called* invariant *(for f) if*

$$\mu(f^{-1}(E)) = \mu(E) ,$$

for every measurable set E.

Recall that if f is not invertible, f^{-1} is the set theoretic inverse, namely

$$f^{-1}(E) = \{x \,|\, f(x) \in E\} . \tag{5.3}$$

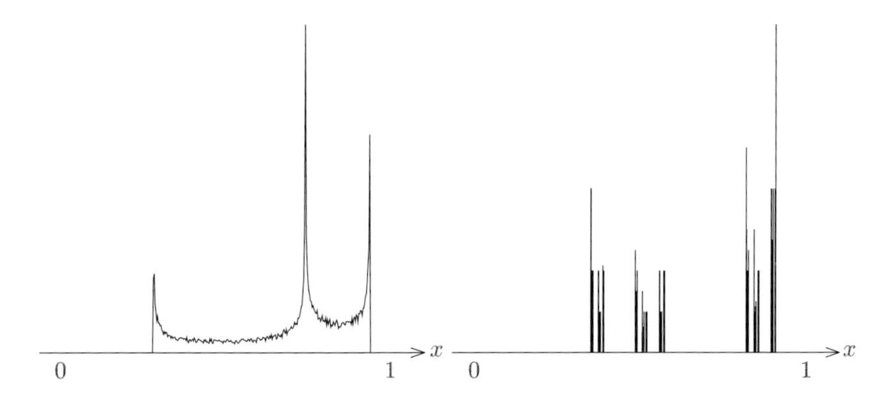

Fig. 5.2. Histograms for the Misiurewicz–Ruelle map (left) and the Feigenbaum map (right)

Exercise 5.4. *Show that the construction of (5.2) defines an invariant measure.*

Remark 5.5. The precise relation between the invariant measure and the histograms will be provided by the ergodic theorem, which says that time averages of observables are equal to space averages.

We need a notion of "not too irregular measure":

Definition 5.6. *A measure is called* absolutely continuous *(with respect to Lebesgue measure) if*

$$\mu(\mathrm{d}x) = h(x)\mathrm{d}x \quad \textit{with} \quad h \in \mathrm{L}^1 ;$$

that is, $\int h(x)\mathrm{d}x < \infty$.

We call such measures *acim*.

Note that by measure we always mean a positive measure, so that $h(x) \geq 0$. Here, we wrote the Lebesgue measure as "$\mathrm{d}x$." A more general definition is

Definition 5.7. *A measure* μ_1 *is called* absolutely continuous *with respect to another measure* μ_2 *if*

$$\mu_2(E) = 0 \quad \Rightarrow \quad \mu_1(E) = 0 ,$$

for every measurable set E. *One writes then* $\mu_1 \prec \mu_2$.

In other words, every set of "volume" 0 for μ_2 also has measure 0 for μ_1.

Remark 5.8. The delta function (δ-measure) is not absolutely continuous with respect to Lebesgue measure. In more detail, consider $\delta_{x_0} = \delta(x - x_0)$. The δ_{x_0}-measure of the set $\{x_0\}$ is 1, but $\lambda(\{x_0\}) = 0$, where λ is Lebesgue measure. Thus, $\delta_{x_0} \not\prec \lambda$. On the other hand, $\lambda([x_0 - 2, x_0 - 1]) = 1$, while $\delta_{x_0}([x_0 - 2, x_0 - 1]) = 0$ so that $\lambda \not\prec \delta_{x_0}$. This also shows that two measures are not in general "comparable."

Remark 5.9. The Feigenbaum measure is neither a δ function, nor absolutely continuous with respect to Lebesgue measure.

Remark 5.10. We discuss here the special case of $\nu = 2$; that is, $f(x) = 1 - 2x^2$. In this case one can compute the invariant measure explicitly, by the following trick. Consider first the map $g(y) = 1 - 2|y|$. This has the (normalized) invariant measure $\frac{1}{2}\mathrm{d}y$. But $g(y) = t\big(f(t^{-1}(y))\big)$, where

$$t(x) = \frac{4}{\pi}\arcsin\sqrt{\frac{x+1}{2}} - 1\,, \quad t^{-1}(y) = 2\sin^2\left((1+y)\frac{\pi}{4}\right) - 1\,,$$

and therefore the invariant density for f equals $\frac{1}{2}\mathrm{d}y = \frac{1}{2}t'(x)\mathrm{d}x = 1/(\pi\sqrt{(1-x^2)})$. (Hint: it is easier to consider the map $\hat{f}(\hat{x}) = 4\hat{x}(1-\hat{x})$ on the interval $[0,1]$ and to compare it to $\hat{g}(\hat{y}) = 1-2|\frac{1}{2}-\hat{y}|$ and then to transform $\hat{x} = 2x-1$, $\hat{y} = 2y-1$. Then, $\hat{t}^{-1}(y) = \sin^2(\hat{y}\frac{\pi}{2})$, and the identification follows from a doubling of the "angle" $\hat{y}$.)

Remark 5.11. Many other invariant measures exist, which are not absolutely continuous with respect to the Lebesgue measure. For example, if we consider the map $f : x \mapsto 2x \pmod 1$, then $\mathrm{d}x$ is an acim, and it is the only absolutely continuous one. But others can be obtained: If $x_1, \dots, x_n$ is a periodic orbit of f (with all points distinct) then

$$\mu(x) = \frac{\delta(x-x_1) + \delta(x-x_2) + \cdots + \delta(x-x_n)}{n}$$

is also an invariant measure. But there are more interesting examples: Let

$$I_{j,k} = \left[\frac{j}{2^k}, \frac{j+1}{2^k}\right)\,, \text{ with } 0 \le j \le 2^k - 1\,.$$

Define $\#j$ as the number of 1's in the binary representation of j. So, $\#0 = 0$, $\#1 = 1$, $\#2 = 1$, $\#3 = 2$, and so on. Give yourself a $p \in [0,1]$ and define a measure μ_p by setting

$$\mu_p(I_{j,k}) = p^{\#j}(1-p)^{k-\#j}\,.$$

For $p = \frac{1}{2}$ this is the Lebesgue measure, but for $p \neq \frac{1}{2}$ this is not an acim, but it is invariant, since for $j < 2^{k-1}$ one has $f^{-1}(I_{j,k}) = I_{j,k+1} \cup I_{j+2^{k-1},k+1}$ and therefore

$$\begin{aligned} \mu_p(f^{-1}(I_{j,k})) &= p^{\#j}(1-p)^{k+1-\#j} + p^{1+\#j}(1-p)^{k+1-1-\#j} \\ &= p^{\#j}(1-p)^{k-\#j} = \mu_p(I_{j,k})\,. \end{aligned} \tag{5.4}$$

Using an approximation of general intervals by dyadic intervals, one can extend this relation to show that for any Borel set A one has $\mu(f^{-1}(A)) = \mu(A)$.

In fact, if $p \neq p'$, the corresponding measures are *mutually singular* (supported on disjoint sets). Although one cannot draw the densities of such singular measures, a representation on 2^{10} bins (these are really 2^{10} cylinder sets) is given in Fig. 5.3.

Remark 5.12. One may wonder why the Lebesgue measure plays such a natural and fundamental role in physics. Perhaps the best argument is that measuring apparatus

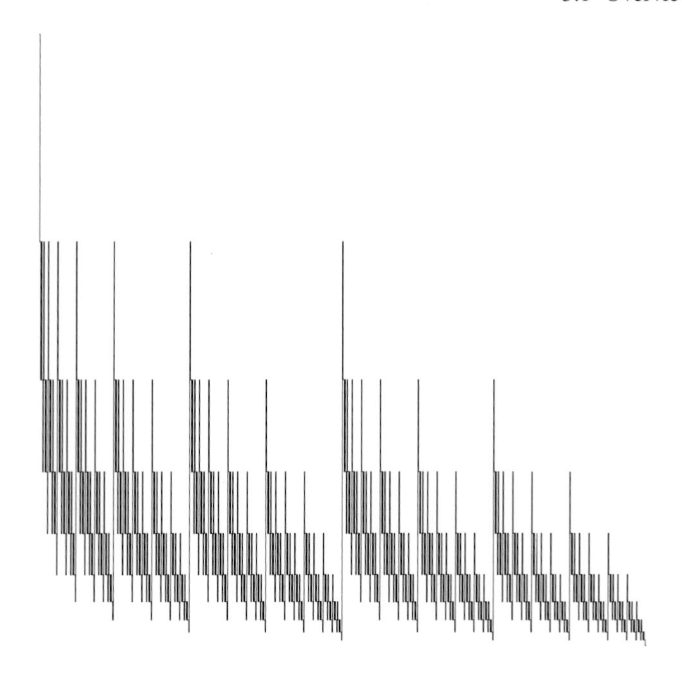

Fig. 5.3. A histogram in 2^{10} bins for $p = 0.6$ as described in (5.4)

are usually based on reading numbers in intervals. A more natural idea would be to use the rationals, but that seems highly impractical. The same question shows up in Sect. 5.7

Another argument goes back to Boltzmann. Consider the problem of choosing a distribution of initial conditions. Partition the phase space into small cubes of equal size. In the absence of any information about the initial condition, all the cubes are equally likely to contain it and this leads to the Lebesgue measure. This argument is connected to the maximum entropy argument of statistical mechanics (the Lebesgue measure has maximal entropy) and to the choice of the uniform prior in Bayesian statistics (we refer to (Jaynes 1982) for more on these questions). Another way to say it is that the choice of a measure different from the Lebesgue measure (singular with respect to the Lebesgue measure) amounts to introducing a bias in the problem, namely pretending that some supplementary information is available. Of course if such supplementary information is indeed available it should be included in the statistical choice of the initial condition.

Examples of rigorous results

i) The Lasota–Yorke theorem (Lasota and Yorke 1973): Let f be a map from $[0, 1]$ to itself which has the form of Fig. 5.4. In mathematical terms: There are a finite number of pieces, and f restricted to each piece is C^2, with $|f'| > \alpha > 2$. Then f has an absolutely continuous invariant measure (with respect to the Lebesgue

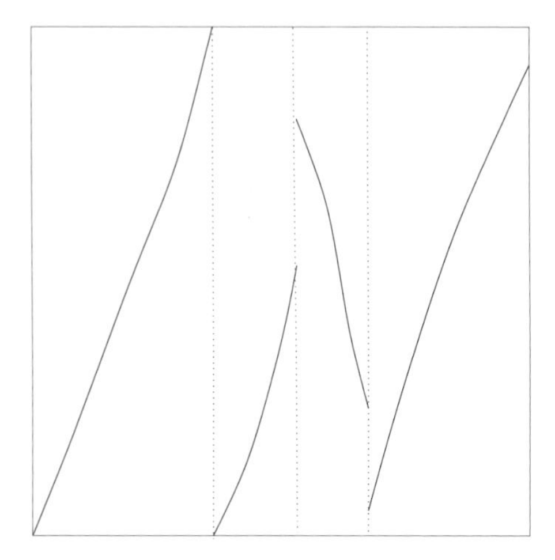

Fig. 5.4. The graph of a map f which satisfies the conditions of the Lasota–Yorke result

measure). It can have other invariant measures, but they would not be absolutely continuous.

ii) The Misiurewicz theorem (Misiurewicz 1981): Let f be a unimodal map with quadratic maximum. If the orbit of the critical point (i.e., the point x_{crit} where f takes its maximum) is periodic, then f has a unique acim. The map $x \mapsto 1-2x^2$ is such a map (see the histogram). Another such example is $f_\nu = x \mapsto 1 - \nu x^2$, where ν is so chosen that $f_\nu^3(0) = f_\nu^4(0)$. In all these cases, the invariant measure has at worst square root singularities $h(x) \sim \text{const}/\sqrt{|x - x_0|}$. See also (Ulam and von Neumann 1967; Ruelle 1977) for an earlier example of this phenomenon. The paper (Misiurewicz 1981) contains a much stronger result, namely that it is sufficient for the orbit of the critical point not to return too close the critical point.

iii) If (Collet and Eckmann 1983)

$$|(f_\nu^n)'(f_\nu(x_{\text{crit}}))| > C\mathrm{e}^{\varepsilon n} ,$$

for all $n \in \mathbb{Z}$ with $\varepsilon > 0$, then f_ν has a unique acim. For a review of extensions (in particular, requiring the condition only for $n \geq 0$) see (Świątek 2001; Avila and Moreira 2005; Nowicki 1998; Nowicki and van Strien 1991).

iv) One can ask for how many ν the map f_ν has an acim. Those of the examples covered by ii) form only a set of (Lebesgue) measure 0. But those satisfying the weaker condition in iii. form a set of positive measure. This set has so-called *Lebesgue points* ν_* by which one means

$$\lim_{\nu \to \nu_*} \frac{\lambda(\text{ good points in } [\nu, \nu_*])}{\lambda([\nu, \nu_*])} = 1 .$$

In the vicinity of these Lebesgue points, the set of parameter values ν for which f_ν has an acim not only has therefore positive measure, but is basically of full measure. The value $\nu_* = 2$ is such a Lebesgue point, and there are many more, in particular all those ν_* for which the map f_{ν_*} has the property that the orbit of the critical point lands on an unstable fixed (or periodic) point, see, for example, (de Melo and van Strien 1993).

Exercise 5.13. *This is a supplement to Exercise 3.26, concerning the dissipative baker's map. Show that there is an invariant measure on the attractor which can be identified as the product of the Lebesgue measure (in the horizontal direction) with the Bernoulli (product) measure of parameters* $(1/2, 1/2)$ *(in the vertical direction).*

Exercise 5.14. *Consider the Definitions 3.23 and 3.24 of attractor and of an attracting set. Show that if we have an attracting set* $\mathscr{A}$*, any invariant measure with support contained in the basin of this attracting set should have support in* $\mathscr{A}$.

5.2 Details

A basic goal of the study of dynamical systems is to understand the effect of time evolution on the state of the system, and primarily the long time behavior of the state (although transient behavior is sometimes very interesting and important in applications). Basically, two approaches to this problem can be found.

The first one can be called topological or differential–geometric, and it studies objects like periodic orbits, attractors, bifurcations, etc.

The second approach can be called ergodic and studies the behavior of trajectories from a measure theoretic point of view. This is interesting for systems with a large variety of dynamical behaviors depending on the initial condition.

In this course we will mostly deal with this second approach. Nevertheless, as we will see below, the interplay between the two approaches (geometric and ergodic) can be important and fruitful. Many observables, for example the Lyapunov exponents, are averages which have their origin in differential quantities of geometric nature. The ergodic approach then allows to study these averages, as well as their fluctuations.

To introduce in more detail the ergodic approach, let us consider the following question which is historically one of the first motivations of ergodic theory. Let A be a subset of the phase space Ω (describing the states with a property of interest, for example, that all the molecules in a room are in its left half). One would like to know, for example, in a long time interval $[0, N]$ (N large) how much time the system has spent in A, namely how often the state has the property described by A. Consider for simplicity a discrete time evolution. If χ_A denotes the characteristic function of the set A, the average time the system has spent in A over an interval $[0, N]$ starting in the initial state x_0 is given by

$$\mathcal{A}_N(x_0, A) = \frac{1}{N} \sum_{j=0}^{N-1} \chi_A\big(f^j(x_0)\big) .$$

It is natural to ask if this quantity has a limit when N tends to infinity. The answer may, of course, depend on A and x_0, but we can already make two important remarks. Assume the limit exists, and denote it by $\mu_{x_0}(A)$.

First, it is easy to check that the limit also exists for $f(x_0)$ and also for any $y \in f^{-1}(x_0) = \{z \mid f(z) = x_0\}$. Moreover, we have

$$\mu_{f(x_0)}(A) = \mu_y(A) = \mu_{x_0}(A) .$$

Second, the limit also exists if A is replaced by $f^{-1}(A)$ and has the same value, namely

$$\mu_{x_0}(f^{-1}(A)) = \mu_{x_0}(A) ;$$

indeed,

$$\chi_A(f^j(x)) = 1 \text{ if and only if } x \in f^{-j}(A) .$$

If one assumes that $\mu_{x_0}(A)$ does not depend on x_0 at least for Borel sets A (or some other sigma algebra but we will mostly consider the Borel sigma algebra below), one is immediately led to the notion of *invariant measure*. We reformulate Definition 5.3 in terms of sigma-algebras (cf. Table 3.1):

Definition 5.15. *A measure μ on a sigma-algebra $\mathscr{B}$ is invariant under the measurable map f if for any measurable set A*

$$\mu(f^{-1}(A)) = \mu(A) . \tag{5.5}$$

A similar definition holds for (semi)flows.

Remark 5.16. If μ_1 is an invariant measure for the dynamical system (Ω_1, f_1) and the map Φ is measurable and a.s. invertible, then the measure $\mu_2 = \mu_1 \circ \Phi^{-1}$ is an invariant measure for the dynamical system (Ω_2, f_2) with $\Omega_2 = \Phi(\Omega_1)$, $f_2 = \Phi \circ f_1 \circ \Phi^{-1}$.

Unless otherwise stated, when speaking below of an invariant measure we will assume that it is a probability measure. We will denote by $(\Omega, f, \mathscr{B}, \mu)$ the dynamical system with state space Ω, discrete time evolution f, $\mathscr{B}$ is a sigma-algebra on Ω such that f is measurable with respect to $\mathscr{B}$ and μ is an f-invariant measure on $\mathscr{B}$. As mentioned above $\mathscr{B}$ will most often be the Borel sigma-algebra and we tacitly assume this throughout.

Exercise 5.17. *Let (Ω, f) be a dynamical system and assume Ω is a metric space. Assume f is continuous, and let μ be a Borel measure. Show that μ is invariant if and only if*

$$\int g \circ f \, \mathrm{d}\mu = \int g \, \mathrm{d}\mu ,$$

for every continuous function g.

The goal of ergodic theory is to study systems $(\Omega, f, \mathscr{B}, \mu)$ and in particular their large time evolution. Consider now an observable g which we recall is a measurable function on the phase space Ω. We have seen above one such observable, namely the function χ_A which takes the value 1 if the state is in A (has the property described by A) and takes the value 0 otherwise. $(\Omega, \mathscr{B}, \mu)$ is a probability space and therefore g is a random variable on this probability space. More generally $\left(g \circ f^n, n \geq 0\right)$ is a discrete time, stationary, stochastic process. Therefore, we can apply all the ideas and results of the theory of stochastic process to dynamical systems equipped with an invariant measure. As mentioned above and as we will see in more detail below, this will be particularly interesting when done in conjunction with questions and concepts coming from the geometric approach.

Although any (stationary) stochastic process can be considered as a dynamical system (with phase space the set of all possible trajectories and transformation given by the time shift), there are however some important differences to mention. First of all, dynamical systems frequently have many invariant measures. This raises the question of whether one of them is a more natural one. This is indeed the case from a physical point of view (the so-called Physical measure when it exists, see below). However, other invariant measures can be interesting from different points of view (measure of maximal entropy for example). In other words, in dynamical system theory sets of measure zero for one invariant measure may be important from some other point of view. Other concepts like Hausdorff dimension and Hausdorff measure can be interesting but may involve sets of measure zero. It is also important to mention that some sets of measure zero can indeed be "observed" (like, for example, in the multifractal formalism). It is worth mentioning that the choice of a particular invariant measure is related to the choice of an initial condition for a trajectory. We will come back to this point when discussing the ergodic theorem. The conclusion is that in dynamical systems, sets of measure zero should not be disregarded as systematically as in probability theory.

We now discuss some simple examples of invariant measures. The simplest situation arises when a system has a *fixed point*, namely a point in phase space which does not move under time evolution. For a discrete time system given by a map f, this is a point ω of phase space such that $f(\omega) = \omega$. For a continuous time system given by a vector field $\mathbf{X}$, a fixed point (also called a stationary state) is a point ω of phase space for which $\mathbf{X}(\omega) = 0$. Such a point satisfies $\varphi_t(\omega) = \omega$ for any $t \in \mathbb{R}$, where φ_t is the flow integrating $\mathbf{X}$. It is easy to verify that if a system has a fixed point, the Dirac mass in this fixed point is an invariant measure; namely, this measure satisfies equation (5.5). More generally, for a (finite) periodic orbit for a discrete time evolution, the average of the Dirac masses at the points of the orbit is an invariant measure.

The Lebesgue measure is also invariant under the map $f\colon x \mapsto 2x \pmod 1$ (Example 2.6). Although we have discussed this in Remark 5.11, we repeat it in slightly different form. We compute $f^{-1}(A)$ for each measurable set $A \subset [0,1]$. It is easy to verify that (if $1 \notin A$)

$$f^{-1}(A) = \bigl(A/2\bigr) \cup \bigl(1/2 + A/2\bigr)$$

where

$$(A/2) = \{x \in [0, 1/2] \,|\, 2x \in A\} ,$$

and

$$(1/2 + A/2) = \{x \in [1/2, 1] \,|\, 2x - 1 \in A\} .$$

If λ denotes the *Lebesgue measure*, we have

$$\lambda(A/2) = \lambda(1/2 + A/2) = \lambda(A)/2$$

and since the intersection of the two sets $(A/2)$ and $(1/2+A/2)$ is empty or reduced to the point $\{1/2\}$, the equality (5.5) follows for $\mu = \lambda$ and any measurable set A. We will see below a generalization of this idea.

Exercise 5.18. *Prove that the Lebesgue measure on $[-1, 1]$ is invariant under the map $g(x) = 1 - 2|x|$.*

Exercise 5.19. *Consider the middle third triadic Cantor set K. Recall that $K = \cap_n K_n$, where each K_n is a disjoint union of 2^n intervals, the intervals of K_{n+1} being obtained from those of K_n by dropping the (open) middle third subinterval. The construction is illustrated in Fig. 5.5.*

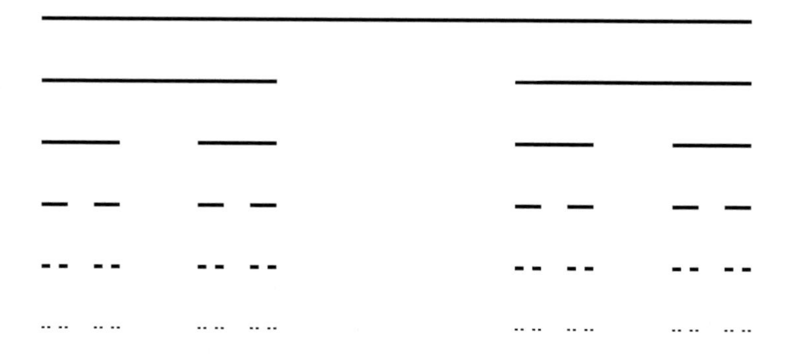

Fig. 5.5. Recursive construction of the Cantor set (top to bottom, omitting each time the middle third of every interval)

Show that this is the set of points whose representation in base 3 does not contain any digit "1." Define a measure μ by giving the weight 2^{-n} to any interval in K_n. Show that μ is invariant under the map $x \mapsto 3x \pmod 1$.

Let p be a Markov transition matrix on a finite alphabet $\mathcal{A}$. Recall that this is a $|\mathcal{A}| \times |\mathcal{A}|$ matrix with nonnegative elements $p_{b,a}$ and such that for any $a \in \mathcal{A}$ we have

$$\sum_{b \in \mathcal{A}} p_{b,a} = 1 .$$

Such matrices are often denoted by $p(b|a)$. There exists an eigenvector q with nonnegative entries and eigenvalue 1, namely for any $b \in \mathcal{A}$

$$q_b = \sum_{a \in \mathcal{A}} p_{b,a} q_a \ . \tag{5.6}$$

Let $\Omega = \mathcal{A}^{\mathbb{Z}}$ and consider the dynamical system on this phase space given by the shift. Given a finite sequence $x_r, \ldots, x_s$ ($r \le s \in \mathbb{Z}$) of elements of $\mathcal{A}$, often denoted by the short hand notation x_r^s, we denote by $C(x_r^s)$ the *cylinder subset* of Ω given by

$$C(x_r^s) = \left\{ \mathbf{y} \in \mathcal{A}^{\mathbb{Z}} \;\middle|\; y_j = x_j \ , \ r \le j \le s \right\} \ .$$

We now define a measure μ on Ω by specifying its value on any (finite) cylinder set:

$$\mu\bigl(C(x_r^s)\bigr) = q_{x_r} \prod_{j=r}^{s-1} p_{x_{j+1}, x_j} \ .$$

It is easy to verify that this defines a measure on Ω, which is invariant under the shift.

Exercise 5.20. *Prove this assertion.*

This is, of course, nothing but a stationary Markov chain with transition probability p. In the particular case where for any $b \in \mathcal{A}$ the numbers $p_{b,a}$ do not depend on a we get a sequence of i.i.d (independent, identically distributed) random variables on $\mathcal{A}$. The measure is an infinite product measure often called a *Bernoulli measure*. This construction can be generalized in various ways and in particular to the construction of Gibbs measures on subshifts of finite type (see (Ruelle 2004)).

Exercise 5.21. *(Another way to look at Fig. 5.3 and earlier calculations.) Let $p \in (0,1)$ and $q = 1 - p$. Consider the phase space $\Omega = \{0,1\}^N$ equipped with the shift operator $\mathcal{S}$. Let μ_p be the infinite product measure with $\mu_p(0) = p$ (coin flipping). Using the conjugation of the shift and the map $f(x) = 2x \pmod 1$ (except for a countable set), show that one obtains an invariant measure (also denoted μ_p) for the map f. Show that the measure of any dyadic interval of length 2^{-n} is of the form $p^r q^{n-r}$, where $0 \le r \le n$ is given by the dyadic coding of the interval. Show that for $p = q = 1/2$ one gets the Lebesgue measure.*

Exercise 5.22. *Prove that the 2-dimensional Lebesgue measure is invariant under the baker's map (Example 2.5), the cat map (Example 2.10), the standard map (Example 2.29), and the Hénon–Heiles map (4.3). Hint: compute the Jacobian.*

5.3 The Perron–Frobenius Operator

It is natural at this point to ask how one finds the invariant measures of a dynamical system. Solving (5.5) is, in general, a nontrivial task, in particular it happens frequently that dynamical systems have an uncountable number of (inequivalent) invariant measures (as in Exercise 5.21). One can then try to determine invariant measures with some special properties. In this section we consider a map f of the unit interval $[0, 1]$ and look for invariant probability measures that are absolutely continuous with respect to the Lebesgue measure, abbreviated *acim*. In other words, we are looking for measures $\mathrm{d}\mu = h\mathrm{d}x$, with h a nonnegative integrable function (of integral one) satisfying (5.5). This is equivalent to requiring

$$\int_0^1 g(x)h(x)\mathrm{d}x = \int_0^1 g(f(x))h(x)\mathrm{d}x \;, \tag{5.7}$$

for any measurable (bounded) function g. Assume next that f is piecewise monotone with finitely many pieces (see Example 2.15), namely that there is a sequence $a_0 = 0 < a_1 < \cdots < a_k = 1$ such that on each interval $[a_j, a_{j+1}]$ the map f is monotone and continuous. We have already seen the case of piecewise expanding maps of the interval, but the piecewise monotone property also holds for the case of the quadratic family of Example 2.18 and more generally for unimodal maps and for continuous maps with finitely many critical points. Let ψ_j denote the inverse of the map $f\big|_{[a_j,a_{j+1}]}$. This inverse exists since f is monotone on $[a_j, a_{j+1}]$, it is also a monotone map from $f([a_j, a_{j+1}])$ to $[a_j, a_{j+1}]$. If the maps ψ_j are differentiable, we can perform a change of variables $y = f(x)$ (restricted to $[a_j, a_{j+1}]$) in the right-hand side of (5.7). More precisely

$$\int_0^1 g(f(x))h(x)\mathrm{d}x = \sum_{j=0}^{k-1} \int_{a_j}^{a_{j+1}} g(f(x))h(x)\mathrm{d}x$$

$$= \sum_{j=0}^{k-1} \int_{f(a_j)}^{f(a_{j+1})} g(y)\psi_j'(y)h(\psi_j(y))\mathrm{d}y \;. \tag{5.8}$$

Since this relation should hold for any bounded and measurable function g, we conclude that h is the density of an acim for the map f if and only if $Ph = h$, where P is the so-called *Perron–Frobenius operator* given by

$$Pg(x) = \sum_j \chi_{f([a_j,a_{j+1}])}(x) g(\psi_j(x)) \left|\psi_j'(x)\right| \;. \tag{5.9}$$

In other words, h is the density of an acim for the map f if and only if it is an eigenvector of eigenvalue one for the operator P. Using the fact that the functions ψ_j are local inverses of f, the Perron–Frobenius operator is also given by

$$Pg(x) = \sum_{y,\, f(y)=x} \frac{g(y)}{\left|f'(y)\right|} \;. \tag{5.10}$$

An immediate consequence of this formula is that for any measurable functions u and g,

$$P(u \cdot g \circ f) = gP(u) \;. \tag{5.11}$$

In d dimensions the Perron–Frobenius operator is

$$P_f \; : \; g \mapsto \sum_{y: f(y)=x} \frac{g(y)}{|\det \mathrm{D}_y f|} ,$$

where $\mathrm{D}_y f$ is the matrix $\partial f_i / \partial x_j$ evaluated at y.

We mention without proof the following result for piecewise expanding maps of the interval. Before we state this result, we recall that a function g on the interval is of *bounded variation* if there is a finite number $C > 0$ such that for any n and any strictly increasing sequence $0 \leq b_0 < b_1 < \ldots < b_n \leq 1$ we have

$$\sum_{j=0}^{n-1} \left| g(b_j) - g(b_{j+1}) \right| \leq C .$$

The space of functions of bounded variation equipped with a norm given by the infimum $C(g)$ of all these numbers C plus the sup norm is a Banach space denoted below by **BV**. We thus write

$$\|g\|_{\mathbf{BV}} = \|g\|_{\mathrm{L}^1} + C(g) .$$

Theorem 5.23. *Any piecewise expanding map of the interval has at least one absolutely continuous invariant probability measure with density in the space of functions of bounded variation.*

This theorem due to Lasota and Yorke is proved by investigating the spectral properties of the Perron–Frobenius operator. We refer to (Hofbauer and Keller 1982) for the details, references, and consequences. We also give some complements to this result in Theorem 5.57.

Equation (5.10) is very reminiscent of (5.6) for the invariant measures of a Markov chain. The next exercise develops this analogy.

Exercise 5.24. *Consider a piecewise expanding Markov map of the interval with affine pieces. Show that there are acim with piecewise constant densities. Using the coding of Exercise 2.24 show that this system is conjugated to a stationary Markov chain. Show that any stationary Markov chain with a finite number of states can be realized by a piecewise expanding Markov map of the interval with affine pieces.*

Exercise 5.25. *Let $p_1, \; p_2, \ldots$ be a sequence of positive numbers summing to 1. Define an infinite sequence of intervals $I_1 = [a_2, a_1]$, $I_2 = [a_3, a_2]$,... , $I_j = [a_{j+1}, a_j]$,..., where $a_1 = 1$ and*

$$a_j = \sum_{\ell=j}^{\infty} p_\ell .$$

Define a map of the unit interval into itself by

$$f(x) = \begin{cases} \frac{x-a_2}{1-a_2} , & \text{if } x \in I_1 \\ a_j + (a_{j-1} - a_j)\frac{x-a_{j+1}}{a_j-a_{j+1}} , & \text{if } x \in I_j \quad \text{for} \quad j > 1 \end{cases} .$$

Choose an initial condition at random with respect to the Lebesgue measure on the interval I_1 and generate the trajectory. Show that this dynamical system can be interpreted as a renewal process *(a process with independent identically distributed waiting times).*

Another version of relation (5.8) will be useful later. Let U be the *Koopman operator* defined on measurable functions by

$$Ug(x) = g(f(x)) \ . \tag{5.12}$$

It is easy to verify that if μ is an invariant measure, then U is an isometry of $\mathrm{L}^2(\mathrm{d}\mu)$. Equation (5.8) can now be written in the following form

$$\int_0^1 g_2 U(g_1) \mathrm{d}x = \int_0^1 P(g_2) g_1 \mathrm{d}x \tag{5.13}$$

for any pair g_1, g_2 of square integrable functions.

Although we have worked here explicitly with the Lebesgue measure, we mention that similar relations can be obtained for other reference measures. Important examples are provided by *Gibbs states* on subshifts of finite type (Example 2.20).

We now discuss these measures. For a subshift of finite type, the phase space Ω is a shift invariant subspace of $\Theta = \mathcal{A}^{\mathbb{Z}}$, where $\mathcal{A}$ is a finite alphabet. We next recall the definition of a metric on symbolic sequences: For two elements $\mathbf{x}$ and $\mathbf{y}$ of Θ, denote by $\delta(\mathbf{x}, \mathbf{y})$ the nearest position to the origin where these two sequences differ, namely

$$\delta(\mathbf{x}, \mathbf{y}) = \min \left\{ |q| \mid x_q \neq y_q \right\} \ .$$

For a given number $0 < \zeta < 1$ we define a distance d_ζ (denoted simply by d when there is no ambiguity in the choice of ζ) by

$$d_\zeta(\mathbf{x}, \mathbf{y}) = \zeta^{\delta(\mathbf{x}, \mathbf{y})} \ . \tag{5.14}$$

Remark 5.26. As we have said earlier, a function is *Lipschitz continuous* (or, equivalently, *Hölder continuous with exponent 1*) if it is continuous and if

$$\sup_{x \neq y} |f(x) - f(y)|/|x - y| = L_f$$

is bounded. The constant L_f is called the *Lipschitz constant* of f.

Exercise 5.27. *Prove that d_ζ is a distance and that Θ is compact in this topology. Show that the phase space Ω (see Example 2.20) of any subshift of finite type with alphabet $\mathcal{A}$ is closed in Θ. Prove that the shift map S on Ω is continuous (and even Hölder), with a continuous inverse.*

Exercise 5.28. *This metric has the* ultrametric *property: Check that every triangle is isosceles.*

Theorem 5.29. *Let φ be a real valued Hölder continuous function on the phase space Ω of a subshift $\mathcal{S}$ of finite type. Assume that the incidence matrix M of the subshift is irreducible and aperiodic (in other words, there is an integer r such that all the entries of the matrix M^r are nonzero). Then there are a unique probability measure μ invariant under the shift and a positive constant $\Gamma > 1$ such that for any cylinder set $C(x_q^p)$ (with $q \le p$) and for any $\mathbf{y} \in C(x_q^p)$, if $\mu(C(x_q^p)) \ne 0$, then*

$$\Gamma^{-1} \le \frac{\mu(C(x_q^p))}{\mathrm{e}^{-(p-q+1)P_\varphi}\mathrm{e}^{\sum_{j=q}^{p}\varphi(\mathcal{S}^j(\mathbf{y}))}} \le \Gamma\,, \tag{5.15}$$

where P_φ is the pressure *defined by*

$$P_\varphi = \lim_{n\to\infty}\frac{1}{2n+1}\log\left(\sum_{x_{-n}^n,\, M_{x_j,x_{j+1}}=1}\mathrm{e}^{\sum_{j=-n}^{n}\varphi(\mathcal{S}^j(\mathbf{x}))}\right).$$

In this last formula, $\mathbf{x}$ *denotes any point of Ω belonging to the cylinder set $C(x_q^p)$.*

Definition 5.30. *The measure whose existence is asserted in Theorem 5.29 is called the* Gibbs state *(on the subshift $\mathcal{S}$, with* potential *φ).*

Exercise 5.31. *Prove that the pressure does not depend on the choice of the point* $\mathbf{x}$ *in the cylinder set $C(x_q^p)$ (use the Hölder continuity of φ).*

We refer to (Bowen 1975) or (Ruelle 2004) for a proof of this result, using the Perron–Frobenius operator and the relation with Gibbs states in statistical mechanics. See also Sect. 7.4.

Remark 5.32. Note that when φ is a constant, the measure does not depend on its value. Furthermore, by (5.15), all cylinders have the same measure up to the factor Γ.

Exercise 5.33. *Under the assumptions of Theorem 5.29, consider the eigenvector v with maximal eigenvalue λ of the matrix M and the left eigenvector w normalized to $\sum_i v_i w_i = 1$. Show that*

$$\mu(C(x_q^p)) = \lambda^{-(p-q)} w_{x_p} M_{x_p,x_{p-1}} \cdots M_{x_{q+1},x_q} v_{x_q} \varphi\,.$$

Exercise 5.34. *Show that all Markov chains with a finite number of states can be realized as a Gibbs measure with a potential $\varphi(\mathbf{x})$ depending only on the first two symbols x_0 and x_1 of* $\mathbf{x}$. *(Start by constructing the incidence matrix.)*

5.4 The Ergodic Theorem

Motivation: One wants to explain *why* the invariant measure plays a certain rôle, and why always the same histogram of Fig. 5.1 appears, somehow independently of the initial point (unless it happens to be a periodic point or some other exceptional point).

Preliminary statement: One is given a transformation f and an *observable* h (an integrable function). Then the *ergodic theorem* tells us

Theorem 5.35 (Ergodic theorem, preliminary version). *If μ is an invariant measure, then*

$$\lim_{n\to\infty} \frac{1}{n} \sum_{j=0}^{n-1} h(f^j(x)) = \int h(y)\mu(\mathrm{d}y) .$$

Note that the l.h.s. is a time average and the r.h.s. is a space average.

Remark 5.36. In the continuous time case, replace the l.h.s. by

$$\lim_{t\to\infty} \frac{1}{t} \int_0^t \mathrm{d}\tau\, h(\varphi^\tau(x)) .$$

Example 5.37. $f(x) = 1 - 2x^2$, $\mu(\mathrm{d}x) = \frac{1}{\pi\sqrt{(1-x^2)}}\mathrm{d}x$. This is the density of Fig. 5.1. That figure is obtained by a normalized histogram, which means that one counts how often an orbit visits a given interval (bin) I divided by the number of iterates. In other words, the observable for each bin is

$$h(x) = \begin{cases} 1\,, & \text{if } x \in I \\ 0\,, & \text{otherwise} \end{cases} .$$

The ergodic theorem tells us that this procedure converges to the histogram of the invariant measure for almost every initial condition x_0.

As already suggested, the statement of Theorem 5.35 cannot be correct as formulated, because it may fail for the periodic points of the example. In general, only a few exceptions appear, if "few" is expressed with respect to the invariant measure. However, this might be "many" with respect to another, perhaps even more naturally seeming measure, see below.

We call (X, μ) (where X is a space and μ is a measure) a *probability space* if $\int_X \mathrm{d}\mu = 1$. And in this case μ is called a *probability measure*.

Theorem 5.38 (Birkhoff ergodic theorem). *Let μ be an invariant probability measure for the map f (on the space X). Let h be an integrable function on X: $\int_X |h|\mathrm{d}\mu < \infty$. Define the partial sums:*

$$S_n(x) = \sum_{i=0}^{n-1} h(f^i(x)) .$$

1. *For μ-almost every x one has*

$$\lim_{n\to\infty} \frac{1}{n} S_n(x) = h^*(x) .$$

2. *The function h^* is f-invariant:*

$$h^*(x) = h^*(f(x)) .$$

3. One has

$$\int h^*(x)\mathrm{d}\mu = \int h(x)\mathrm{d}\mu .$$

Definition 5.39. *The notion* for μ-almost every x *means* for a set of x of μ-measure 1. *It is also written μ-a.e. x.*

The most interesting case is of course when the limit in the ergodic theorem is independent of the initial condition (except for a set of μ-measure zero). This leads to the definition of ergodicity (recall that μ is a probability):

Definition 5.40. *Given a map f on a space X, an invariant probability measure μ is called* ergodic *with respect to f if $f^{-1}(E) = E$ implies $\mu(E) = 1$ or $\mu(E) = 0$. In other words, the only invariant sets E are those of measure 0 or 1, and in particular, it is not possible that a part of the space is invariant.*

The definition can be formulated in several other ways:

Lemma 5.41. *Consider a dynamical system (f, X) with invariant measure μ.*

i) If μ is ergodic, then for any integrable function h the limit function h^* given by the ergodic theorem (Theorem 5.38) is almost surely constant, and furthermore

$$h^*(x) = \lim_{N\to\infty} \frac{1}{N+1} \sum_{j=0}^{N} h\bigl(f^j(x)\bigr) = \int h(y)\mathrm{d}\mu(y) . \tag{5.16}$$

ii) If, for all integrable h, the function $h^*(x)$ is almost surely constant, then μ is ergodic.
iii) If μ is ergodic, then any invariant function is μ-almost surely constant.
iv) If every invariant function is almost surely constant, then μ is ergodic.

Remark 5.42. This is an **important remark**. All the statements of the ergodic theorem are *only* with respect to the measure μ. Thus, they say nothing about sets of initial points whose μ-measure is 0, but which could have positive measure (for another measure ν).

For example, in Example 5.37, a possible invariant measure is the one given there, $\mu(\mathrm{d}x) = \frac{1}{\pi\sqrt{(1-x^2)}}\mathrm{d}x$, but one can also consider the example of

$$\nu(\mathrm{d}x) = \delta(x - x_0)\mathrm{d}x ,$$

where $x_0 = 1/2$ is an unstable fixed point of the map $x \mapsto 1 - 2x^2$. One should note that the ergodic theorem holds for both measures, but while the μ-variant says something about a.e. x with respect to the Lebesgue measure (i.e., almost all points in $[-1, 1]$) the ν-variant talks only about the point $x = 1/2$.

Proof of Lemma 5.41. We first prove i). Let

$$c_M = \{x \in X \mid h^*(x) \text{ exists and } h^*(x) < M\} .$$

This set is f-invariant, since the function $x \mapsto h^*(x)$ is f-invariant. But, by ergodicity, this implies that the measure $\mu(c_M)$ is either 0 or 1. Define now the constant $\tilde{h}$ by

$$\tilde{h} = \inf\{M \mid \mu(c_M) = 1\} .$$

Then

$$\tilde{h} \int d\mu = \int h^*(x) d\mu = \int h(x)\, d\mu .$$

This implies that $h^*(x) = \tilde{h}$, μ-almost surely. To prove ii), observe that if E is an invariant set, then its characteristic function χ_E is invariant. Taking $h(x) = \chi_E(x)$, it follows from the definition of h^* that $h = h^*$, which implies, by i), that $\chi_E(x)$ is almost surely constant. This is only possible if $\mu(E)$ equals 0 or 1. The statement iii) follows from i) by observing that if h is an invariant function, then $h^* = h^*(x) = h(x)$, so h must be constant. The proof of iv) follows like ii), by taking $h = \chi_E$ for an invariant set. □

Proof of Theorem 5.38. We may suppose $h \geq 0$; otherwise, we just decompose h into its (measurable) positive and negative parts $h = h^+ - h^-$ and deal with each of them separately. We define

$$\bar{h}(x) = \limsup_{n\to\infty} \frac{1}{n} S_n(x) , \qquad \underline{h}(x) = \liminf_{n\to\infty} \frac{1}{n} S_n(x) .$$

We shall prove

$$\int \bar{h}(x)\mu(dx) \leq \int h(x)\mu(dx) \leq \int \underline{h}(x)\mu(dx) , \tag{5.17}$$

from which we can conclude $\bar{h}(x) \leq \underline{h}(x)$ for μ-a.e. x. We only show the second inequality of (5.17); the proof of the first inequality is very similar.

We will present a proof which is based on ideas of nonstandard analysis (see (Kamae 1982; Katznelson and Weiss 1982)). We begin by a wrong, but illuminating argument. It is based on the wrong assumption that the integrable function h must be bounded: $|h| \leq M < \infty$. (In another approach, one shows directly that it suffices to prove (5.17) for bounded measurable functions to deduce the result for any $h \in L^1$.)

Clearly, this implies

$$\bar{h}(x) = \limsup_{n\to\infty} \frac{1}{n} S_n(x) \leq M .$$

Then, by the definition of $S_n(x)$,

$$\bar{h}(f(x)) - \bar{h}(x) = \lim_{n\to\infty} \frac{1}{n}\big(h(f^n(x)) - h(x)\big) = 0 .$$

So, *$\bar{h}$ is f-invariant.*

We now *fix* $\varepsilon > 0$. For every x there is a smallest n for which

$$\frac{1}{n} S_n(x) \geq \bar{h}(x) - \varepsilon . \tag{5.18}$$

We call this $n(x)$; note that $n(x)$ also depends on ε. We make a second wrong assumption:

$$\sup_x n(x) = N < \infty ; \tag{5.19}$$

again, $N = N_\varepsilon$ depends on ε. We define recursively:

$$\begin{aligned} n_0(x) &= 0 , \quad n_1(x) = n(x) , \\ n_2(x) &= n_1(x) + n(f^{n_1(x)}(x)) , \\ &\cdots \\ n_{j+1}(x) &= n_j(x) + n(f^{n_j(x)}(x)) . \end{aligned}$$

Fix now $L \geq M \cdot N_\varepsilon/\varepsilon$. Let $J(x)$ be the last j for which $n_j(x) \leq L$. We write (in a complicated way, using the f-invariance of $\bar{h}$),

$$L\bar{h}(x) = \sum_{i=0}^{L-1} \bar{h}(f^i(x)) = \sum_{j=1}^{J(x)} \sum_{i=n_{j-1}(x)}^{n_j(x)-1} \bar{h}(f^i(x)) + \sum_{i=n_{J(x)}(x)}^{L-1} \bar{h}(f^i(x)) .$$

By (5.19), the last sum is bounded by

$$\sum_{i=n_{J(x)}(x)}^{L-1} \bar{h}(f^i(x)) \leq N \cdot M .$$

We also get from (5.18),

$$S_{n(y)}(y) \geq \bar{h}(y)n(y) - n(y)\varepsilon ,$$

so that (taking $y = f^{n_{j-1}(x)}(x)$ below), and using $h \geq 0$ we get

$$\begin{aligned} L\bar{h}(x) &\leq \sum_{j=1}^{J(x)} \Big(S_{n_j(x)-n_{j-1}(x)}\big(f^{n_{j-1}(x)}(x)\big) + \big(n_j(x) - n_{j-1}(x)\big)\varepsilon \Big) + N \cdot M \\ &\leq \sum_{i=0}^{n_{J(x)}(x)-1} h(f^i(x)) + L\varepsilon + N \cdot M \\ &\leq \sum_{i=0}^{L-1} h(f^i(x)) + 2L\varepsilon . \end{aligned}$$

Dividing by L, we get

$$\bar{h}(x) \leq \sum_{i=0}^{L} h(f^i(x)) + 2\varepsilon .$$

Integrating, and using the invariance of the measure, we get

$$\int \bar{h}(x)\mu(\mathrm{d}x) \le \int h(x)\mu(\mathrm{d}x) + 2\varepsilon \int \mu(\mathrm{d}x) .$$

Since this holds for all $\varepsilon > 0$, and $\int \mu(\mathrm{d}x) = 1$, we get (5.17).

Second, more correct proof. The first observation is that if $h \in \mathrm{L}^1$, this implies

$$\lim_{M\to\infty} \mu(\{(x) \mid h(x) > M\}) = 0 ,$$

in other words, while we cannot guarantee that h is bounded, the set on which it is large has small μ-measure. We fix an $M < \infty$, and we define $\bar{h}_M(x) = \min(M, \bar{h}(x))$. Note that $\bar{h}_M(x) = \bar{h}_M(f(x))$. We also fix $\varepsilon > 0$ and define, as before, but for $\bar{h}_M$:

$$n(x) = \inf\{n \mid \bar{h}_M(x) - \varepsilon \le S_n(x)/n\} .$$

Note that $n(x)$ is μ-almost everywhere finite, and therefore

$$\lim_{N_0\to\infty} \mu(\{x \mid n(x) > N_0\}) = 0 .$$

So, in summary, the "wrong" assumptions are wrong only on a set of small μ-measure. It will suffice to modify the wrong proof correspondingly, by omitting small sets, and arriving at the end at a bound of 3ε instead of 2ε. More precisely, fix $\varepsilon > 0$ and choose M such that $\mu(\{x \mid h(x) > M\}) < \varepsilon$, and then N_0 large enough such that $\mu(\mathcal{A}) < \varepsilon/M$, where $\mathcal{A} = \{x \mid n(x) > N_0\}$. We need a little gymnastics since the orbit of x can enter and leave $\mathcal{A}$. For this, we let

$$\check{h}(x) = \begin{cases} h(x) , & \text{if } x \notin \mathcal{A} \\ \max(M, h(x)) , & \text{otherwise} \end{cases} ,$$

and

$$\check{n}(x) = \begin{cases} n(x) , & \text{if } x \notin \mathcal{A} \\ 1 , & \text{otherwise} \end{cases} .$$

So, $\mathcal{A}^{\mathrm{c}}$ is the set where things are as before, while $\mathcal{A}$ is the set where the wrong assumptions fail. Then, we get

$$\sum_{i=0}^{\check{n}(x)-1} \bar{h}_M(f^i(x)) \le \sum_{i=0}^{\check{n}(x)-1} \check{h}(f^i(x)) + \check{n}(x)\varepsilon , \tag{5.20}$$

since if $x \in \mathcal{A}$, we have $\check{n}(x) = 1$ and $\bar{h}_M(x) \le M \le \check{h}(x)$, while if $x \notin \mathcal{A}$ the inequality follows from the definition of $n(x)$ and the inequality $h(f^i(x)) \le \check{h}(f^i(x))$.

As before, one inductively defines

$$n_0(x) = 0 ,$$

$$\dots$$

$$n_{j+1}(x) = n_j(x) + \check{n}(f^{n_j(x)}x) .$$

Now we can sum, as before

$$\begin{aligned}\sum_{i=0}^{L} \bar{h}_M(f^i(x)) &= \sum_{j=1}^{J(x)} \sum_{i=n_{j-1}(x)}^{n_j(x)-1} \bar{h}_M(f^i(x)) + \sum_{i=n_{J(x)}(x)}^{L-1} \bar{h}_M(f^i(x)) \\ &\leq \sum_{i=0}^{L-1} \check{h}(f^i(x)) + L\varepsilon + N_0 \cdot M\ ,\end{aligned}$$

where $J(x)$ is the largest integer such that $n_{J(x)}(x) \leq L$, and the last inequality follows from (5.20) (one distinguishes the cases $f^{n_{J(x)}(x)}(x) \in \mathcal{A}^c$, where $L - n_{J(x)}(x) \leq N_0$, and the complementary case, where $L - n_{J(x)}(x) = 1$ by definition of $\check{n}$).

On the other hand, integrating, we get

$$\int \check{h}(x)\mu(\mathrm{d}x) \leq \int_{\mathcal{A}^c} h(x)\mu(\mathrm{d}x) + \int_{\mathcal{A}} h(x)\mu(\mathrm{d}x) + M \int_{\mathcal{A}} \mu(\mathrm{d}x)\ .$$

The last term is bounded by ε by the construction of N_0 (and hence $\mathcal{A}$), and thus, we get by letting L tend to infinity

$$\int \bar{h}_M d\mu \leq \int \check{h} d\mu + 3\varepsilon \leq \int h d\mu + 4\varepsilon\ ,$$

and the proof is completed by letting M tend to infinity and then ε tend to 0. □

Apart from the (strong) Birkhoff ergodic theorem, there are other variants. The ergodic theorem of von Neumann (von Neumann 1932) applies in an L^2 context, while the Birkhoff ergodic theorem (Birkhoff 1931) applies almost surely (we refer to (Krengel 1985; Kalikov 2000; Keane 1991; Petersen 1983; Keane and Petersen 2006) for proofs, extensions and applications).

We add here a list of comments and precautions about the ergodic theorem.

i) The set of μ-measure zero where nothing is claimed depends on both f and μ. One can often use a set independent of f (for example if $\mathrm{L}^1(\mathrm{d}\mu)$ is separable). We comment below on the dependence on μ.
ii) The theorem is often remembered as saying that the time average is equal to the space average. This has to be taken with a grain of salt. As we will see below, changing the measure may change drastically the exceptional set of measure zero and this can lead to completely different results for the limit.
iii) In (5.16), the initial state x does not appear on the right-hand side, but it is hidden in the fact that the formula is only true outside a set of measure zero.
iv) The ergodic theorem tells us that $\lim_{n\to\infty} \frac{1}{n} S_n(x)$ exists for μ-almost every x. This does *not* mean that the limit exists for Lebesgue (λ) almost every x if μ is not absolutely continuous. We will see later that for very chaotic systems, this desirable additional property indeed will hold for some μ which are not absolutely continuous (the so-called Physical measures which will be defined later).

Example: Suppose f is the cat map, for $x \in [0,1) \times [0,1)$. Every rational point is periodic. Take one of them, x_0, and define $x_i = f^i(x_0)$ for $i = 0, \ldots, p-1$, where p is the period of x_0. Then $\mu = \frac{1}{p}\sum_{i=0}^{p-1} \delta_{x_i}$ is an invariant measure, and the ergodic theorem applies. But a property which is true for μ-almost every x is true *only* on the set $\{x_0, \ldots, x_{p-1}\}$ and a "few" other points (the stable manifold of the periodic orbit, see later). But for a set of λ-measure 1, the sum $S_n(x)/n$ converges to $\int h(x)\lambda(\mathrm{d}x)$, and *not* to $\int h(x)\mu(\mathrm{d}x)$.

v) The set of initial conditions where the limit does not exist, although small from the point of view of the measure μ may be big from other points of views (see (Barreira and Schmeling 2000)).

vi) It often happens that a dynamical system (X, f) has several ergodic invariant (probability) measures. Let ν and μ be two different ones. It is easy to verify that they must be disjoint, namely one can find a set of ν-measure one which is of μ-measure zero and vice versa. This explains why the ergodic theorem applies to both measures leading in general to different time averages.

vii) For nonergodic measures, one can use an ergodic decomposition (disintegration). We refer to (Krengel 1985) for more information. However, in concrete cases this often leads to rather complicated sets.

viii) In probability theory, the ergodic theorem is usually called the law of large numbers for stationary sequences.

ix) Birkhoff's ergodic theorem holds for semiflows (continuous time average). It also holds of course for the map obtained by sampling the semiflow uniformly in time. However, nonuniform sampling may spoil the result (see (Reinhold 2000) and references therein).

x) Simple cases of nonergodicity come from Hamiltonian systems with the invariant Liouville measure. First of all since the energy is conserved (the Hamiltonian is a nontrivial invariant function), the system is not ergodic. One has to restrict the dynamics to each energy surface. More generally if there are other independent constants of the motion one should restrict oneself to lower dimensional manifolds. For completely integrable systems, one is reduced to a constant flow on a torus which is ergodic if the frequencies are incommensurable. It is also known that generic Hamiltonian systems are neither integrable nor ergodic (see (Markus and Meyer 1974)).

xi) Proving ergodicity of a measure is sometimes a very hard problem. Note that it is enough to prove (5.16) for a dense set of functions, for example the continuous functions.

Some of the above remarks are illustrated in Figs. 5.6–5.8. The dynamical system is the map $3x \pmod 1$ of the unit interval and the function h is the characteristic function of the interval $[0, 1/2]$. The figure shows S_n/n as a function of n for two initial conditions. The Lebesgue measure is invariant and ergodic and we expect that if we choose an initial condition uniformly at random on $[0, 1]$, then S_n/n should converge to $1/2$ (the average of h with respect to the Lebesgue measure). This is what we see in Fig. 5.6, where two different initial conditions were chosen at random according to the Lebesgue measure. Note that the convergence to the limit $1/2$ is not

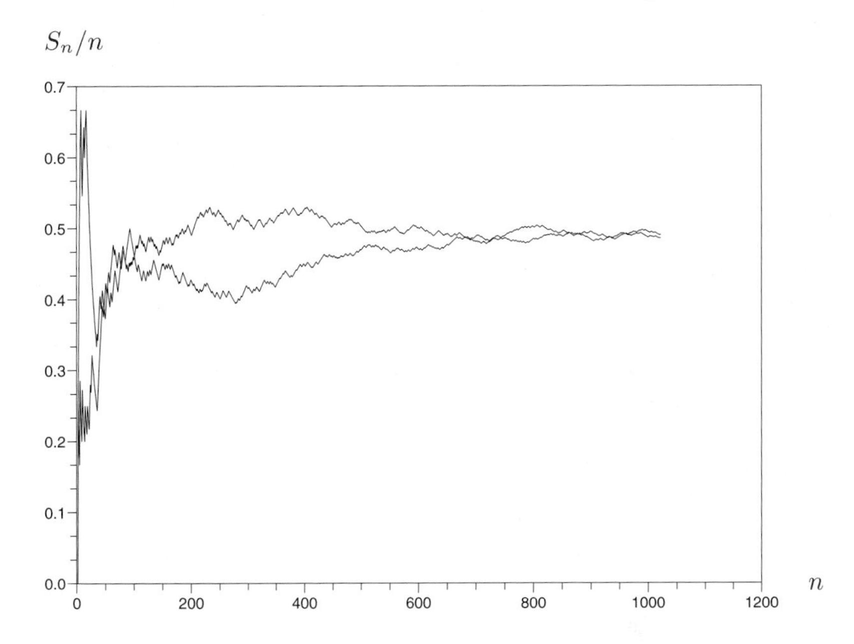

Fig. 5.6. Evolution of S_n/n for two random initial conditions for the map $x \mapsto 3x \pmod 1$

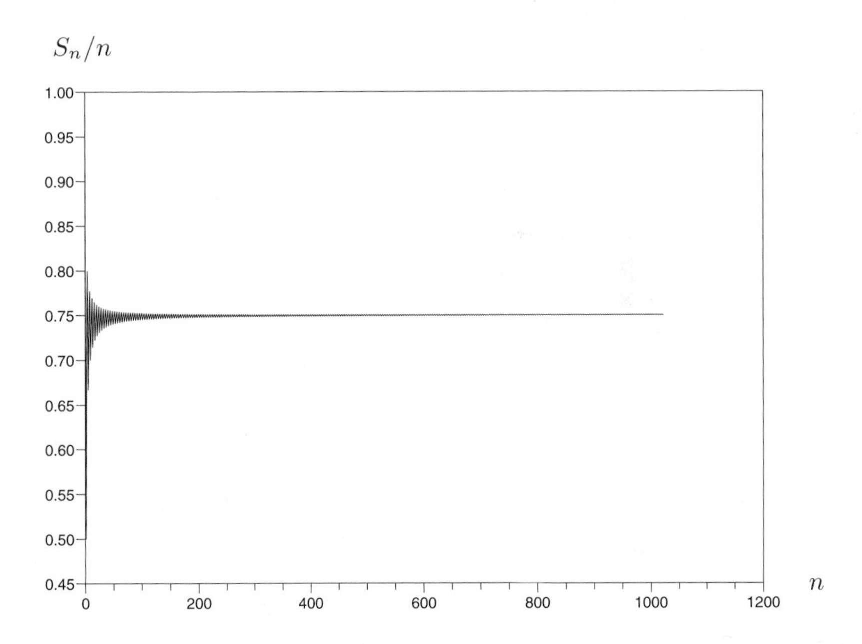

Fig. 5.7. Evolution of S_n/n for a periodic initial condition for the map $x \mapsto 3x \pmod 1$

very fast. We come back to this question later. Figure 5.7 shows S_n/n as a function of n for the initial condition $x_0 = 7/80$. This point is periodic of period 4, and its orbit has three points in the interval $[1/2, 1]$ (Exercise: verify this statement). The invariant measure (the sum of the Dirac measures on the 4 points of the orbit) is ergodic and therefore, S_n/n converges to $3/4$ (the average of f with respect to the discrete invariant measure supported by the orbit).

Figure 5.8 was drawn using an atypical initial condition for which S_n/n will oscillate forever between two extreme values, although the oscillations will slow down as n increases.

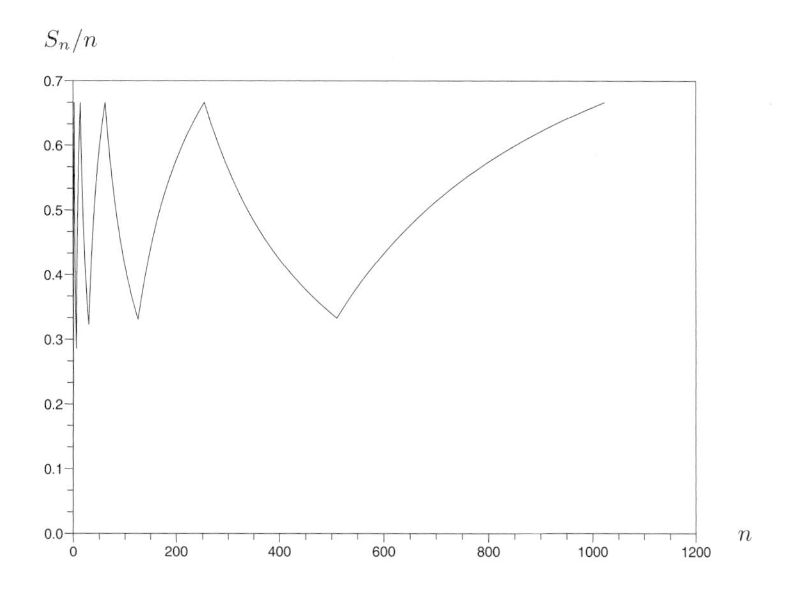

Fig. 5.8. Evolution of S_n/n for the map $x \mapsto 3x \pmod 1$ for an atypical initial condition. This initial condition has the representation $x = \sum_{i=0}^{\infty} s_i 3^{-i}$, with $s_i = 1$ for $i \in [2^{2k}, 2^{2k+1} - 1]$, $k \in \mathbb{N}$ and 0 otherwise

Let $p \in (0, 1)$ with $p \neq 1/2$. Consider the probability measure μ_p on $\Omega = \{0, 1\}^{\mathbb{N}}$ of Exercise 5.21. This is equivalent to flipping a coin with probability p to get head and probability $q = 1 - p$ to get tail. It is easy to prove that the measure μ_p is invariant and ergodic for the shift S. In this case, ergodicity follows from the law of large numbers and the ergodic theorem. We discuss later some other methods of proving ergodicity using mixing.

An extension of the ergodic theorem is the so-called *subadditive ergodic theorem* which has many useful applications.

Theorem 5.43. *Let $(\Omega, \mathscr{B}, f)$ be a measurable dynamical system and μ an ergodic invariant measure. Let $\{h_n\}$ be a sequence of integrable functions satisfying for each $x \in \Omega$ and for any pair of integers m and n the inequality*

$$h_{n+m}(x) \le h_n(x) + h_m(f^n(x)) , \tag{5.21}$$

and such that

$$\liminf_{n\to\infty} \frac{1}{n} \int h_n \mathrm{d}\mu > -\infty .$$

Then μ*-almost surely*

$$\lim_{n\to\infty} \frac{h_n}{n} = \lim_{n\to\infty} \frac{1}{n} \int h_n \mathrm{d}\mu .$$

We refer to (Krengel 1985) for a proof.

5.5 Convergence Rates in the Ergodic Theorem

It is natural to ask about the speed of convergence of the ergodic average to its limit in Theorem 5.38. Not much can be said in general, and obviously, this is bad for practical applications, since one can never really know whether one has reached the limit or not. There are in fact two problems. The first is that one cannot know whether the initial point is really typical. See, for example, Fig. 5.8, which shows what happens for an atypical point for the map $x \mapsto 3x \pmod 1$. The second problem is that even if the initial condition is typical for the measure, the convergence can be very slow as one sees in Fig. 5.6.

We enumerate some of the results dealing with these questions.

i) At the full level of generality any kind of velocity above $1/n$ can occur. Indeed Halász and Krengel have proved the following result (see (Kachurovskiĭ 1996) for a review):

Theorem 5.44. *Consider a (measurable) automorphism f of the unit interval $\Omega = [0,1]$, leaving the Lebesgue measure $\lambda = \mathrm{d}x$ invariant.*

i) For any increasing, diverging sequence $a_1, a_2, \ldots$ with $a_1 \ge 2$, and for any number $\alpha \in (0,1)$, there is a measurable subset $A \in \Omega$ such that $\lambda(A) = \alpha$, and

$$\left| \frac{1}{n} \sum_{j=0}^{n-1} \left(\chi_A \circ f^j - \lambda(A)\right) \right| \le \frac{a_n}{n} ,$$

λ-almost surely, for all n.

ii) Let $\{b_n\}$ be a sequence of numbers diverging arbitrarily slowly to ∞. Then, there is a measurable subset $B \in \Omega$ with $\lambda(B) \in (0,1)$ such that almost surely

$$\lim_{n\to\infty} \left| \frac{b_n}{n} \sum_{j=0}^{n-1} \left(\chi_B \circ f^j - \lambda(B)\right) \right| = \infty .$$

Remark 5.45. The first part of the theorem tells us that although one expects in general a deviation from the mean of order $n^{-1/2}$ (see e.g. Theorem 7.4), there are subsets where the deviation is quite close to $1/n$. The second part tells us that on some other subset, the mean deviation can go to 0 arbitrarily slowly.

One can construct rotations on infinite adding machines that have arbitrary prescribed convergence properties (see (O'Brien 1983)).

ii) In spite of these negative results, there is however a class of somewhat surprising general theorems dealing with aspects of "constructive analysis" (Bishop 1967). They show the following result: Assume $S_n/n \to h^*$. Consider the interval $I_\varepsilon = [h^* - \varepsilon, h^* + \varepsilon]$, and ask how often the sequence $\{S_n(x)/n\}$ *down-crosses* (traverses from top to bottom) the interval I_ε. Let Ω_Q be those points x for which the sequence $\{S_n(x)/n\}$ has exactly Q down-crossings. Then the probability $\mu(\Omega_Q)$ is bounded by

$$\mu(\Omega_Q) \le \left(\frac{h^* - \varepsilon}{h^* + \varepsilon}\right)^Q .$$

Astonishingly, all this does depend neither on the map nor on the observable. To formulate the result more precisely, we first define the sequence of down-crossings for a sequence $\{u_n\}_{n\in\mathbb{N}}$. (This sequence will be given later as $u_k = h(f^k(x))$, $k = 0, \ldots$.) Let a and b be two numbers such that $0 < a < b$. If $u_k \le a$, for some $k \ge 0$, we define the first down-crossing from b to a after k as the smallest integer $n_d > k$ (if it exists) for which

i) $u_{n_d} \le a$,
ii) There exists at least one integer $k < j < n_d$ such that $u_j \ge b$.

Let now $\{n_\ell\}$ be the sequence of successive down-crossings from b to a (this sequence may be finite and even empty). We denote by $N\big(a, b, p, \{u_n\}\big)$ the number of successive down-crossings from b to a occurring before time p for the sequence $\{u_n\}$, namely

$$N(a, b, p, \{u_n\}) = \sup\left\{\ell \,\middle|\, n_\ell \le p\right\} .$$

Theorem 5.46. *Let $(\Omega, \mathscr{B}, f, \mu)$ be a dynamical system. Let h be a nonnegative observable with $\mu(h) > 0$. Let a and b be two positive real numbers such that $0 < a < \mu(h) < b$, then for any integer r*

$$\mu\left(\left\{x \,\middle|\, N\Big(a, b, \infty, \{h(f^n(x))\}\Big) > r\right\}\right) \le \left(\frac{a}{b}\right)^r .$$

The interesting fact about this theorem is that the bound on the right-hand side is explicit and independent of the map f and of the observable h. We refer to (Ivanov 1996a;b; Collet and Eckmann 1997; Kalikow and Weiss 1999) for proofs and extensions.

To get more precise information on the rate of convergence in the ergodic theorem, one has to make some hypotheses on the dynamical system and on the observable.

5.6 Mixing and Decay of Correlations

If one considers the numerator of the ergodic average, namely the *ergodic sum*

$$S_n(h)(x) = \sum_{j=0}^{n-1} h\bigl(f^j(x)\bigr) , \tag{5.22}$$

this can be considered as a sum of random variables, which in general are not independent. It is still natural to ask if there is something similar to the central limit theorem in probability theory. We will treat this question in detail in Sect. 7. In any case, one has first to obtain information about the limiting variance, which we discuss now. Assuming for simplicity that the average of h is zero, we are faced with the question of convergence of the sequence

$$\begin{aligned} &\frac{1}{n}\int \Bigl(S_n(h)(x)\Bigr)^2 \,\mathrm{d}\mu(x) \\ &\quad = \int h^2(x)\,\mathrm{d}\mu(x) + 2\sum_{j=1}^{n-1} \frac{n-j}{n} \int h(x)\, h\bigl(f^j(x)\bigr)\,\mathrm{d}\mu(x) . \end{aligned} \tag{5.23}$$

Exercise 5.47. *Use the invariance of the measure μ to prove this formula.*

Here we restrict of course the discussion to observables h which are square integrable. This sequence may diverge when n tends to infinity. It may also tend to zero. This is for example the case if $h = u - u \circ f$ with $u \in \mathrm{L}^2(\mathrm{d}\mu)$. Indeed, in that case S_n is a telescopic sum, and so $S_n(h) = u - u \circ f^n$ is of order one in L^2 and not of order $\sqrt{n}$ (see (Kachurovskiĭ 1996) for more details and references), and the sequence (5.23) converges to 0.

A quantity which arises naturally from the above formula is the *auto-correlation function* $C_{h,h}$ of the observable h. This is the sequence (function of the integer valued variable j) defined (for h of integral $\int h \mathrm{d}\mu = 0$) by

$$C_{h,h}(j) = \int h(x)\, h\bigl(f^j(x)\bigr)\,\mathrm{d}\mu(x) . \tag{5.24}$$

If this sequence belongs to ℓ^1, the limiting variance exists and is given by

$$\sigma_h^2 = C_{h,h}(0) + 2\sum_{j=1}^{\infty} C_{h,h}(j) . \tag{5.25}$$

This shows that the decay of the auto-correlation function (5.24) is an important quantity (if we want the limiting variance to be finite).

A natural generalization of the auto-correlation is the cross correlation (often called simply the *correlation*) between two square integrable observables g and h. This function (sequence) is defined by

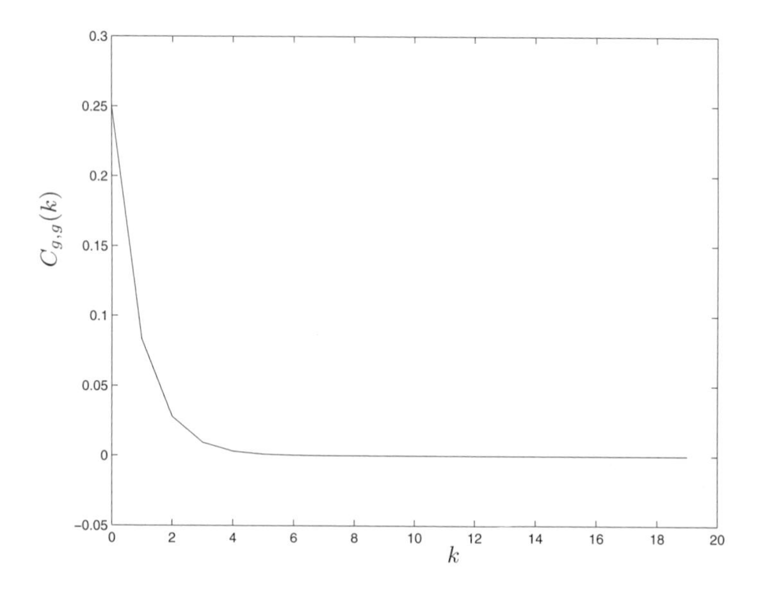

Fig. 5.9. The correlation function for the observable g of Exercise 5.56 for the map $x \mapsto 3x$ (mod 1)

$$C_{g,h}(j) = \int g(x)\, h\big(f^j(x)\big)\, \mathrm{d}\mu(x) - \int g\, \mathrm{d}\mu \int h\, \mathrm{d}\mu\,. \tag{5.26}$$

The second term appears in the general case when neither g nor h has zero average.

The case in which g and h are characteristic functions is particularly interesting. If $g = \chi_A$ and $h = \chi_B$, we have

$$\begin{aligned} C_{\chi_A,\chi_B}(n) = &\int_A \chi_B \circ f^n \mathrm{d}\mu - \mu(A)\mu(B) \\ =&\mu\big(A \cap f^{-n}(B)\big) - \mu(A)\mu(B) \\ =&\mu\big(\{x \mid f^n(x) \in B, x \in A\}\big) - \mu(A)\mu(B)\,. \end{aligned}$$

If for large time the system loses memory of its initial condition, it is natural to expect that $\mu\big(f^n(x) \in B \big| x \in A\big)$ converges to $\mu(B)$ as $n \to \infty$. This leads to the definition of mixing. In this case the correlation function will converge to zero as $n \to \infty$. We say that for a dynamical system $(\Omega, \mathscr{B}, f)$ the f invariant measure μ is *mixing* if for any measurable subsets A and B of Ω, we have

$$\lim_{n\to\infty} \mu\big(A \cap f^{-n}(B)\big) = \mu(A)\mu(B)\,. \tag{5.27}$$

Note that this can also be written (for $\mu(B) > 0$)

$$\lim_{n\to\infty} \mu\big(f^{-n}(B)\big|A\big) = \mu(B)\,,$$

and if the map f is invertible

$$\lim_{n\to\infty} \mu\big(B\big|f^n(A)\big) = \mu(B) .$$

In other words, if we have the information that at some time the state of the system is in the subset A, then if we wait long enough (more precisely asymptotically in time), this information is lost in the sense that we get the same statistics as if we had not imposed this condition. Many other mixing conditions can be formulated, including various aspects of convergence, for example the velocity of convergence. We refer to (Doukhan 1994) for details. See also the recent work in (Bradley 2005).

Example 5.48. In Fig. 5.10, we illustrate how mixing occurs in the cat map.

Exercise 5.49. *Show that mixing implies ergodicity.*

Exercise 5.50. *Let f be a map with a finite periodic orbit $x_0, \dots, x_{p-1}$ ($f(x_r) = x_{r+1 \pmod p}$ for $0 \le r < p$). Consider the invariant probability measure obtained by putting Dirac masses with weight $1/p$ at the points of the orbit. Show that this measure is ergodic but not mixing.*

Exercise 5.51. *Consider the map $x \mapsto 2x \pmod 1$ and the ergodic invariant measure μ_p of Exercise 5.21. Show that this dynamical system is mixing (hint: show (5.27) for any cylinder set and conclude by approximation).*

As explained before, to show that the ergodic sum has a (normalized) limiting variance, we need to estimate the convergence rate in (5.27). For general measurable sets (or for square integrable functions in (5.26)) this is often a hopeless task. Furthermore, it frequently happens that the rate of convergence in (5.26) depends on the class of functions one considers.

Note that the correlation function (5.26) can also be written

$$C_{g_1,g_2}(j) = \int g_1(x)\, U^j g_2(x)\, \mathrm{d}\mu(x) - \int g_1 \,\mathrm{d}\mu \int g_2 \,\mathrm{d}\mu ,$$

where U is the Koopman operator (5.12). In the case of maps of the interval, when the invariant probability measure μ is absolutely continuous with respect to the Lebesgue measure with density h, the correlations can be expressed in terms of the Perron–Frobenius operator using equality (5.13), namely

$$C_{g_1,g_2}(j) = \int \mathcal{L}^j g_1(x)\, g_2(x)\, \mathrm{d}\mu(x) - \int g_1 \,\mathrm{d}\mu \int g_2 \,\mathrm{d}\mu , \tag{5.28}$$

with

$$\mathcal{L}g = \frac{1}{h} P\big(hg\big) . \tag{5.29}$$

Note that since $P(h) = h$, this formula is well-defined Lebesgue-almost everywhere—even if h vanishes—by setting $\mathcal{L}g = 0$, where $h = 0$. It is now obvious from (5.28) that information about the spectrum of $\mathcal{L}$ translates into information on the decay of correlations. Since $\mathcal{L}$ is conjugated to the Perron–Frobenius operator P, one often studies in more detail this last operator. We recall that the relation between

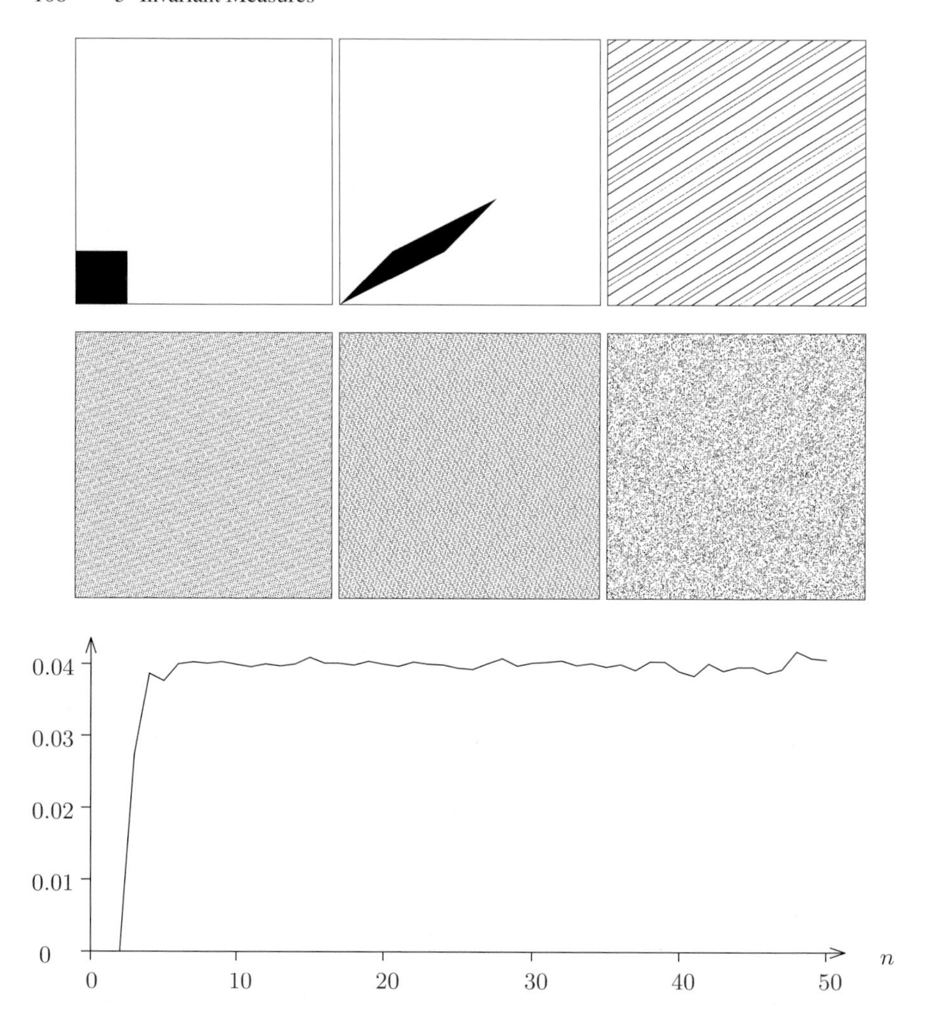

Fig. 5.10. The mixing in the cat map. Starting from $22'500$ equally spaced points in the region $[0, 0.2] \times [0, 0.2]$ we show in the top six panels (top left to right, bottom left to right), where these points lie after $n = 0, 1, 5, 10, 25, 50$ iterations, respectively. The bottom panel shows the relative number of points that are in the diagonally opposite square $[0.8, 1] \times [0.8, 1]$ after n iterations. The theoretical limit is 0.04

decay of correlations and spectrum of the transition matrix is also a well-known fact in the theory of Markov chains.

Consider, for example, the map $f(x) = 2x \pmod 1$ of the interval $[0, 1)$ (Example 2.6) with the Lebesgue measure as invariant measure. If g and h belong to L^2, one can use their Fourier series $\{g_p\}$ and $\{h_p\}$, defined by

$$g(x) = \sum_{p \in \mathbb{Z}} g_p \mathrm{e}^{2\pi i p x}$$

where

$$g_p = \int_0^1 \mathrm{e}^{-2\pi \mathrm{i} p y} g(y) \mathrm{d}y \,. \tag{5.30}$$

One gets

$$C_{h,g}(j) = \sum_{p,q} h_p g_q \int_0^1 \mathrm{e}^{2\pi \mathrm{i} p x} \mathrm{e}^{2\pi \mathrm{i} q 2^j x} \mathrm{d}x - h_0 g_0 = \sum_{q \neq 0} h_{q 2^j} g_{-q} \,. \tag{5.31}$$

A first brutal estimate using the Schwarz inequality gives

$$\left| C_{h,g}(j) \right| \leq \|g\|_2 \left(\sum_{q \neq 0} \left| h_{q 2^j} \right|^2 \right)^{1/2} \,. \tag{5.32}$$

It follows easily from $h \in \mathrm{L}^2$ using Parseval's inequality that the last term tends to zero when j tends to infinity. In other words, the map $2x \pmod 1$ is mixing when the invariant measure is the Lebesgue measure. Note that it is crucial in the argument above that the sum does not contain the term $q = 0$, this is exactly the rôle of the subtraction of the last term in definition (5.26), see also (5.31).

To simplify the discussion, we view the observable as being defined on the circle: $h : [0, 1) \to \mathbb{R}$ is of the form $h(\varphi) = \sum_n h_n \mathrm{e}^{2\pi \mathrm{i} n \varphi}$. Let $H(\mathrm{e}^{2\pi \mathrm{i} \varphi}) \equiv h(\varphi)$. If $z \mapsto H(z)$ extends to an analytic function in a neighborhood of $|z| = 1$, then the Fourier coefficients h_n decay exponentially fast in $|n|$. This is what we assume now. In this case, for any $g \in \mathrm{L}^2$ we find a *doubly* exponential decay of the correlations. See also Sect. 9.2.

Exercise 5.52. *Prove the assertions made in the last two paragraphs, the first one using formula (5.30), the second one using formula (5.32).*

If we only have a power law bound on the decay of the Fourier coefficients of h, we can conclude only that the correlations have only an exponential and not a doubly exponential decay. To see this, note that the behavior of the Fourier coefficients is related to the regularity properties of the function. For example, if h is only C^1 (on the circle), then for any $p \neq 0$,

$$|h_p| \leq \frac{1}{2\pi p} \left\| \frac{\mathrm{d}h}{\mathrm{d}x} \right\|_2 \,.$$

Exercise 5.53. *Prove this estimate using integration by parts in formula (5.30).*

Using the estimate (5.32), we conclude that for any $g \in \mathrm{L}^2$ and any $h \in C^1$ one has

$$\left| C_{h,g}(j) \right| \leq \Gamma \|g\|_2 \|h'\|_2 2^{-j} \,, \tag{5.33}$$

with

$$\Gamma = \frac{1}{2\pi} \left(\sum_{q \neq 0} \frac{1}{q^2} \right)^{1/2} \,.$$

By comparing the cases of analytic h and $h \in C^1$, we see that the bounds we have obtained depend on the regularity properties of the function h. This is not an artifact of our crude estimation method. To see that the decay indeed depends on the regularity, we consider a particular example. For $p \neq 0$, let

$$h_p = \frac{1}{|p|^\alpha} ,$$

for some fixed $\alpha \in (1/2, 1)$. Using Parseval's inequality, it follows at once that the function h whose Fourier series is $\{h_p\}$ belongs to L^2. We take $g_p = 0$ for all p except that $g_1 = g_{-1} = 1$. We get from (5.31)

$$C_{h,g}(n) = 2 \cdot 2^{-n\alpha} .$$

We conclude by comparison with (5.33) that the function h does not belong to C^1 (in fact h' is not in L^2) and its correlation decays indeed more slowly.

Exercise 5.54. *Let $u(x) = \sin(2\pi x)$, and let $f(x) = 2x \pmod 1$ on the interval $[0,1]$. Although for the Lebesgue measure this map is very chaotic, we now construct observables in L^2 with very slow decay of correlations. Notice that for any $n > 0$*

$$\int_0^1 u(x)\,\mathrm{d}x = 0 , \qquad \textit{and} \qquad \int_0^1 u(x)\,u\big(f^n(x)\big)\,\mathrm{d}x = 0 .$$

Let $\{\alpha_k\}$ be the sparse sequence defined by

$$\alpha_k = \begin{cases} \frac{1}{p} , & \textit{if } k = 2^p \\ 0 , & \textit{otherwise} \end{cases} .$$

Let

$$g(x) = \sum_k \alpha_k u \circ f^k .$$

Show that there is a constant $C > 0$ such that for infinitely many values of j we have

$$C_{g,g}(j) \geq \frac{C}{(\log j)^2} .$$

Show that the function g is continuous. This function belongs to the well-studied family of examples of continuous nondifferentiable functions (see (Weierstrass 1886, p. 97) and (Hardy 1917)).

Exercise 5.55. *Use the same technique of Fourier series to discuss the decay of correlations for the cat map of Example 2.10.*

Exercise 5.56. *Consider the map of the interval $f(x) = 3x \pmod 1$ and the Lebesgue measure. Compute the autocorrelation function for the observable $g(x) = \chi_{[0,1/3]}$. Let $h(x) = \chi_{[0,1/2]} - 1/2$ and show that this function is an eigenvector of the Perron–Frobenius operator. Compute its autocorrelation function. Compute the limiting variance (5.25). See Fig. 5.9.*

The situation described above relating the rate of decay of correlations with some kind of regularity properties of the observables is quite frequent in the theory of dynamical systems. One looks for two Banach spaces $\mathcal{B}_1$ and $\mathcal{B}_2$ of functions on the phase space and a function $C_{\mathcal{B}_1,\mathcal{B}_2}(n)$ tending to zero when n tends to infinity such that for any $g \in \mathcal{B}_1$, for any $h \in \mathcal{B}_2$ and for any integer n,

$$\left|C_{g,h}(n)\right| \leq C_{\mathcal{B}_1,\mathcal{B}_2}(n)\, \|g\|_{\mathcal{B}_1} \|h\|_{\mathcal{B}_2} \ . \tag{5.34}$$

Note that it is not necessary for $\mathcal{B}_1$ and $\mathcal{B}_2$ to be contained in $\mathrm{L}^2(\mathrm{d}\mu)$, they only need to be in duality in some sense (see (Liverani 1995)).

As explained above, this kind of estimate may follow from adequate information on the spectral theory of a Perron–Frobenius operator. We mention here without proof the case of piecewise expanding maps of the interval (Example 2.15).

Theorem 5.57. *Let f be a piecewise expanding map of the interval. The associated Perron–Frobenius operator P maps the space $\mathbf{BV}$ of functions of bounded variation into itself. In this space, the spectrum is composed of the eigenvalue 1 with finite multiplicity, a finite number (maybe there are none) of other eigenvalues of modulus 1 with finite multiplicity, and the rest of the spectrum is contained in a disk centered at the origin and of radius $\sigma < 1$. Any eigenvector of eigenvalue 1 is a linear combination of a finite number of nonnegative eigenvectors. Their supports have disjoint interiors, which are a finite union of intervals permuted by the map. The eigenvalues on the unit circle different from 1 are rational multiples of 2π and therefore do not occur in the spectrum of a sufficiently high power of P.*

We refer to (Hofbauer and Keller 1982) for a proof of this theorem, and to (Collet and Isola 1991; Baladi 2000) for more results and extensions.

We can now consider some iterate of f and choose one of the absolutely continuous invariant measures with positive density h and support an interval $[a, b]$ (recall that a finite iterate of a piecewise expanding map of the interval is also a piecewise expanding map of the interval).

In that case, in the space $\mathbf{BV}([a, b])$ of functions of bounded variation with support equal to the support of h, the Perron–Frobenius operator P has 1 as a simple eigenvalue (with eigenvector h), and the rest of the spectrum is contained in a disk centered at the origin and of radius $\sigma < 1$. In particular, for any $1 > \varrho > \sigma$, there is a constant $\Gamma_\varrho > 0$ such that for any $g \in \mathbf{BV}([a, b])$ with

$$\int_a^b g(x)\, h(x)\, \mathrm{d}x = 0 \ ,$$

and any integer n, we have

$$\left\|P^n g\right\|_{\mathbf{BV}([a,b])} \leq \Gamma_\varrho \varrho^n \left\|g\right\|_{\mathbf{BV}([a,b])} \ .$$

This immediately implies that for any $g_1 \in \mathbf{BV}([a, b])$ and any $g_2 \in \mathrm{L}^1(\mathrm{d}x)$, we have

$$\left|C_{g_1,g_2}(n)\right| \le \Gamma_\varrho \varrho^n \|g_1\|_{\mathbf{BV}([a,b])} \|g_2\|_{\mathrm{L}^1} \,. \tag{5.35}$$

In other words, we have the estimate (5.34) with $\mathcal{B}_1 = \mathbf{BV}([a,b])$, $\mathcal{B}_2 = \mathrm{L}^1$, and $C_{\mathcal{B}_1,\mathcal{B}_2}(n) = \Gamma_\varrho \varrho^n$.

Perron–Frobenius operators have also been used to prove the decay of correlations for Gibbs states on subshifts of finite type (see (Bowen 1975) and (Ruelle 2004)), and for some nonuniformly hyperbolic dynamical systems (see (Young 1998) and references therein). The relation between the rate of decay of correlations and other quantities is not simple. We refer to (Collet and Eckmann 2004) for some results and examples.

Another important method is the method of coupling borrowed from probability theory. We refer to (Barbour, Gerrard, and Reinert 2000) for the case of piecewise expanding maps of the interval, (Bressaud, Fernández, and Galves 1999a;b) for the case of chains with complete connections which generalize the case of Gibbs states on subshifts of finite type, to (Bressaud and Liverani 2002), and to (Young 1999) for nonuniformly hyperbolic dynamical systems.

Let f be a piecewise expanding map on the interval with an acim μ with a nonvanishing density h. In this situation we can define a Markov chain X on the interval by the following formula for its transition probability

$$p(y|x) = \frac{h(y)}{h(x)} \delta\big(x - f(y)\big) = \frac{1}{h(x)} \sum_{z,\, f(z)=x} \frac{h(z)}{|f'(z)|} \delta(y - z) \,. \tag{5.36}$$

Exercise 5.58. *Show that the integral over y is equal to 1.*

This transition probability is nothing but the kernel of the operator $\mathcal{L}$ defined in (5.29). This is a rather irregular kernel, it only charges a finite number of points, namely the points y such that $f(y) = x$. These are called the *preimages* of x under the map f. Nevertheless, for piecewise expanding maps of the interval the following result has been proven in (Barbour, Gerrard, and Reinert 2000).

To formulate the result, we first recall that if we have two measures ν on a space X and μ on a space Y, a *coupling* between ν and μ is a measure on the product space such that its marginals (projections onto X resp. Y) are ν and μ respectively.

Theorem 5.59. *For any piecewise expanding map on the interval with a mixing acim μ with a nonvanishing density, there are three positive constants C, $\varrho_1 < 1$ and $\varrho_2 < 1$, and for any pair of points x_1 and x_2 in the interval, there is a coupling $\mathbb{P}_{x_1,x_2}$ between the two Markov chains (X_n^1) and (X_n^2) defined by the kernel (5.36) starting in x_1 and x_2 respectively and such that for any $n \in \mathbf{N}$*

$$\mathbb{P}_{x_1,x_2}\left(\left|X_n^1 - X_n^2\right| > \varrho_1^n\right) \le C\varrho_2^n \,.$$

Here, $\mathbb{P}$ denotes probability.

The statement in the original paper (Barbour, Gerrard, and Reinert 2000) is in fact stronger but we will use only the above form.

Note that the constants C, ϱ_1 and ϱ_2 do not depend on x_1 and x_2. Note also the difference with the coupling for Markov chains on finite state space. Here, because of the singularity of the kernel, the realizations of $\left(X_n^1\right)$ and $\left(X_n^2\right)$ will in general not meet. However, they can approach each other fast enough. The condition that h does not vanish is not restrictive. We have seen already that this is first connected with ergodicity (see Theorem 5.57). Once ergodicity is ensured, the vanishing of h can be discussed in detail. We refer to (Barbour, Gerrard, and Reinert 2000; Buzzi 1997) for the details.

We now apply Theorem 5.59 to prove the decay of correlations. Let g be a function of bounded variation. We have from the definition of the Markov chains $\left(X_n^1\right)$ and $\left(X_n^2\right)$ and from the definition of the coupling (here and below we define $\mathbb{E}$ to denote the *expectation*):

$$\begin{aligned}(\mathcal{L}^n g)(x_1) - (\mathcal{L}^n g)(x_2) &= \mathbb{E}_{x_1}\left(g(X_n^1)\right) - \mathbb{E}_{x_2}\left(g(X_n^2)\right) \\ &= \mathbb{E}_{x_1,x_2}\Big(\left(g(X_n^1)\right) - \left(g(X_n^2)\right)\Big) ,\end{aligned}$$

where in the last expression, the expectation is with respect to any coupling between $\left(X_n^1\right)$ and $\left(X_n^2\right)$. We now use the coupling of Theorem 5.59 and assume that g is a Lipschitz function with Lipschitz constant L_g. Then we get by applying Theorem 5.59,

$$\begin{aligned}&\left|\mathbb{E}_{x_1,x_2}\Big(g(X_n^1) - g(X_n^2)\Big)\right| \\ &\le \mathbb{E}_{x_1,x_2}\Big(\chi_{\left|X_n^1 - X_n^2\right| \le \varrho_1^n}\left|g(X_n^1) - g(X_n^2)\right|\Big) \\ &+\mathbb{E}_{x_1,x_2}\Big(\chi_{\left|X_n^1 - X_n^2\right| > \varrho_1^n}\left|g(X_n^1) - g(X_n^2)\right|\Big) \\ &\le L_g \varrho_1^n + 2\, C\, L_g \varrho_2^n .\end{aligned}$$

We now consider a correlation $C_{u,v}(n)$ between a Lipschitz function u and an integrable function v. Using formula (5.28) and the invariance of μ, we conclude that

$$C_{u,v}(n) = \int\!\!\int \Big(\mathcal{L}^n u(x_1) v(x_1) - \mathcal{L}^n u(x_1) v(x_2)\Big)\, \mathrm{d}\mu(x_1)\, \mathrm{d}\mu(x_2) .$$

We can now use the above estimate to get

$$\left|C_{u,v}(n)\right| \le \left(L_u \varrho_1^n + 2\, C\, L_u \varrho_2^n\right) \int |v|\, \mathrm{d}\mu ,$$

which is the exponential decay of correlations as in (5.35).

We explain here very shortly the idea of coupling on the particularly simple case of the map $f(x) = 2x \pmod 1$. In that case, the transition probability is given by

$$p(y|x) = \frac{1}{2} \sum_{z\,,\,2z=x \pmod 1} \delta(y-z) = \delta(2y-x) . \tag{5.37}$$

Let $\mathcal{P}$ be the partition of the interval $[0,1)$ given by $\mathcal{P} = \{[0,1/2),[1/2,1)\}$. It is easy to verify that, except for a countable number of points x, any of the 2^n atoms of the partition $\vee_{j=1}^{n} f^{-j}\mathcal{P}$ contains one and only one preimage of order n of a point x (the points y such that $f^n(y)=x$).

Exercise 5.60. *Use the coding with the full unilateral shift on two symbols to prove this assertion (see Example 2.15).*

In other words, for two points x and x' we have a coupling at the topological level of their preimages by putting together those preimages which are in the same atom of $\vee_{j=1}^{n} f^{-j}\mathcal{P}$. We can now write the coupling on the trajectories of the Markov chain. For any pair of points x and x' in $[0,1)$, for any integer n, and for any pair of n-tuples $(y_1,\ldots,y_n)$ and $(y'_1,\ldots,y'_n)$ of $[0,1)^n$ we define

$$\mu_{x,x'}\left((y_1,\ldots,y_n),(y'_1,\ldots,y'_n)\right)$$

$$= \sum_{I\in\vee_{j=1}^{n} f^{-j}\mathcal{P}} \chi_I(y_n)\chi_I(y'_n)\prod_{\ell=1}^{n}\delta(2y_\ell - y_{\ell-1})\prod_{\ell=1}^{n}\delta(2y'_\ell - y'_{\ell-1})$$

where for convenience of notations we defined $y_0 = x$ and $y'_0 = x'$.

Exercise 5.61. *Show that the above formula defines a measure on $\left([0,1)\times[0,1)\right)^{\mathbb{N}}$ (use Kolmogorov's theorem, see, for example, (Gikhman and Skorokhod 1996)). Show that except for a countable number of points x and x' of the interval, this measure is a coupling between the two Markov chains with transition kernel (5.37) and initial points x and x'. Show that this coupling satisfies the conclusions of Theorem 5.59.*

The more general cases use basically the same ideas with the complication that the transition probability is not constant. We refer to the above-mentioned references for the details.

5.7 Physical Measures

As we have seen already earlier, dynamical systems have frequently several ergodic invariant measures. It is therefore natural to ask if there is one of these measures which in some sense is more important than the others. From a physical point of view that we do not discuss here (see however Remark 5.12), the Lebesgue measure of the phase space is singled out. However, if the dynamical system has a rather small attractor, for example a set of Lebesgue measure zero as it often happens in dissipative systems, this does not look very helpful. In such cases there is no invariant measure which is absolutely continuous with respect to the Lebesgue measure.

The study of Axiom A systems by Sinai, Ruelle, and Bowen (SRB)—which we introduce in Sect. 6.2—leads naturally to a generalization of the notion of ergodicity which makes use of the Lebesgue measure on the whole phase space and which we discuss now.

Definition 5.62. *Consider a dynamical system given by a map f on a phase space Ω which is a compact Riemannian manifold. An ergodic invariant measure μ for a map f is called a* Physical measure *if there is a subset V of Ω of positive Lebesgue measure such that for any continuous function g on Ω and any initial point $x \in V$ we have*

$$\lim_{n\to\infty} \frac{1}{n} \sum_{j=0}^{n-1} g\big(f^j(x)\big) = \int g \, \mathrm{d}\mu \, .$$

Remark 5.63. The point of the definition is that the set of "good points" has positive *Lebesgue* measure, even if the invariant measure is concentrated on a very small set (and not absolutely continuous with respect to the Lebesgue measure).

For Axiom A systems it is known that Physical measures always exist. They have some more properties and are particular cases of SRB measures (see Sect. 6.2).

Example 5.64. Let f have a stable fixed point at x_0. The δ function at x_0 is an invariant measure, and since a neighborhood of x_0 is attracted to it, and has positive Lebesgue measure, δ_{x_0} is a Physical measure. We will see later that even intrinsically unstable systems can have Physical measures.

Example 5.65. Consider the map $f : x \mapsto 2x \pmod 1$. The fixed point at $x_0 = 0$ is unstable. The δ function at x_0 is an invariant measure. On the other hand, the measure $\mathrm{d}x$ is invariant and ergodic. Since, by the ergodic theorem, one has for any continuous function g and Lebesgue-almost every x convergence of the ergodic average to $\int g(x)\,\mathrm{d}x$, it is not possible that for a set of positive measure this limit equals $g(0)$ (except if $\int g = g(0)$, which is not the case in general). Therefore, δ_{x_0} is not a Physical measure.

Example 5.66. In the preceding example, the Physical measure was absolutely continuous. In Fig. 4.5, the δ function at the hyperbolic fixed point P_0 is a Physical measure which is not absolutely continuous (even in the unstable direction). But the flow is not Axiom A. See Example 4.54.

The idea behind the definition is that if the dynamical system has an attractor, this set should be transversally attracting. An orbit will therefore rapidly approach the attractor, and a continuous function will take almost the same value for a point on the attractor and a point nearby. One would then be able to apply the ergodic theorem for the ergodic measure μ (supported by the attractor) provided exceptional sets of μ-measure zero would in some sense be projections of sets of Lebesgue measure zero in the phase space. This requires an additional technical property (the so-called absolute continuity of the projection along the stable manifolds). We refer to (Young 2002) for details and references.

We have already seen examples of Physical measures. The simplest case is of course when the system has an invariant measure which is absolutely continuous with respect to the Lebesgue measure. For the quadratic family of Example (2.18) it is known that there is a unique Physical measure for all the parameters and a subset of positive measure of the parameter space (the interval $[0, 2]$) for which such a measure is absolutely continuous. We refer to (Avila and Moreira 2005) for recent work and references. Similarly, for the Hénon map of example (2.31) it has been shown that for a set of positive Lebesgue measure of parameter values, Physical measures exist (see (Benedicks and Carleson 1991; Benedicks and Young 1993)). We refer to (Eckmann and Ruelle 1985a) for a general discussion.

Exercise 5.67. *Show that the invariant measure of the dissipative baker's map constructed in Exercise 3.26 is a Physical measure.*

We mention that there are dynamical systems for which there is no Physical measure. A simple example is given by the map of the cylinder $\Omega = S^1 \times \mathbb{R}$ given by $T(\vartheta, x) = (\vartheta, x/2)$. The circle $S^1 \times \{0\}$ is invariant and globally attracting. Any point is a fixed point and the ergodic invariant measures are the Dirac masses $\delta_{(\vartheta,0)}$. Obviously none of them "attract" a set of positive 2-dimensional Lebesgue measure. We refer to (Young 2002) for more information.

5.8 Lyapunov Exponents

An important application of the subadditive ergodic theorem (Theorem 5.43) is the proof of the Oseledec theorem about Lyapunov exponents. We will see here the interplay between ergodic theory and differentiable structure on the phase space and the time evolution. We first explain the ideas in an informal way.

Complicated dynamical behavior is often called chaotic, but one would like to have a more precise definition of this notion. Frequently, one observes in these systems that the trajectories depend in a very sensitive way on the initial conditions (see Sect. 4.4.1). If one starts with two very nearby generic initial conditions, the distance between the orbits grows (at small distances) exponentially fast in time. A very different behavior takes place for simple systems, such as for a harmonic oscillator which is an integrable (nonchaotic) system. In that case, for two nearby initial conditions with the same energy, the distance does not grow at all, and, even if the energies are different, the distance only grows linearly in time. The phenomenon of growth of small errors in the initial condition is called *sensitive dependence on initial conditions* (see Sect. 4.4.1). The precise definition of this and similar notions varies in the literature; for a possible set of definitions, see (Eckmann and Ruelle 1985a). In this section, we discuss its strongest form, namely the positivity of Lyapunov exponents. This phenomenon is illustrated in Fig. 5.11 for the Lorenz model (see (2.7)). We show a plot of the x-component as a function of time for two "very" nearby initial conditions. After staying together for a while, they suddenly separate macroscopically. Since it is so important and so common in chaotic systems, one would like to have a quantitative measurement of the sensitive dependence on initial conditions. Assume that the time evolution map f is differentiable (and the phase space Ω is a differentiable manifold, we will assume $\Omega = \mathbb{R}^d$ for simplicity). If we start with two nearby initial conditions x and $y = x + \mathbf{h}$ with $\mathbf{h}$ small, we would like to estimate the size of

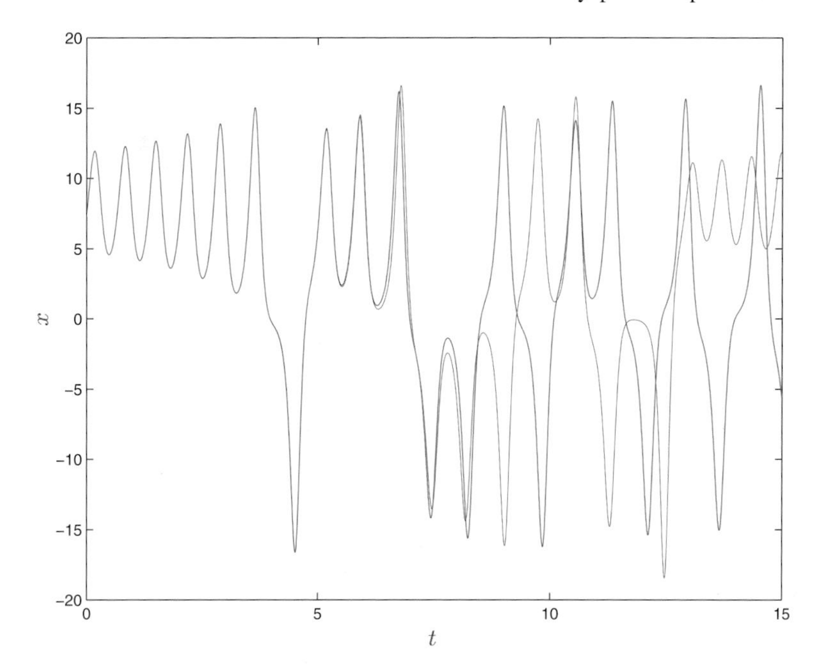

Fig. 5.11. Sensitive dependence with respect to initial condition in the Lorenz model. The plot shows the x-coordinate as a function of time. The two orbits start with a separation of 0.01, which reaches order 1 after about $t = 8$

$$f^n(x + \mathbf{h}) - f^n(x) ,$$

namely how the initially small error $\mathbf{h}$ evolves in time. As long as the error has not grown too much, we can expect to have a reasonable answer from the first-order Taylor formula, namely

$$f^n(x + \mathbf{h}) - f^n(x) = \mathrm{D}_x f^n \mathbf{h} + \mathcal{O}(\mathbf{h}^2) ,$$

where $\mathrm{D}_x f^n$ denotes the differential of the map f^n at the point x. We recall that from the chain rule one finds

$$\mathrm{D}_x f^n = \mathrm{D}_{f^{n-1}(x)} f \cdots \mathrm{D}_x f ,$$

where in general the matrices in this product do not commute. As this is a product of n terms, this suggests that we look (test) for an exponential growth. In other words, we take the logarithm, divide by n, and take the limit $n \to \infty$. Since the matrices do not commute, the formulation of the Oseledec theorem for the general case will be slightly more sophisticated. We will warm up with some examples.

First (easy) case: dimension 1, the phase space is an interval ([–1,1]), the dynamics is given by a regular map f. One gets by Taylor's formula

$$f^n(x + h) = f^n(x) + f^{n\prime}(x)h + \mathcal{O}(h^2) .$$

By the chain rule ($f^n = f \circ f \circ \cdots \circ f$, n times), we have

$$f^{n\prime}(x) = \prod_{j=0}^{n-1} f'\left(f^j(x)\right) .$$

As explained above, this naturally suggests an exponential growth in n, and to look for the exponential growth rate per step (unit of time), namely

$$\frac{1}{n} \log \left| f^{n\prime}(x) \right| = \frac{1}{n} \sum_{j=0}^{n-1} \log \left| f'\left(f^j(x)\right) \right| .$$

On the right-hand side appears a temporal average and we can apply Birkhoff's ergodic theorem (Theorem 5.38). Let μ be an ergodic invariant measure and assume the modulus of $\log \left| f' \right|$ is μ-integrable. Then, on a set of full μ-measure, we have convergence of the temporal average, and moreover

$$\lim_{n\to\infty} \frac{1}{n} \log \left| f^{n\prime}(x) \right| = \int \log \left| f'(\cdot) \right| \mathrm{d}\mu .$$

This limit is called the Lyapunov exponent of the transformation f for the measure μ. Here one should stress again the importance of the initial condition. For many initial conditions the limit does not exist. For many other initial conditions, the limit exists, but takes another value. For example, the initial conditions typical of a different ergodic measure.

Remark 5.68. Note that the Lyapunov exponent does not measure the separation of close-by initial conditions, but really the rate of separation of infinitesimally close initial points. Indeed, there cannot be any indefinitely long exponential separation of the orbits of close-by points in a compact system, since the orbits can only separate some finite amount.

Exercise 5.69. *Consider the continuous skew map given by (2.4). Show that the Lebesgue measure is f invariant and ergodic, and that its Lyapunov exponent is equal to* $\frac{1}{3} \log 3 + \frac{2}{3} \log(3/2)$. *Show that the transformation f has a fixed point $x = 3/5$, that the Dirac measure in this point is also invariant and ergodic, and its Lyapunov exponent is equal to* $\log(3/2)$. *Show that for any ergodic invariant measure the Lyapunov exponent belongs to the interval* $[\log(3/2), \log 3]$. *Show that there is an uncountable set of invariant ergodic measures, all with different Lyapunov exponents (taking all the values between* $\log(3/2)$ *and* $\log 3$*). Hint: use a conjugacy (coding) to the unilateral shift on two symbols, and then consider the infinite product Bernoulli measures.*

We now consider the next level of difficulty: When the dimension of the phase space is larger than 1. As a simple example we will use the dissipative baker's map f given in (2.6), whose attractor has been studied in Exercise 3.26.

Example 5.70. We consider a variant of the dissipative baker's map given by

$$f(x,y) = \begin{cases} (3x, y/4)\,, & \text{if } 0 \le x < 1/3 \\ (3x/2 - 1/2, 1 - y/3)\,, & \text{if } 1/3 \le x \le 1 \end{cases}.$$

If we start from the initial point (x, y), with an initial error $\mathbf{h}$, we have after n iteration steps (using the chain rule) an error given (up to order $\mathcal{O}(\mathbf{h}^2)$) by

$$\mathrm{D}_{(x,y)} f^n \mathbf{h} = \mathrm{D}_{f^{n-1}(x,y)} f \cdot \mathrm{D}_{f^{n-2}(x,y)} f \cdots \mathrm{D}_{(x,y)} f \mathbf{h}\,.$$

To estimate $\mathrm{D}_{(x,y)} f^n$ we take a product of matrices. In general matrices do not commute (hence one should be careful with the order in the product). However, here they do commute (and they are even diagonal). Therefore, to obtain the product matrix, it is enough to take the product of the diagonal elements:

$$\mathrm{D}_{f^{n-1}(x,y)} f \cdot \mathrm{D}_{f^{n-2}(x,y)} f \cdots \mathrm{D}_{(x,y)} f = \begin{pmatrix} \prod_{j=0}^{n-1} u\big(f^j(x,y)\big) & 0 \\ 0 & \prod_{j=0}^{n-1} v\big(f^j(x,y)\big) \end{pmatrix}$$

where

$$u(x,y) = 3\chi_{[0,1/3]}(x) + (3/2)\chi_{[1/3,1]}(x)$$

and

$$v(x,y) = (1/4)\chi_{[0,1/3]}(x) - (1/3)\chi_{[1/3,1]}(x)\,.$$

We can now take the logarithm of the absolute value of each diagonal entry of the product and apply Birkhoff's ergodic theorem as in the 1-dimensional case. Since the functions u and v do not depend on y, the integral with respect to the Physical measure reduces to the integration with respect to the 1-dimensional Lebesgue measure. Therefore, for Lebesgue-almost any (x, y) (2-dimensional Lebesgue measure, recall the definition of Physical measure)

$$\lim_{n\to\infty} \frac{1}{n} \sum_{j=0}^{n-1} \log \big|u\big(f^j(x,y)\big| = (1/3)\log 3 + (2/3)\log(3/2) = \log\big((27/4)^{1/3}\big)\,.$$

Similarly

$$\lim_{n\to\infty} \frac{1}{n} \sum_{j=0}^{n-1} \log \big|v\big(f^j(x,y)\big| = -(1/3)\log 4 - (2/3)\log(3) = \log\big((36)^{-1/3}\big)\,.$$

We conclude that

$$\mathrm{D}_{f^{n-1}(x,y)} f \mathrm{D}_{f^{n-2}(x,y)} f \cdots \mathrm{D}_{(x,y)} f \approx \begin{pmatrix} (27/4)^{n/3} & 0 \\ 0 & (36)^{-n/3} \end{pmatrix},$$

and we can now interpret this result.

If we take a vector $\mathbf{h} = (h_1, h_2)$, with $h_1 \neq 0$, we see that the first component of $\mathrm{D}_{(x,y)} f^n \mathbf{h}$ grows exponentially at the rate $\log\left((27/4)^{1/3}\right)$, and this component dominates the other one which decreases exponentially at the rate $\log\left(36^{-1/3}\right)$ (<0). In other words, almost any error grows exponentially fast with rate $\log\left((27/4)^{1/3}\right)$. This quantity is the maximal Lyapunov exponent. However, there is another Lyapunov exponent $\log\left(36^{-1/3}\right)$ (<0) which corresponds to special initial error vectors satisfying $h_1 = 0$. These errors do not grow but decay exponentially fast. This distinction is similar to the diagonalization of matrices but the interpretation is slightly more involved. We have to distinguish two subspaces. The first is the entire space $E_0 = \mathbb{R}^2$. The second is the subspace of codimension one $E_1 = \left\{(0, h_2)\right\} \subset E_0$. If $\mathbf{h} \in E_0 \backslash E_1$, (and $\mathbf{h} \neq \mathbf{0}$) the initial error grows exponentially fast with rate $\log\left((27/4)^{1/3}\right)$. If $\mathbf{h}$ is in E_1 the initial error decays exponentially fast with rate $\log\left(36^{-1/3}\right)$. This is somewhat analogous to the case of a hyperbolic fixed point although here one considers general points, not necessarily periodic ones. This ends Example 5.70.

We now come to the study of Lyapunov exponents in the general case of dynamical systems. We assume for simplicity that f is a diffeomorphism of $\mathbb{R}^n$ with a compact attracting set. The result also holds on manifolds and even in infinite dimensions (under suitable hypotheses).

This general case combines the two preceding ideas (product of matrices and ergodic theorem) together with a new fact: the subspaces E_0, $E_1, \dots$ now depend on the initial condition, and they vary from point to point. Moreover, the matrices appearing in the product do not commute.

For the moment let us first consider the behavior of the norm and define

$$h_n(x) = \log \left\| \mathrm{D}_x f^n \right\| .$$

It is easy to verify that this sequence of functions satisfies the subadditive inequality (5.21).

Exercise 5.71. *Prove it.*

Assume now that an ergodic invariant measure μ for f is given. Then we can apply Theorem 5.43 (provided we can check the second assumption) and conclude that

$$\lim_{n \to \infty} \frac{1}{n} \log \left\| \mathrm{D}_x f^n \right\|$$

exists μ-almost surely. If this quantity is positive, we know that some initially small errors are amplified exponentially. If this quantity is negative or zero, we know that initial errors cannot grow very rapidly.

More generally, if we have a fixed vector $\mathbf{h}$, recall that

$$\left\| \mathrm{D}_x f^n \mathbf{h} \right\|^2 = \left\langle \mathbf{h} \middle| \left(\mathrm{D}_x f^n\right)^{\mathrm{t}} \mathrm{D}_x f^n \mathbf{h} \right\rangle$$

where $\langle \cdot | \cdot \rangle$ denotes the scalar product in $\mathbb{R}^d$ and A^{t} is the transpose of the (real) matrix A. This suggests to study the asymptotic exponential growth of the matrix

$(\mathrm{D}_x f^n)^{\mathrm{t}} \mathrm{D}_x f^n$. The answer is provided by the following theorem due initially to Oseledec, see also (Ruelle 1979; Johnson, Palmer, and Sell 1987):

Theorem 5.72. *Let μ be an ergodic invariant measure for a diffeomorphism f of a compact manifold Ω. Then for μ-almost every initial condition x, the sequence of symmetric nonnegative matrices*

$$\left((\mathrm{D}_x f^n)^{\mathrm{t}} \mathrm{D}_x f^n\right)^{1/2n}$$

converges to a symmetric nonnegative matrix Λ (independent of x). Denote by $\lambda_0 > \lambda_1 > \ldots > \lambda_k$ the strictly decreasing sequence of the logarithms of the eigenvalues of the matrix Λ (some of them may have nontrivial multiplicity).

These numbers are called the Lyapunov exponents *of the map f for the ergodic invariant measure μ. For μ-almost every point x there is a decreasing sequence of subspaces*

$$\Omega = E_0(x) \supsetneq E_1(x) \supsetneq \cdots \supsetneq E_k(x) \supsetneq E_{k+1}(x) = \{\mathbf{0}\} ,$$

satisfying (μ-almost surely)

$$\mathrm{D}_x f E_j(x) = E_j\big(f(x)\big)$$

and for any $j \in \{0, \ldots, k\}$ and any $\mathbf{h} \in E_j(x) \backslash E_{j+1}(x)$ one has

$$\lim_{n \to \infty} \frac{1}{n} \log \left\| \mathrm{D}_x f^n \mathbf{h} \right\| = \lambda_j .$$

We refer to (Barreira and Pesin 2001; 2002; Barreira 1997) for reviews on the Oseledec theorem.

Remark 5.73.

i) Note that the theorem says that

$$\left\| \mathrm{D}_x f^n \mathbf{h} \right\| \sim \mathrm{e}^{n\lambda_j} .$$

There may be large or small (subexponential) prefactors which can depend on $\mathbf{h}$ and x.

ii) Positive Lyapunov exponents are obviously responsible for sensitive dependence on initial conditions. Their corresponding "eigen" directions are tangent to the attractor.

iii) Transversally to the attractor one gets contracting directions, namely negative Lyapunov exponents.

iv) One of the astonishing facts about the theorem is that the noncommutativity of the matrices does not really destroy the reasoning we gave in the commutative case.

Exercise 5.74. *Prove that if we conjugate the map f to a map f_* by a diffeomorphism, and consider an image measure, then we get the same Lyapunov exponents.*

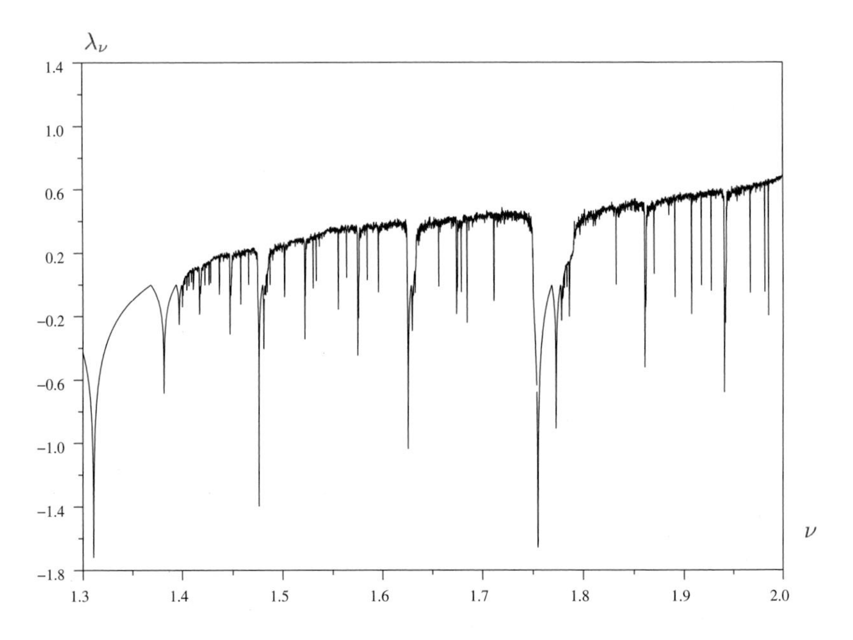

Fig. 5.12. Lyapunov exponent as a function of the parameter for the quadratic family $f_\nu(x) = 1 - \nu x^2$

If the map depends on a parameter, the Lyapunov exponents depend also in general on the parameter (one has also to decide on an ergodic invariant measure for each parameter). This is illustrated in Fig. 5.12, where the Lyapunov exponent of the quadratic family of Example 2.18 is drawn as a function of the parameter ν for the Physical measure.

6

Entropy

In Sect. 3.2 we already addressed the subject of fuzzy knowledge from a purely topological point of view. Here, we come back to this idea, but now, we also connect it to the notion of invariant measure. That is, we account (in particular in the natural case of the Physical measure) for how often (or how long) a trajectory stays in a particular region. One can ask how much information this gives us about the long-time dynamics of the system and the variability of the set of orbits. The main difference with the topological entropy is that we are not going to consider all trajectories but only those "typical" for a given measure.

For the convenience of the reader, we repeat some issues from Sect. 3.1. Often one does not have access to the points of phase space but only to some fuzzy approximation. For example, if one uses a real apparatus which always has a finite precision. One can formalize this idea in several ways. We present two of them.

i) The phase space Ω is a metric space with metric d. For a given precision $\varepsilon > 0$, two points at distance less than ε are not distinguishable.
ii) One gives a (measurable) partition $\mathcal{P}$ of the phase space, $\mathcal{P} = \{A_1, \ldots, A_k\}$ (k finite or not) , $A_j \cap A_\ell = \emptyset$ for $j \neq \ell$ and

$$\Omega = \bigcup_{j=1}^{k} A_j \ .$$

Two points in the same atom of the partition $\mathcal{P}$ are considered indistinguishable. If there is a given measure μ on the phase space, it is often useful to use partitions modulo sets of μ-measure zero.

We also recall from Sect. 3.1 that the notion of partition leads naturally to a coding of the dynamical system. This is a map Φ from Ω to $\{1, \ldots, k\}^{\mathbb{Z}^+}$ given by

$$\Phi_n(x) = \ell \qquad \text{if} \qquad f^n(x) \in A_\ell \ .$$

If the map is invertible, one can also use a bilateral coding. If S denotes the shift on sequences, it is easy to verify that $\Phi \circ f = S \circ \Phi$. In general $\Phi(\Omega)$ is a complicated subset of $\{1,\ldots,k\}^{\mathbb{N}}$; i.e., it is difficult to say which codes are admissible. There are however some examples of very nice codings like for Axiom A attractors (see (Bowen 1975; Keane 1991) and (Ruelle 2004)), as we have seen in Sect. 4.5.

Let $\mathcal{P}$ and $\mathcal{P}'$ be two partitions, the partition $\mathcal{P} \vee \mathcal{P}'$ is defined as before by

$$\mathcal{P} \vee \mathcal{P}' = \{A \cap B \mid A \in \mathcal{P} \,,\ B \in \mathcal{P}'\} \,.$$

If $\mathcal{P}$ is a partition,

$$f^{-1}\mathcal{P} = \{f^{-1}(A)\}$$

is also a partition. Recall that (even in the noninvertible case) $f^{-1}(A) = \{x \mid f(x) \in A\}$.

A partition $\mathcal{P}$ is said to be *generating* if (modulo sets of measure zero when a particular measure has been chosen)

$$\bigvee_{n=0}^{\infty} f^{-n}\mathcal{P} = \varepsilon$$

with ε the partition into points. In this case the coding is injective (modulo sets of measure zero).

Exercise 6.1. *For the map of the interval* $3x \pmod 1$, *show that*

$$\mathcal{P} = \{[0,1/3),[1/3,2/3),[2/3,1)\}$$

is a generating partition (modulo sets of Lebesgue measure zero).

We have already seen the definition of topological entropy in Sect. 3.2, we discuss now the so-called Kolmogorov–Sinai entropy. Both measure how many different orbits one can detect through a fuzzy observation.

What is new in this chapter is that we now allow for an invariant measure to come into play; the reader might think of a Physical measure. When an ergodic invariant measure μ is considered, the disadvantage of the topological entropy is that it measures the total number of (distinguishable) trajectories, including trajectories which have an anomalously small probability to be chosen by μ. It even often happens that these trajectories are much more numerous than the ones favored by μ. The Kolmogorov–Sinai entropy is then more adequate.

If $\mathcal{P}$ is a (measurable) partition, its entropy $H_\mu(\mathcal{P})$ with respect to the measure μ is defined by

$$H_\mu(\mathcal{P}) = -\sum_{A\in\mathcal{P}} \mu(A)\log\mu(A) \,.$$

This is Shannon's formula, and in communication theory one often uses the logarithm base 2 (see (Shannon 1948), reprinted in (Shannon 1993)). This quantity measures the average information obtained by observing a long stretch of the orbit of a μ-randomly chosen initial point. More precisely, the longer one observes the sequence of atoms an orbit visits, the better one knows the initial point. The Shannon entropy quantifies the asymptotic average gain of this information about the initial point.

The entropy of the dynamical system with respect to the partition $\mathcal{P}$ and the (invariant, ergodic, probability) measure μ is defined by

$$H_\mu(f,\mathcal{P}) = \lim_{n\to\infty} \frac{1}{n} H_\mu(\vee_{j=0}^{n-1} f^{-j}\mathcal{P}) \,.$$

One can prove that the limit always exists. The interpretation is the average information per unit time we get by specifying longer and longer pieces of code of a trajectory. Indeed, a longer code gives a more precise information about the initial condition (see Sect. 3.1).

Finally the *Kolmogorov–Sinai entropy* is defined by

$$h_\mu(f) = \sup_{\mathcal{P}\text{ finite or countable}} H_\mu(f,\mathcal{P}) \,.$$

If $\mathcal{P}$ is generating, and $h_\mu(f) < \infty$, then

$$h_\mu(f) = H_\mu(f,\mathcal{P}) \,.$$

One has also the general inequality

$$h_\mu(f) \le h_{\text{top}}(f) \,, \tag{6.1}$$

and moreover

$$h_{\text{top}}(f) = \sup_{\mu\,:\,f\text{ ergodic}} h_\mu(f) \,.$$

However, the maximum may not be reached. We refer to (Sinai 1976) or (Petersen 1983; Keane and Petersen 2006) for proofs and more results on the Kolmogorov Sinai entropy.

An important application of the entropy is Ornstein's isomorphism theorem for Bernoulli shifts, we refer to (Ornstein 1974) for more information.

6.1 The Shannon–McMillan–Breiman Theorem

Another important application of the entropy is the Shannon–McMillan–Breiman theorem which in a sense counts the number of typical orbits for the measure μ.

Theorem 6.2. *Let $\mathcal{P}$ be a finite generating partition for a map f with an ergodic invariant measure μ. For any $\varepsilon > 0$ there is an integer $N(\varepsilon)$ such that for any $n > N(\varepsilon)$ one can separate the atoms of the partition $\mathcal{P}_n = \vee_{j=0}^{n-1} f^{-j}\mathcal{P}$ into two disjoint subsets*

$$\mathcal{P}_n = \mathcal{B}_n \cup \mathcal{G}_n$$

such that

$$\mu\Big(\bigcup_{A\in\mathcal{B}_n} A\Big) \le \varepsilon$$

and for any $A \in \mathcal{G}_n$

$$e^{-n(h_\mu(f)+\varepsilon)} \le \mu(A) \le e^{-n(h_\mu(f)-\varepsilon)} .$$

In other words, the atoms in $\mathcal{B}_n$ are in some sense the "bad" atoms (for the measure μ), but their total mass is small. On the other hand, the good atoms (those in $\mathcal{G}_n$) all have almost the same measure, and of course their union gives almost the total weight. An immediate consequence is that

$$\mathrm{card}\, \mathcal{G}_n \approx e^{n\, h_\mu(f)} ,$$

where $\mathrm{card}\, \mathcal{G}_n$ denotes the cardinality of the set $\mathcal{G}_n$. This is similar to the well-known formula of Boltzmann in statistical mechanics relating the entropy to the logarithm of the (relevant) volume in phase space. Another obvious connection is given by the equivalence of ensembles in statistical mechanics. We refer to (Krengel 1985) and (Lanford 1973) for more information.

As mentioned before, it often happens that $\mathrm{card}\, \mathcal{G}_n \ll \mathrm{card}\, \mathcal{B}_n$.

Example 6.3. We illustrate these notions with a simple example, the Bernoulli shift on two symbols. Let $p \in (0,1)$ with $p \neq 1/2$. Consider the probability measure μ_p on $\Omega = \{0,1\}^{\mathbb{N}}$ of Exercise 5.21. This is equivalent to flipping a coin with probability p to get head and probability $q = 1-p$ to get tail. It is easy to prove that the measure μ_p is invariant and ergodic for the shift $\mathcal{S}$. We use the definition of a distance d on Ω (and in fact a one-parameter family of distances) in (5.14). It is also easy to prove that $h_{\mathrm{top}}(\mathcal{S}) = \log 2$.

Exercise 6.4. *Prove this assertion.*

The partition

$$\mathcal{P} = \Big\{ \{x_0 = 0\}, \{x_0 = 1\} \Big\}$$

is generating. An atom of $\bigvee_{j=0}^{n-1} \mathcal{S}^{-j}\mathcal{P}$, called a cylinder set in this context, is given by specifying the first n symbols of the sequences. For the entropy one has $h_{\mu_p}(\mathcal{S}) = -p\log p - q\log q$. Therefore, for large n we get (since $p \neq 1/2$)

$$\big|\mathcal{G}_n\big| \approx e^{n\, h_{\mu_p}(\mathcal{S})} \ll 2^n \approx \big|\mathcal{B}_n\big| .$$

One can use this classical example and ask for a more precise asymptotics of the set of typical sequences among the sequences of zeros (tail) and ones (head) of length n. To count them, we classify them according to the number m of ones (head). If m is given, there are $\binom{n}{m}$ sequences of length n with m ones and $n-m$ zeros. Of course

$$2^n = \sum_{m=0}^{n} \binom{n}{m}$$

but also

$$1=\sum_{m=0}^{n}\binom{n}{m}p^m q^{n-m}\ .$$

Let us look more carefully at this well-known identity. If a sum of $(n+1)$-positive terms is equal to 1, there are two extreme cases: all the terms are equal (to $1/(n+1)$), or one is equal to 1 and the others are 0. For large n, the present case is more like this second situation. Indeed, by Stirling's formula

$$\log\left(\binom{n}{m}p^m q^{n-m}\right)$$

$$\approx n\log n - m\log m-(n-m)\log(n-m)+m\log p+(n-m)\log q\ ,$$

and this quantity is maximal when m equals $m_* = p\,n$. It decreases very fast when m deviates from m_*. For example, for $|m-m_*|>2.6\sqrt{pqn}$ we have

$$\sum_{|m-m_*|>2.6\sqrt{pqn}}\binom{n}{m}p^m q^{n-m}\le 0.01\ .$$

Hence, there is an extremely small probability to observe a sequence with an m differing from m_* by more than $2.6\sqrt{pqn}$. When m equals m_*, then $\log\left(p^m q^{n-m}\right)$ is asymptotically equal to e^{-nh} with

$$h=-p\log p-q\log q\ ,$$

the entropy per flip. For m near m_*

$$\binom{n}{m}p^m q^{n-m}\approx\frac{1}{2\pi\sqrt{pqn}}\mathrm{e}^{-(m-m_*)^2/(2pqn)}\ .$$

Note that $p=1/2$ gives the maximal entropy which is equal to $\log 2$. In this case all possible sequences of outcome have the same probability. This measure is called the measure of maximal entropy (its entropy is equal to the topological entropy and this is the maximum as follows from inequality (6.1)). All the trajectories are equally typical. This is why we assumed before $p\neq 1/2$.

Let us summarize (for $p\neq 1/2$). Out of the 2^n possible outcomes, the observed ones belong with overwhelming probability to a much smaller subset ($m\simeq m_*$). In this subset, all sequences have about the same probability $p^{m_*}q^{n-m_*}\approx\mathrm{e}^{-nh}$. Since the set of these sequences has almost full measure, their number must satisfy $N_p(n)\approx\mathrm{e}^{nh}$. For the Kolmogorov–Sinai entropy of the independent coin flipping this leads to

$$\lim_{n\to\infty}\frac{1}{n}\log N_p(n)=h=-p\log p-q\log q\ .$$

and

$$N_p(n)\approx\mathrm{e}^{nh}\ll 2^n$$

for $p\neq 1/2$. Said differently, one can separate the possible outcomes of length n in two categories: the good ones and the bad ones. This discrimination depends heavily on the measure. All bad outcomes together form a set of small measure (not only each one separately). The good outcomes have all about the same probability e^{-nh} and together account for almost all the weight (one). Their number is about e^{nh}. This ends the discussion of Example 6.3.

Another way to formulate the Shannon–McMillan–Breiman theorem is to look at the measure of cylinder sets. For a point $x \in \Omega$, let $C_n(x)$ be the atom of $\vee_{j=0}^{n-1} f^{-j}\mathcal{P}$, which contains x. In other words, $C_n(x)$ is the set of $y \in \Omega$ such that for $0 \le j \le n-1$, $f^j(x)$ and $f^j(y)$ belong to the same atom of $\mathcal{P}$ (the trajectories are indistinguishable up to time $n-1$ from the fuzzy observation defined by $\mathcal{P}$, they have the same code). Then for μ-almost every x we have

$$h_\mu(f) = -\lim_{n\to\infty} \frac{1}{n} \log \mu\big(C_n(x)\big) .$$

A similar result holds using a metric instead of a partition. It is due to Brin and Katok, and uses the so-called *Bowen balls* defined for $x \in \Omega$, the transformation f, $\delta > 0$ and an integer n by

$$B(x, f, \delta, n) = \Big\{ y \,|\, d\big(f^j(x), f^j(y)\big) < \delta \text{ for } j = 0, \ldots, n-1 \Big\} .$$

These are again the initial conditions leading to trajectories indistinguishable (at precision δ) from that of x up to time $n-1$.

Theorem 6.5. *If μ is f ergodic, then for μ-almost any initial condition*

$$h_\mu(f) = \lim_{\delta\searrow 0} \liminf_{n\to\infty} -\frac{1}{n} \log \mu\big(B(x, f, \delta, n)\big) .$$

We refer to (Brin and Katok 1983) for proofs and related results.

Another way to get the entropy was obtained by Katok (Katok 1980) in the case of a phase space Ω with a metric d and a continuous transformation f. The formula is analogous to the formula for the topological entropy except that it refers only to typical orbits for a measure. Let A be a set, we denote by $N_n(A, \varepsilon)$ the maximal number of ε-different orbits before time n with initial condition in A. Recall that two orbits of initial condition x and y respectively are ε-different before time n if

$$\sup_{0\le k\le n} d\big(f^k(x), f^k(y)\big) > \varepsilon .$$

Note that the orbits are not required to stay in A but only to start in A. Let μ be an ergodic invariant probability measure for f. We define for $\delta \in (0, 1)$ the sequence of numbers $\tilde{N}_n(\mu, \delta, \varepsilon)$ by

$$\tilde{N}_n(\mu, \delta, \varepsilon) = \inf_{A,\ \mu(A)\ge 1-\delta} N_n(A, \varepsilon) .$$

In this definition, the trajectories that are "not seen" by the measure μ are discarded by taking the infimum. Note that there might be many more trajectories that are discarded than trajectories which are kept.

Theorem 6.6. *For any $\delta > 0$ we have*

$$h(\mu) = \lim_{\varepsilon \to 0} \limsup_{n \to \infty} \frac{\log \tilde{N}_n(\mu, \delta, \varepsilon)}{n} = \lim_{\varepsilon \to 0} \liminf_{n \to \infty} \frac{\log \tilde{N}_n(\mu, \delta, \varepsilon)}{n} .$$

We refer to (Katok 1980) for the proof. We refer to (Keane 1991; Katok and Hasselblatt 1995; Cornfeld, Fomin, and Sinaĭ 1982), and references therein for more results.

A way to measure the entropy in coded systems was discovered by Ornstein and Weiss using return times to the first cylinder. The result was motivated by the investigation of the asymptotic optimality of Ziv's compression algorithms which have very popular implementations (gzip, zip, etc.)

In more technical terms, let q be a finite integer, and assume the phase space of the dynamical system is a shift invariant subset of $\{1, \ldots, q\}^{\mathbb{N}}$. As before, we denote the shift by $\mathcal{S}$. Let μ be an ergodic invariant measure. Let n be an integer and for $x \in \Omega$, define $R_n(x)$ as the smallest integer such that the first n symbols of x and of $\mathcal{S}^{R_n(x)}(x)$ are identical.

Theorem 6.7. *For μ-almost every x we have*

$$\lim_{n \to \infty} \frac{1}{n} \log R_n(x) = h_\mu(\mathcal{S}) .$$

We refer to (Ornstein and Weiss 1993) for the proof.

A metric version was recently obtained by Downarowicz and Weiss, using again Bowen balls, in the context of dynamical systems with a metric phase space. Define for a number $\delta > 0$, an integer n, and a point $x \in \Omega$ the integer

$$R(x, f, \delta, n) = \inf \left\{ \ell > 0 \mid f^\ell(x) \in B(x, f, \delta, n) \right\} .$$

In other words,

$$R(x, f, \delta, n) = \inf \left\{ \ell > 0 \mid d\big(f^{\ell+j}(x), f^j(x)\big) < \delta \text{ for } j = 0, \ldots, n-1 \right\} .$$

Namely, starting at position $R(x, f, \delta, n)$, the orbit imitates its first chunk of length n with precision δ.

Theorem 6.8. *For μ-almost every x, we have*

$$\lim_{\delta \searrow 0} \limsup_{n \to \infty} \frac{1}{n} \log R(x, f, \delta, n) = h_\mu(f) .$$

We refer to (Downarowicz and Weiss 2004) for the proof and related results.

6.2 Sinai–Bowen–Ruelle Measures

It is natural to ask if there is a connection between entropy and Lyapunov exponents. The first relation is an inequality due to Ruelle.

Theorem 6.9. *Let μ be an ergodic invariant measure. The entropy of μ is less than or equal to the sum of the positive Lyapunov exponents of μ.*

We give only a rough idea of the proof. Take a ball B of radius ε (small) such that $\mu(B) > 0$. Iterate k times (k not too large such that $f^k(B)$ is still small). One gets an ellipsoid $f^k(B)$, and we cover it with balls of the same radius ε. To cover $f^k(B)$ by balls of radius ε, we need at most $N = \mathrm{e}^{k\Theta}$ balls, where Θ is the sum of the positive exponents. Indeed, in the direction of exponent j with $\lambda_j > 0$, we stretched from a size ε to a size $\mathrm{e}^{k\lambda_j}\varepsilon$, and we need now $\mathrm{e}^{k\lambda_j}$ balls of radius ε to cover "in this direction."

We now interpret the construction. Let $b_1, \ldots, b_N$ be the balls of radius ε used to cover $f^k(B)$. If two initial conditions in B manage to ε-separate before time k, they will land in two different balls b_j and b_l (forget re-encounters). If they do not separate, they will land in the same ball b_k. Therefore

$$N_n(\varepsilon, A) \le \mathrm{e}^{k\Theta} N_\varepsilon(A) ,$$

where $N_\varepsilon(A)$ is the smallest number of balls of radius ε needed to cover A (recall A is precompact), and using the result of Katok, Theorem 6.6, we get

$$h(\mu) \le \lim_{k\to\infty} \frac{\log N}{k} \le \Theta = \sum_{\lambda_i(\mu)>0} \lambda_i(\mu) .$$

One would like to know when equality holds. To explain the answer to this question, we first need some definitions.

Consider a negative Lyapunov exponent for an ergodic invariant measure μ and let x be a typical point for this measure. Then from Oseledec's theorem we know that at the linear level, there are directions at x which are contracted exponentially fast. One may ask if such a fact also holds for the complete dynamics and not just for its linearization. This is the concept of local invariant manifolds with exponents introduced in Sect. 4.2.

We repeat here Definition 4.27.

Definition 6.10. *An (open) submanifold $\mathcal{W}$ of the phase space Ω is called a local stable manifold with exponent $\gamma > 0$ if there is a constant $C > 0$ such that for any x and y in $\mathcal{W}$ and any integer n we have*

$$d\big(f^n(x), f^n(y)\big) \le C\mathrm{e}^{-\gamma n} .$$

Note that it follows immediately from the definition that if $\mathcal{W}$ is a local stable manifold with exponent γ, then for any integer p, $f^p(\mathcal{W})$ is also a local stable manifold with exponent γ. When f is invertible, one can define the local unstable manifolds as the local stable manifolds of the inverse. We refer to (Hirsch, Pugh, and Shub 1977) for the general case and more details. A local stable manifold containing a point x is called a local stable manifold of x.

Given a measure μ, we can consider stable manifolds except on sets of measure zero. The idea being that the size of the local stable manifold depends on the point x (for example, on the distance of x to the boundary of the manifold). The nice picture is when points without local stable manifolds (interpreted as local stable manifolds of width zero) are of measure zero, and when there is some coherence among the local stable manifolds through the dynamics. This leads naturally to the following definitions.

Definition 6.11. *Given a map f with an invariant measure μ, a coherent field of local stable manifolds for f with exponent $\gamma > 0$ is a collection $\big(\mathcal{W}(x)\big)_{x\in\mathcal{A}}$ of local stable manifolds for f with exponent γ indexed by a subset $\mathcal{A}$ of full measure and such that*

i) For any x in $\mathcal{A}$, $x \in \mathcal{W}(x)$.
ii) For any x in $\mathcal{A}$, $f\big(\mathcal{W}(x)\big) \subset \mathcal{W}\big(f(x)\big)$.

Definition 6.12. *Given an invertible map f with an invariant measure μ, a coherent field of local unstable manifolds for f with exponent $\gamma > 0$ is a collection $\big(\mathcal{W}(x)\big)_{x\in\mathcal{A}}$ of local unstable manifolds for f with exponent γ indexed by a subset $\mathcal{A}$ of full measure and such that*

i) For any x in $\mathcal{A}$, $x \in \mathcal{W}(x)$.
ii) For any x in $\mathcal{A}$, $f^{-1}\big(\mathcal{W}(x)\big) \subset \mathcal{W}\big(f^{-1}(x)\big)$.

These definitions generalize the notions of local stable/unstable manifolds for uniformly hyperbolic systems introduced in Sect. 4.1.1.

We now state an "integrated version" of the Oseledec theorem 5.72.

Theorem 6.13. *Let μ be an ergodic invariant measure for a diffeomorphism f of a compact manifold Ω. Let $\lambda_1 > \ldots > \lambda_k > 0$, be the collection of positive Lyapunov exponents, and $\Omega = E_1 \supsetneq E_1 \supsetneq \cdots \supsetneq E_k$ the sequence of associated linear bundles in the Oseledec theorem. Then there is a set B of full measure such that for any ε satisfying*

$$0 < \varepsilon < \inf_{1\le j\le k-1} \left(\lambda_j - \lambda_{j+1}\right) ,$$

and for each $x \in B$, there exist k nested submanifolds

$$\mathcal{W}_k(x) \subsetneq \mathcal{W}_{k-1}(x) \subsetneq \cdots \subsetneq \mathcal{W}_1(x) ,$$

where for any j $(1 \le j \le k)$, $\mathcal{W}_j(x)$ is a local unstable manifold for f with exponent $\lambda_j - \varepsilon$.

A similar result holds for the negative exponents.

We refer to (Pesin 1976) and (Ruelle 1979) for the proof and extensions. One of the big differences with the case of uniformly hyperbolic systems like Axiom A systems (see Sect. 4.1.1) is that there is, in general, no uniform lower bound on the size of the local stable or unstable manifolds.

We now state the definition of SRB measures.

Definition 6.14. *Let the phase space Ω of a dynamical system be a compact Riemannian manifold. Let the map f be a C^2 diffeomorphism. An ergodic invariant measure μ is said to be an* SRB measure *if it has a positive Lyapunov exponent and the conditional measures on the local unstable manifolds are absolutely continuous.*

In other words, one can disintegrate μ as follows, where g is any continuous function

$$\int g(x)\mathrm{d}\mu(x) = \int_{\mathcal{W}^u} \mathrm{d}\mu_{\mathcal{W}}(w) \int_w g(y)\varrho_w(y)\mathrm{d}_w y \tag{6.2}$$

where $\mathcal{W}^u$ is the set of local unstable manifolds, $\mathrm{d}_w y$ is the Lebesgue measure on the local unstable manifold w, $\varrho_w(y)$ is a nonnegative density supported by w, and $\mu_{\mathcal{W}}$ is called the *transverse measure*.

Remark 6.15. The set of local unstable manifolds is not uniquely defined. Indeed, a piece of local unstable manifold is also a local unstable manifold.

Example 6.16. The δ function on a stable fixed point of a map can also be considered an SRB measure. The condition on the unstable manifolds is redundant in this case.

We can now state an important theorem due initially to Pesin and generalized by Ruelle which complements the inequality in Theorem 6.9.

Theorem 6.17. *Let the phase space Ω of a dynamical system be a compact Riemannian manifold. Let the map f be a C^2 diffeomorphism. Let μ be an ergodic invariant measure. Then μ is an SRB measure if and only if the entropy of μ is equal to the sum of its positive Lyapunov exponents.*

In particular, in dimension 1, an ergodic invariant measure is absolutely continuous if and only if it has a positive Lyapunov exponent and its entropy is equal to this exponent. In Axiom A systems, Physical and SRB measures are the same. They have further properties connected with stochastic perturbations that we discuss in Sect. 8.5. We mention that there are dynamical systems which do not possess SRB measures. We refer to (Young 2002) for more information.

6.3 Dimensions

Because of the sensitive dependence on initial conditions and the compactness of the attractors, invariant sets have often a complicated (fractal) structure. One way to capture (measure) this property is to determine their dimension. Many definitions for the dimension of a set exist. Two among them are most often used in dynamical system theory: the Hausdorff dimension and the box counting dimension (also called *capacity*). They are not equivalent, the Hausdorff dimension is more satisfactory from a theoretical point of view, the box counting dimension (or box dimension) is easier to compute numerically.

The *Hausdorff dimension* of a set A is defined as follows. Let $B_r(x)$ denote the ball of radius r centered at x. For $d > 0$ and $\varepsilon > 0$, we define

$$\mathcal{H}_d(\varepsilon, A) = \inf_{A \subset \cup_j B_{r_j}(x_j)\,,\, \sup_j r_j \leq \varepsilon} \sum_j r_j^d \,. \tag{6.3}$$

The infimum is taken over all coverings of A by sequences of balls $B_{r_j}(x_j)$ of radius $r_j \leq \varepsilon$ (the radii can depend on j):

$$A \subset \bigcup_j B_{r_j}(x_j) \,.$$

Note that $\mathcal{H}_d(\varepsilon, A)$ is a nonincreasing function of ε, which may diverge when ε tends to zero. Moreover, if the limit when $\varepsilon \searrow 0$ is finite for some d, it is equal to zero for any larger d. This limit is nonincreasing in d. Moreover, if it is finite and nonzero for some d, it is infinite for any smaller d. The Hausdorff dimension of A, denoted below by $d_{\mathrm{H}}(A)$, is defined as the infimum of the positive numbers d such that the limit vanishes (for this special d the limit may be zero, infinite, finite or does not exist), namely

$$d_{\mathrm{H}}(A) = \inf \left\{ d \,\Big|\, \lim_{\varepsilon \searrow 0} \mathcal{H}_d(\varepsilon, A) = 0 \right\} \,. \tag{6.4}$$

This is also the supremum of the set of positive numbers d such that the limit is infinite.

The *box counting dimension* of a subset A of $\mathbb{R}^n$ is defined as follows. For a positive number r, let $N_A(r)$ be the smallest number of balls of radius r needed to cover A. If for (all) small r we have $N_A(r) \approx r^{-d}$, we say that the *box dimension* A is d, and we denote it by $d_{\mathrm{Box}}(A)$.

Exercise 6.18. *Show that for any (bounded) set A, $d_{\mathrm{H}}(A) \leq d_{\mathrm{Box}}(A)$ (there are sets for which the inequality is strict).*

Exercise 6.19. *Show that for an interval A (a square, a disk, a cube, etc.), $d_{\mathrm{H}}(A) = d_{\mathrm{Box}}(A) = 1$ (respectively 2, 2, 3, etc.). Show that for the triadic Cantor set, see Fig. 5.5, one has $d_{\mathrm{H}}(K) = d_{\mathrm{Box}}(K) = \log 2/\log 3$.*

We refer to (Kahane 1985; Falconer 1986) and (Mattila 1995) for other definitions of dimensions and the main properties.

A difficulty is that in general one does not have access to the whole attractor. As explained at the beginning of this chapter, frequently, one knows only one trajectory which is not necessarily uniformly spread over all the attractor, even though it may be dense. One can however assume that this trajectory is typical for a certain measure μ (for example, an SRB measure), and this leads to the definition of the dimension of a probability measure.

Definition 6.20. *The* dimension *of a probability measure is the infimum of the dimensions of the (measurable) sets of measure* 1.

Remark 6.21. In practice one should, of course, say which dimension is used (Hausdorff, box counting, and so on). Note also that the infimum may not be attained. If a measure is supported by an attractor, its dimension is of course at most the dimension of the attractor (the dimension of a measure is at most the dimension of its support). This inequality may be strict.

The dimension of a measure is related to the behavior of the measure of small balls as a function of the diameter. We start by a somewhat informal argument. Let μ be a measure, and let $d_{\text{Box}}(\mu)$ be its box counting dimension. This roughly means that we need about $N_A(r) \approx r^{-d_{\text{Box}}(\mu)}$ balls of radius r to cover (optimally) the "smallest" set A of full measure. Let $B_1, B_2, \ldots, B_{N_A(r)}$ be these balls. To have as few balls as possible, they should be almost disjoint, and to simplify the argument, we will indeed assume that they are disjoint. Since the set A is of full measure, we find

$$1 = \mu(A) = \mu\left(\bigcup_{j=1}^{N_A(r)} B_j\right) = \sum_{j=1}^{N_A(r)} \mu(B_j) \; .$$

In the simplest, homogeneous, case, all the balls have the same measure, and we get for $1 \le j \le N_A(r)$,

$$\mu(B_j) = \frac{1}{N_A(r)} \approx r^{d_{\text{Box}}(\mu)} \; .$$

A rigorous version of this argument leads to a lower bound for the Hausdorff dimension.

Lemma 6.22 (Frostman Lemma). *Assume that there are two constants $C > 0$ and $\delta > 0$ such that a probability measure μ with compact support on $\mathbb{R}^n$ satisfies for any $x \in \mathbb{R}^n$ and any $r > 0$ the inequality*

$$\mu\left(B_r(x)\right) \le C r^{\delta} \; .$$

Then

$$d_{\text{H}}(\mu) \ge \delta \; .$$

Proof. Let A be a set of measure one, and let $B_{r_j}(x_j)$ be any sequence of balls covering A. Since

$$A \subset \bigcup_j B_{r_j}(x_j) \; ,$$

we have (since the balls are not necessarily disjoint)

$$1 = \mu(A) \le \sum_j \mu(B_{r_j}(x_j)) \; .$$

Therefore, from the assumption of the lemma, we conclude that

$$1 \le C \sum_j r_j^\delta .$$

In other words, for any $\varepsilon > 0$ and any set A of full measure we have

$$\mathcal{H}_\delta(\varepsilon, A) \ge \frac{1}{C} .$$

The claim follows immediately from the definition (6.4) of the Hausdorff dimension. □

We refer to (Kahane 1985) for more on the Frostman lemma, and in particular its converse.

We will need the following corollary of the Frostman lemma.

Corollary 6.23. *Let μ be a probability measure on a metric space Ω and assume that for some constant $d > 0$ we have almost surely*

$$\liminf_{r \to 0} \frac{\log \mu(B_r(x))}{\log r} \ge d .$$

Then $d_{\mathrm{H}}(\mu) \ge d$.

In the opposite direction, one has a similar result: If, μ-almost surely,

$$\limsup_{r \to 0} \frac{\log \mu(B_r(x))}{\log r} \le d ,$$

then $d_{\mathrm{H}}(\mu) \le d$, (see (Young 1982)).

Proof (of Corollary 6.23). From the definition of the $\liminf$ it follows that for any $\varepsilon > 0$ there is a (measurable) set F_ε with $\mu(F_\varepsilon) \ge 1/2$ and a number $r(\varepsilon) > 0$ such that for any $x \in F_\varepsilon$ and any $r \in (0, r(\varepsilon))$ we have

$$\mu(B_r(x)) \le r^{d-\varepsilon} .$$

Let A_ε be a measurable set such that $\mu(A_\varepsilon) = 1$ and $d_{\mathrm{H}}(\mu) + \varepsilon \ge d_{\mathrm{H}}(A_\varepsilon) \ge d_{\mathrm{H}}(\mu)$. The existence of such a set follows directly from the definition (6.20) of $d_{\mathrm{H}}(\mu)$. Consider now the probability measure $\tilde{\mu}$ defined by

$$\tilde{\mu}(B) = \frac{\mu(A_\varepsilon \cap F_\varepsilon \cap B)}{\mu(A_\varepsilon \cap F_\varepsilon)} .$$

For any $x \in A_\varepsilon \cap F_\varepsilon$, and any $r \in (0, r(\varepsilon))$ we have

$$\tilde{\mu}(B_r(x)) \le \frac{\mu(B_r(x))}{\mu(A_\varepsilon \cap F_\varepsilon)} \le 2\, r^{d-\varepsilon} ,$$

since $x \in F_\varepsilon$ and $\mu(A_\varepsilon \cap F_\varepsilon) \geq 1/2$. For some $x \notin A_\varepsilon \cap F_\varepsilon$ and some $r \in (0, r(\varepsilon)/2)$ we might have $\mu(B_r(x) \cap A_\varepsilon \cap F_\varepsilon) > 0$. In that case, there exists $x' \in A_\varepsilon \cap F_\varepsilon$ such that $B_r(x) \subset B_{2r}(x')$. We conclude that for any $r \in (0, r(\varepsilon)/2)$ and any x we have

$$\tilde{\mu}(B_r(x)) \leq 2^{1+d}\, r^{d-\varepsilon} .$$

We then apply the Frostman Lemma 6.22 to conclude that $d_{\mathrm{H}}(\tilde{\mu}) \geq d - \varepsilon$. Since $\tilde{\mu}(A_\varepsilon) = 1$ we conclude from the definition of the Hausdorff dimension of a measure that $d_{\mathrm{H}}(A_\varepsilon) \geq d - \varepsilon$. This implies from $d_{\mathrm{H}}(A_\varepsilon) \leq d_{\mathrm{H}}(\mu) + \varepsilon$ the inequality

$$d_{\mathrm{H}}(\mu) + \varepsilon \geq d - \varepsilon .$$

Since this inequality holds for any $\varepsilon > 0$ the corollary is proved. □

Exercise 6.24. *Let μ be a probability measure on a compact manifold Ω, and let f be a diffeomorphism of Ω. Let $d_{\mathrm{sup}}(x)$ be defined by*

$$d_{\mathrm{sup}}(x) = \limsup_{r \to 0} \frac{\log \mu(B_r(x))}{\log r} .$$

Show that this quantity is constant along the orbit of x under f. Hint: show that for some constant $1 > \alpha > 0$, for any r small enough and for any $x \in \Omega$,

$$B_{\alpha r}(f(x)) \subset f(B_r(x)) \subset B_{\alpha^{-1} r}(f(x))$$

Show that if μ is ergodic, the quantity d_{sup} is almost surely constant. Show the same result for the lim inf. *Show that the subset of Ω where the limit exists is of measure 0 or 1.*

Remark 6.25. One can construct examples where the limit in the above exercise exists only on a set of measure zero, see (Ledrappier and Misiurewicz 1985).

It is natural to ask if there are relations between the dimension of a measure and the other quantities like entropy and Lyapunov exponents. In dimension 2, L.-S. Young (Young 1982) found an important relation, which was conjectured earlier in (Kaplan and Yorke 1979).

Theorem 6.26. *For a regular (C^∞) invertible map of a compact manifold and an ergodic invariant probability measure μ with two exponents satisfying $\lambda_1 > 0 > \lambda_2$ we have*

$$d_{\mathrm{H}}(\mu) = h(\mu) \left(\frac{1}{\lambda_1} + \frac{1}{|\lambda_2|} \right) .$$

In the case of an SRB measure, we have under the hypothesis of the above theorem and using Theorem 6.17, $h(\mu) = \lambda_1$, hence

$$d_{\mathrm{H}}(\mu) = 1 + \frac{\lambda_1}{|\lambda_2|} .$$

In the particular case of the Hénon map (see Example 2.31), $\lambda_1 + \lambda_2 = \log b$ and hence for an SRB measure

$$d_{\mathrm{H}}(\mu) = 1 + \frac{\lambda_1}{|\log b - \lambda_1|} \,.$$

We refer to (Young 1982) for a proof of the theorem, but we now give an intuitive argument for this result. It may be useful to look at Fig. 6.1. From the Katok result Theorem 6.6, we should count the maximal number of trajectories which are ε-separated over a time interval of length k. For that purpose, we will make the counting at an intermediate time $0 < k_1 < k$ (k_1 optimally chosen later). At time k_1, we cover a set of positive measure by balls of radius $\varepsilon' \ll \varepsilon$ (ε' will be optimally chosen below). From k_1 to k the balls of radius ε' are stretched. Therefore,

$$\varepsilon = \varepsilon' \mathrm{e}^{k_2 \lambda_1} ,$$

where $k_2 = k - k_1$ (one looks at the worst case: separation takes place at the last time k). From k_1 to 0, the balls of radius ε' are stretched by the inverse map, hence

$$\varepsilon = \varepsilon' \mathrm{e}^{-k_1 \lambda_2} \,.$$

Therefore,

$$\varepsilon = \varepsilon' \mathrm{e}^{-k_1 \lambda_2} = \varepsilon' \mathrm{e}^{k_2 \lambda_1} ,$$

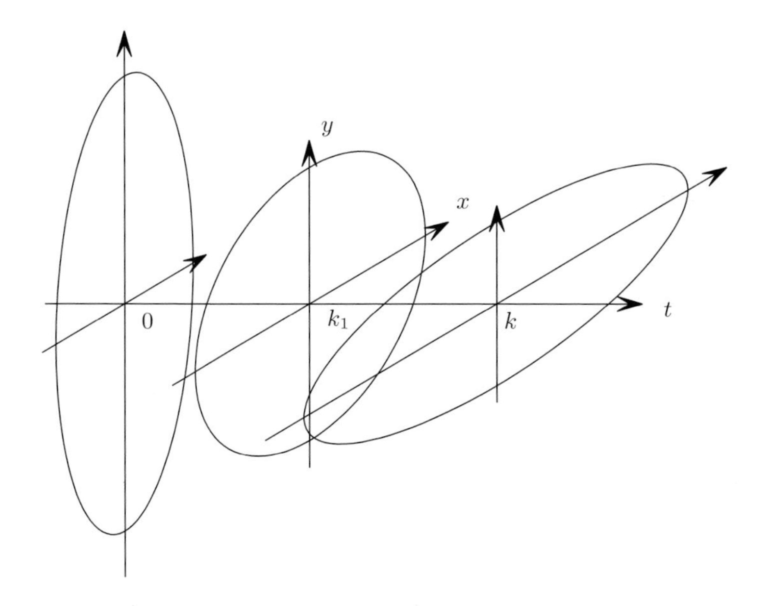

Fig. 6.1. An illustration of how an ellipse deforms from time 0 to time k. At time k_1 it is a circle. The y-axis is squeezed and the x-axis is stretched as time advances (from left to right)

which implies $k_1 = k\lambda_1/(\lambda_1 - \lambda_2)$, from which we derive ε'. However, the number of trajectories we are looking for is the minimal number of balls of radius ε' necessary to make a set of (almost) full measure. Since the balls have measure $\varepsilon'^{d_{\text{Box}}(\mu)}$, we need $\varepsilon'^{-d_{\text{Box}}(\mu)}$ of them. The maximum number of pairwise ε-separated trajectories is therefore

$$\varepsilon'^{-d_{\text{Box}}(\mu)} = \varepsilon^{-d_{\text{Box}}(\mu)} \mathrm{e}^{d_{\text{Box}}(\mu) k_2 \lambda_1} \approx \mathrm{e}^{kh(\mu)} ,$$

and therefore

$$h(\mu) = \frac{d_{\text{Box}}(\mu)\, k_2\, \lambda_1}{k} = \frac{d_{\text{Box}}(\mu)\, \lambda_1\, |\lambda_2|}{\lambda_1 + |\lambda_2|} .$$

Another relation between dimension and Lyapunov exponents is the so-called *Kaplan–Yorke formula* (Kaplan and Yorke 1979). Let $\lambda_1 \geq \lambda_2 \geq \ldots \geq \lambda_d$ be the decreasing sequence of Lyapunov exponents **with multiplicity** (in other words the number of terms is equal to the dimension). Let

$$k = \max\left\{ i \,\middle|\, \lambda_1 + \cdots + \lambda_i \geq 0 \right\} .$$

Kaplan and Yorke defined the *Lyapunov dimension* by

$$d_{\text{L}}(\mu) = k + \frac{|\lambda_{k+1}|}{\lambda_1 + \cdots + \lambda_k} .$$

They conjectured that $d_{\text{L}}(\mu) = d_{\text{H}}(\mu)$ (the *Kaplan–Yorke formula*) but counterexamples were discovered later. Note however that under the hypothesis of Theorem 6.26 the equality holds. In the general case, Ledrappier proved the following inequality.

Theorem 6.27. *For any ergodic invariant measure μ we have $d_{\text{H}}(\mu) \leq d_{\text{L}}(\mu)$.*

We now give a rough idea for the “proof” of the Kaplan–Yorke formula. Take a ball B of radius ε (small) in the attractor. Its measure is about $\approx\varepsilon^{d_{\text{Box}}(\mu)}$. If one iterates this ball k times, (ε small, k not too large), one obtains an ellipsoid $\mathcal{E} = f^k(B)$, elongated in the unstable directions, contracted in the stable directions. We now cover this ellipsoid $\mathcal{E}$ by balls of radius

$$\varepsilon' = \mathrm{e}^{k\lambda_j}\varepsilon$$

with j chosen (optimally) later on but such that $\lambda_j < 0$. A ball of radius ε' has measure $\varepsilon'^{d_{\text{Box}}(\mu)}$, and since the measure μ is invariant, we get

$$\varepsilon^{d_{\text{Box}}(\mu)} = \mu(B) = \mu(\mathcal{E}) = N\, \varepsilon'^{d_{\text{Box}}(\mu)}$$

where N is the number of balls of radius ε' necessary to cover the ellipsoid $\mathcal{E} = f^k(B)$. We now evaluate N as in the proof of Theorem 6.9. In the direction of the exponent λ_1, the original ball is stretched ($\lambda_1 > 0$), and its size becomes $\varepsilon \exp(\lambda_1 k)$. We need therefore

$$\frac{\varepsilon \mathrm{e}^{\lambda_1 k}}{\varepsilon'}$$

balls to cover $\mathcal{E}$ in that direction. The same argument holds in any direction with $\ell < j$ (with multiplicities), even if $\lambda_\ell < 0$. In the direction of λ_j one needs only one ball, which also covers in all directions ($\ell > j$) since they are contracted. Summarizing, since

$$N \approx \prod_{\ell=1}^{j} \frac{\varepsilon \mathrm{e}^{\lambda_\ell k}}{\varepsilon'}$$

and using $\varepsilon^{d_{\mathrm{Box}}(\mu)} = N\varepsilon'^{d_{\mathrm{Box}}(\mu)} = N\varepsilon^{d_{\mathrm{Box}}(\mu)}\mathrm{e}^{k\lambda_j d_{\mathrm{Box}}(\mu)}$, we obtain

$$\varepsilon^{d_{\mathrm{Box}}(\mu)-j} = \varepsilon'^{d_{\mathrm{Box}}(\mu)-j}\mathrm{e}^{kS_{j-1}} = \varepsilon^{d_{\mathrm{Box}}(\mu)-j}\mathrm{e}^{k\lambda_j(d_{\mathrm{Box}}(\mu)-j)}\mathrm{e}^{kS_{j-1}}$$

with $S_p = \lambda_1 + \cdots + \lambda_p$. This must be true for any integer k, hence

$$S_{j-1} + \lambda_j(d_{\mathrm{Box}}(\mu) - j) = 0\ .$$

It seems that there is only one equation for two unknowns j and $d_{\mathrm{Box}}(\mu)$. Nevertheless there is only one solution! To see this, it is convenient to plot the piecewise linear interpolation of S_p as a function of p. Between $p = k$ and $p = k + 1$ the linear interpolation is $S_k + \lambda_{k+1}(x - k)$. Therefore, d_{L} is the unique zero of this function (see Fig. 6.2).

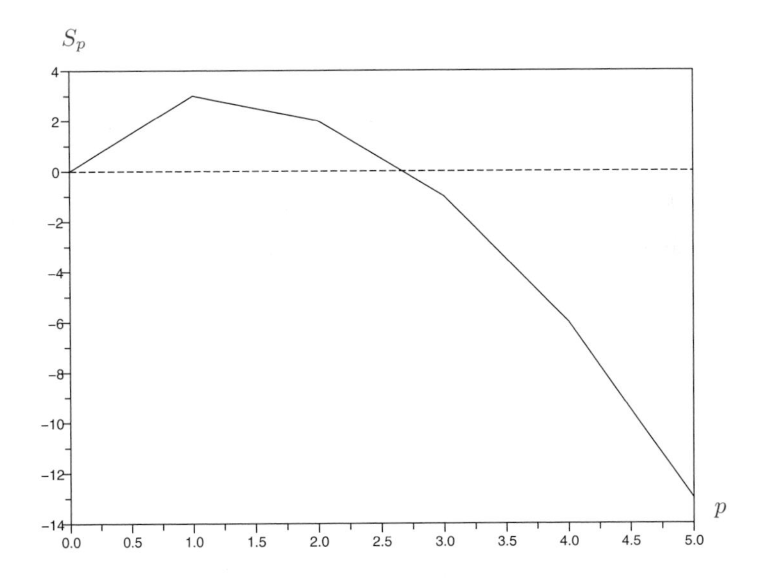

Fig. 6.2. Piecewise linear interpolation of the numbers S_p as a function of p

There are extensions of these ideas to physics, for example, for the Navier–Stokes equations. For example, it is shown in (Constantin, Foias, and Temam 1988) that for the Navier–Stokes equation in two dimensions, with periodic boundary conditions, one has the bound,

$$d_{\mathrm{H}} \leq \mathcal{O}\big(G^{2/3}(\log G)^{1/3}\big) ,$$

where G is the (generalized) *Grashof number*, defined by

$$G = |f| l^2 / \nu^2 ,$$

where f is the body force, l the side of the periodic box, and ν the viscosity. Such bounds are obtained by bounding the lowest nontrivial eigenvalue of the Laplacian in the domain in question; see e.g. (Eckmann and Ruelle 1985b).

7

Statistics and Statistical Mechanics

This chapter is somewhat more technical than the earlier ones. Its aim is to discuss some more recent results in dynamical systems which refine our knowledge of statistical properties. These results follow from a combination of methods from statistics and statistical mechanics.

One of the difficulties one encounters in experimental contexts is that one has basically only one orbit of the dynamical system at one's disposal, namely the one which is being measured (see Chap. 9). Therefore, one of the important questions in dynamical systems is whether single sample paths have any chance of being typical. We will illustrate this question in Sect. 7.1 where we describe various, successively more refined statements about the convergence to the central limit.

7.1 The Central Limit Theorem

Here, we consider a map f, and an observable g (of zero mean). As in (5.22), we define the sum

$$S_n(g)(x) = \sum_{j=0}^{n-1} g\big(f^j(x)\big) ,$$

and as in (5.25) we define the standard deviation σ_g by

$$\sigma_g^2 = C_{g,g}(0) + 2\sum_{j=1}^{\infty} C_{g,g}(j) ,$$

where

$$C_{g,g}(j) = \int g(x)\, g\big(f^j(x)\big)\, \mathrm{d}\mu(x) ,$$

and μ is an invariant measure. We will assume throughout that $\infty > \sigma_g > 0$.

The successive terms in the sum defining S_n are *not* independent, since they are all determined uniquely by the initial point x, because we consider a deterministic system. But, when that system is "chaotic," the terms in the sum should still behave like independent random variables, because somehow enough decay of correlations makes them almost independent. We will see that the more we integrate over the initial point x, the easier it is to get good results.

We therefore ask to which extent the sum S_n converges to its mean like a sum of independent random variables with standard deviation σ_g. More precisely, we say that the *central limit theorem* holds for g if

$$\lim_{n\to\infty} \mu\left(\left\{x \,\middle|\, \frac{S_n(g)(x)}{\sigma_g\sqrt{n}} \le t\right\}\right) = \frac{1}{\sqrt{2\pi}}\int_{-\infty}^{t} e^{-u^2/2}du\,. \tag{7.1}$$

We emphasize that this kind of result has been established only for certain classes of dynamical systems and observables. We refer to (Denker 1989) and (Luzzatto 2005) for reviews and (Young 1999) for recent results.

The central limit theorem is numerically illustrated in Figs. 7.1 & 7.2 for the map $3x \pmod 1$ of the interval and with the observable g being the characteristic function of the interval $[0, 1/2]$.

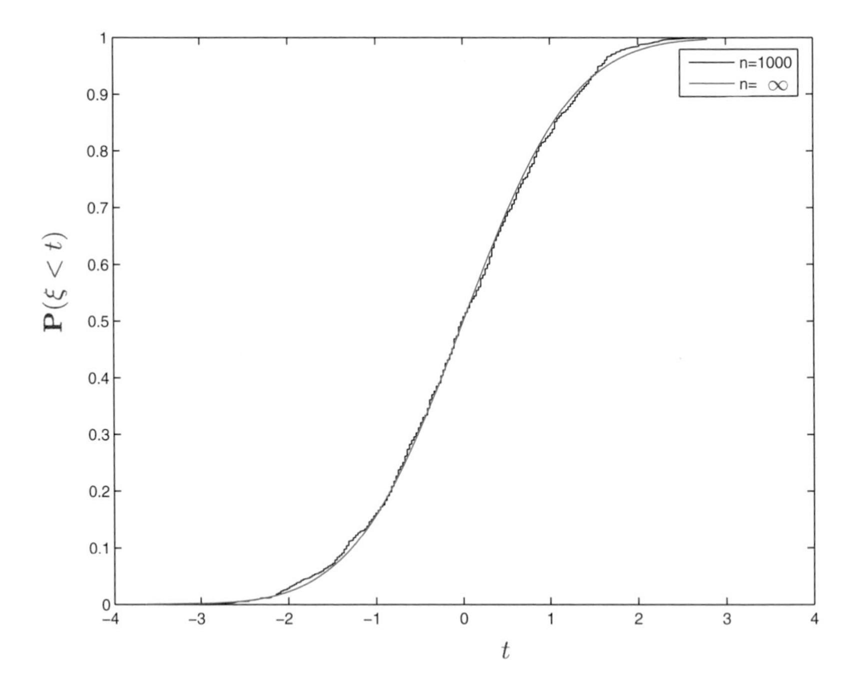

Fig. 7.1. Numerical illustration of the central limit theorem Eq. (7.1) for the map $f : x \mapsto 3x \pmod 1$. The observable g is $g = \chi_{[0,1/2]} - 1/2$. For 1000 initial points x_i, $i = 1, \ldots, 1000$, we computed $\xi_i = S_{3000}(g)(x_i)/\sqrt{3000}$. The vertical axis shows the normalized number of $\xi_i \le t$. The theoretical curve is $y(x) = \frac{1}{\sigma\sqrt{2\pi}}\int_{-\infty}^{x} du\, e^{-u^2/(2\sigma^2)}$, with $\sigma = 1/\sqrt{2}$ (see Exercise 5.56)

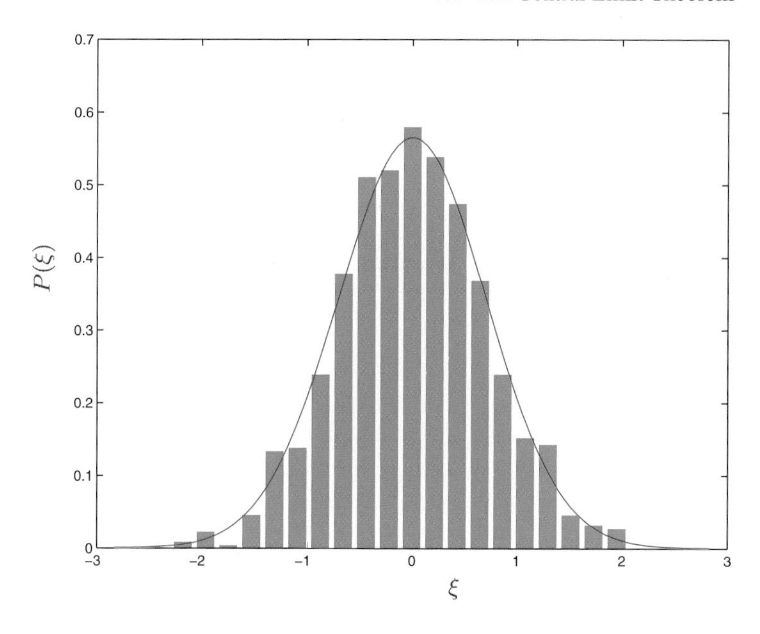

Fig. 7.2. Numerical illustration of the central limit theorem with the same parameters as in Fig. 7.1. The histogram shows the distribution of the ξ_i, normalized by the theoretical variance (which is $\sqrt{1/2}$ in this case)

Two methods were mostly used up to now for proving (7.1). One is based on a result of Gordin (see (Gordin 1969; 1993)), which reduces the problem to the known central limit theorem for martingales. The other method can be used when a Perron–Frobenius operator with adequate spectral properties is available. We illustrate the second idea for the case of piecewise expanding maps of the interval (see Example 2.15 for the definition and Theorem 5.57 for the spectral result). Let g be a real function on the unit interval with bounded variation and zero average with respect to an ergodic and mixing acim with density h. According to a theorem of Paul Lévy (see, for example, (Billingsley 1999) or (Feller 1957, 1966)), if $\sigma_g > 0$, in order to prove (7.1), it is enough to prove convergence of the Fourier transforms of the laws, namely that for any real ϑ,

$$\lim_{n\to\infty} \int e^{i\vartheta S_n(g)(x)/\sqrt{n}} h(x)dx = e^{-\sigma_g^2 \vartheta^2/2} . \tag{7.2}$$

To establish this result, we first define a family of operators P_v ($v \in \mathbf{BV}$) by

$$P_v u(x) = P\left(e^v u\right)(x) , \tag{7.3}$$

where P is the Perron–Frobenius operator. Using repeatedly relations (5.11) and (5.13), it follows that the integral in the left-hand side of (7.2) is equal to

$$\int P^n_{ig\vartheta/\sqrt{n}} h dx .$$

For large n, the function $\mathrm{i}g\vartheta/\sqrt{n}$ is small and therefore we expect the operator $P_{\mathrm{i}g\vartheta/\sqrt{n}}$ to be in some sense near the operator P. This is indeed the case in the following sense. In the space of functions of bounded variation, it is easy to verify, when working on the space BV, that for any $v \in \mathbf{BV}$, one has

$$\|P - P_v\|_{\mathbf{BV}} \le \mathcal{O}(1)\|v\|_{\mathbf{BV}} \,.$$

Therefore, if $\|v\|_{\mathbf{BV}}$ is small enough, using Theorem 5.57 and the analytic perturbation theory around simple eigenvalues (see (Kato 1984)), one can establish the following result.

Theorem 7.1. *There are a positive number η and two positive numbers Γ_1 and $\varrho_1 < 1$ such that for any complex valued function v of bounded variation satisfying $\|v\|_{\mathbf{BV}} < \eta$ the operator P_v has a unique simple eigenvalue $\lambda(v)$ of modulus larger than ϱ_1 with eigenvector h_v and eigenform α_v (satisfying $\alpha_v(h_v) = 1$). $\lambda(v)$, h_v and α_v are analytic in v in the ball of functions of bounded variation centered at the origin and of radius η. Moreover, $\lambda(0) = 1$, $h_0 = h$, and α_0 is the integration against the Lebesgue measure. The rest of the spectrum of P_v (in the space $\mathbf{BV}$) is contained in the disk of radius ϱ_1 and we find for any positive integer m the estimate*

$$\|P_v^m \Pi_v\|_{\mathbf{BV}} \le \Gamma_1 \varrho_1^m \,,$$

where Π_v is the spectral projection of $\lambda(v)$ given by

$$\Pi_v(w) = \alpha_v(w)\, h_v \,.$$

We can now apply this result with $v = \mathrm{i}g\vartheta/\sqrt{n}$ and $m = n$ because the estimates are uniform. Choosing n large enough so that $|\vartheta|/\sqrt{n} \ll \eta$, one gets

$$P^n_{\mathrm{i}\vartheta g/\sqrt{n}} w = \lambda(\mathrm{i}\vartheta g/\sqrt{n})^n h_{\mathrm{i}\vartheta g/\sqrt{n}}\, \alpha_{\mathrm{i}\vartheta/\sqrt{n}}(w) + \mathcal{O}(1)\varrho_1^n \,.$$

A simple perturbation theory computation (see (Kato 1984) or (Collet 1996)) shows that since g has zero average,

$$\lambda(zg) = 1 + \frac{z^2}{2\sigma_g^2} + \mathcal{O}\big(|z|^3\big) \,,$$

and the result (7.2) follows. □

For some particular classes of dynamical systems more precise results have been established, with an n-dependent bound on the error in (7.1). For Gibbs states over subshifts of finite type (see Example 2.20 and 5.15), a *Berry–Esseen inequality* has been obtained in (Coelho and Parry 1990). They proved that for such dynamical systems, for any Hölder continuous observable g (of zero average and with $\infty > \sigma_g > 0$), there is a constant $C > 0$ such that for any integer $n > 0$,

$$\left| \mu\left(\left\{ x \,\middle|\, \frac{S_n(g)(x)}{\sigma_g\sqrt{n}} \le t \right\}\right) - \frac{1}{\sqrt{2\pi}} \int_{-\infty}^{t} \mathrm{e}^{-u^2/2} du \right| \le \frac{C}{\sqrt{n}} \,.$$

Analogous results have been obtained in the nonuniformly hyperbolic case in (Gouëzel 2005).

The next step of refinement is to study the *sequence* of random variables $S_n(g)/\sqrt{n}$ (with g of zero average and $\infty > \sigma_g > 0$). For each fixed x, define a sequence $\{\xi_n\}$ of functions of t by

$$\xi_n(t)(x) = \frac{1}{\sigma_g\sqrt{n}}\sum_{j=0}^{[nt]} g\circ f^j(x) + \frac{1}{\sigma_g\sqrt{n}}\big(nt-[nt]\big)g\circ f^{[nt]+1}(x)\ , \tag{7.4}$$

where [ ] denotes the integer part. This defines a random sequence of continuous functions, the randomness coming from the choice of the initial condition x with respect to the invariant measure. If we drop the last term in (7.4), we obtain a random sequence of piecewise constant functions.

Exercise 7.2. *Prove that for any fixed integer n the function ξ_n defined in (7.4) is continuous.*

We next need a definition:

Definition 7.3. Brownian motion $t \mapsto B_t$ *is the unique continuous time Gaussian process with independent increments, defined for $t \geq 0$, with $B_0 = 0$, with B_t of zero average for any $t > 0$ and such that*

$$\mathbb{E}\big(B_t\, B_s\big) = \min\big\{t, s\big\}\ .$$

For several classes of dynamical systems one can show that the process $\{\xi_n\}$ converges weakly to a Brownian motion. We state here the case of piecewise expanding map of the interval. We refer to (Denker 1989) for more results and references.

Theorem 7.4. *Let f be a piecewise expanding map of the interval and μ a mixing absolutely continuous invariant probability measure with a nonvanishing density h. Let g be a Lipschitz continuous function on the interval with zero μ average and with nonzero standard deviation σ_g. Then the sequence of processes $\big\{\xi_n(\,\cdot\,)\big\}$ defined in (7.4) converges weakly to the Brownian motion.*

Of course, the weak convergence means that one averages over initial conditions, which is difficult to realize in an experiment.

This theorem also holds without the last term in equation (7.4). The first step in the proof is to show that for any fixed finite sequence of real numbers $0 < t_1 < \ldots < t_k$, we have

$$\lim_{n\to\infty}\int\prod_{j=1}^{k}\xi_n(t_j, x)\,d\mu(x) = \mathbb{E}\big(\prod_{j=1}^{k} B_{t_j}\big)\ .$$

For piecewise expanding maps of the interval, this is proven by extending the arguments leading to the central limit theorem which we described above. The second step is the proof of tightness (suitable compactness in the space of stochastic processes). We refer to the literature (for example (Denker 1989; Ziemian 1985; Broise 1996; Melbourne and Nicol 2005)) for the details. From this compactness one concludes existence of limiting processes, and the uniqueness (hence convergence) follows from the uniqueness of the finite dimensional moments proved in the first step.

The case of the map $x \mapsto 3x \pmod 1$ of the unit interval with the Lebesgue invariant measure and the observable $g = \chi_{[0,1]-1/2}$ is illustrated in Fig. 7.3.

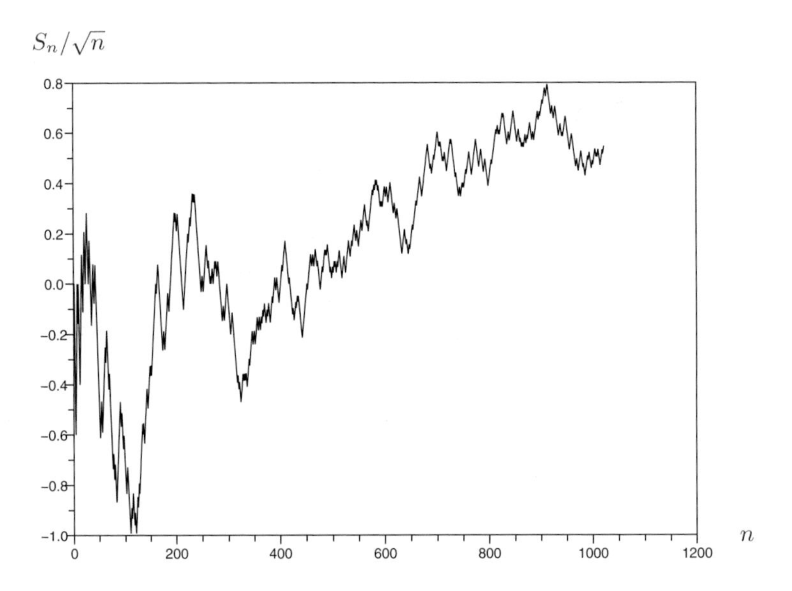

Fig. 7.3. Evolution of $S_n(g)(x_0)/\sqrt{n}$ as a function of n for the map $x \mapsto 3x \pmod 1$ with observable $g(x) = \chi_{[0,1]-1/2}$, and some random initial condition x_0

We now come to a more refined class of results, which hold almost surely with respect to the invariant measure, and which are therefore more adequate in experimental contexts. But they are also harder to prove and usually hold under more stringent conditions. We just mention here a few; they have been proven under various hypotheses, and we refer the reader to the literature for the details. For a general review of invariance principles, see e.g. (Merlevède, Peligrad, and Utev 2006).

We consider again the ergodic sum S_n of an observable g with zero average and finite (nonzero) variance σ_g, and satisfying some adequate hypothesis.

The *law of iterated logarithm* says that almost surely,

$$\limsup_{n\to\infty} \frac{S_n(g)}{\sigma_g \sqrt{2n \log\log n}} = 1 \, .$$

We refer to (Denker 1989) and (Broise 1996) for precise hypotheses and proofs.

One can also prove *almost sure central limit theorems*, for example that for any real s, the sequence of random variables

$$\frac{1}{\log n}\sum_{k=1}^{n}\frac{1}{k}\,\vartheta\left(s-\frac{S_k}{\sigma_g\sqrt{k}}\right)$$

converges almost surely to

$$\frac{1}{\sqrt{2\pi}}\int_{-\infty}^{s}\mathrm{e}^{-u^2/2}du\ ,$$

as $n\to\infty$. We refer to (Chazottes and Collet 2005) for the hypotheses, proofs, and references. An analog of the Erdös–Renyi asymptotics is also proven in (Chazottes and Collet 2005). Moderate deviations have been treated in (Dembo and Zeitouni 1997).

One would like to say that $S_n(g)(x)$ converges to the random walk B_n not only in the weak sense, but, hopefully, for μ-almost all x. This is in general not the case. One can avoid this problem by using the so-called *almost sure invariance principle*. By the classical *invariance principle*, one means the weak convergence of the normalized sum of random variables to the normal distribution $\mathcal{N}_{0,1}$ with mean 0 and variance 1. More precisely,

$$\frac{S_n}{\sqrt{n}}\to\eta\mathcal{N}_{0,\sigma_g}\qquad\text{as } n\to\infty\ .$$

The almost sure invariance principle (see (Philipp and Stout 1975) for a general approach) establishes (through a renormalization procedure) that there exists a sequence of random variables $\tilde{S}_n$ having the *same* distribution as S_n and such that there are an (almost surely finite) integer valued random variable M and a constant $\delta>0$ such that for any $n>M$,

$$\left|\tilde{S}_n-\sigma_g B_n\right|\le n^{-\delta+1/2}\ . \tag{7.5}$$

See in particular (Denker 1989; Ziemian 1985; Broise 1996; Melbourne and Nicol 2005) for the hypotheses and the proofs. If Eq. (7.5) holds, then, if one measures the variance of the process S_n, one can conclude that—up to corrections as close to $n^{-1/2}$ as one wishes—over long stretches of time, one will see a Brownian motion with that variance.

7.2 Large Deviations

Beyond the central limit theorem and its variants and corrections, one can study the probability that an ergodic sum deviates from its limit. More precisely, if μ is an ergodic invariant measure and g an integrable observable with zero average, then $\lim_{n\to\infty}S_n(g)/n=0$ (almost surely). One can ask for the probability of a large deviation, namely for $t>0$ this is the quantity

$$\mu\left(\{x \mid S_n(g)(x)/n > t\}\right) .$$

One can look for a similar quantity for $t < 0$ (or equivalently change g to $-g$). This quantity often decays exponentially fast with n, and it is therefore convenient to define the large deviation function $\varphi(t)$ by

$$\varphi(t) = -\lim_{n\to\infty} \frac{1}{n} \log \mu\left(\{x \mid S_n(g)(x)/n > t\}\right) . \tag{7.6}$$

Of course, the limit may not exist and one can take the limsup or the liminf. To investigate this quantity, one can use the analogy with statistical mechanics (see (Lanford 1973) and (Ruelle 2004)). There, one first computes another quantity called the *pressure function* which is defined in the present context (for $z \in \mathbb{C}$) by

$$\Xi(z) = \lim_{n\to\infty} \frac{1}{n} \log\left(\int \mathrm{e}^{zS_n(g)}\mathrm{d}\mu\right) . \tag{7.7}$$

Of course, this limit may not exist either. We now explain informally the relation between the large deviation function φ and the pressure Ξ. From the definition of the pressure (7.7), we find

$$\int \mathrm{e}^{zS_n(g)}\mathrm{d}\mu \approx \mathrm{e}^{n\Xi(z)} .$$

On the other hand, if z is real, the definition of φ in (7.6) and *Chebyshev's inequality* imply

$$\int \mathrm{e}^{zS_n(g)}\mathrm{d}\mu \geq \mathrm{e}^{nzt}\mu\left(\left\{x \,\middle|\, S_n(g) > nt\right\}\right) \approx \mathrm{e}^{nzt}\mathrm{e}^{-n\varphi(t)} .$$

We conclude immediately from these two inequalities that for any real z

$$\varphi(t) \geq zt - \Xi(z) .$$

Since this should hold for any real z, we conclude that

$$\varphi(t) \geq \sup_{z\in\mathbb{R}} \left(zt - \Xi(z)\right) . \tag{7.8}$$

The right-hand side of this inequality is called the *Legendre transform of the pressure*. This argument can be made rigorous under certain hypotheses on the pressure (basically differentiability which in statistical mechanics corresponds to the absence of phase transitions). One can also establish under some hypotheses that equality holds in (7.8). We refer to (Plachky and Steinebach 1975; Lanford 1973) and (Ellis 1985) for the details. We now explain briefly how to prove the existence of the limit in (7.7) in the simple case of piecewise expanding maps of the interval. We consider the case of μ the acim with density h and an observable g of bounded variation. Using the properties of the Perron–Frobenius operator, in particular, formulas (5.11) and (5.12) we get

$$\int \mathrm{e}^{zS_n(g)}\mathrm{d}\mu = \int \mathrm{e}^{zS_n(g)} h\mathrm{d}x = \int P_{zg}^n(h)\,\mathrm{d}x ,$$

where P_{zg} is the operator defined in (7.3).

Exercise 7.5. *Prove this equality.*

Intuitively the integral should behave like the eigenvalue $\lambda(z)$ of largest modulus of the operator P_{zg}. For $|z|$ not too large, Theorem 7.1 shows that the peripheral spectrum is reduced to a simple eigenvalue and we get immediately

$$\Xi(z) = \log \lambda(z) .$$

Differentiability of the pressure can be derived similarly, and as explained before, this allows one to obtain the large deviation function in a neighborhood of the origin (even analyticity of the pressure follows from the analyticity properties of $\lambda(z)$ and the fact that this function does not vanish near the origin). For the case of (piecewise regular) Markov maps of the interval and for Gibbs states over subshifts of finite type (with Hölder potentials), one can show that for Hölder continuous functions g, the limit in (7.7) exists for any real z and is regular. In those cases one can construct the large deviation function for all t (more precisely between the supremum and the infimum of the observable g). We refer to (Bowen 1975) and (Ruelle 2004) for a detailed study of these cases. We refer to (Young 1990) for a general approach to the large deviations of dynamical systems and to (Kifer 1990) for more details.

7.3 Exponential Estimates

The large deviation results are very precise but, as we saw above, they require proving that the pressure exists. This is often a difficult problem and one would sometimes prefer to obtain less precise estimates at a lower cost.

Several results in the nonindependent case have been obtained and it turns out that one can prove such results for some classes of dynamical systems as well. To be more precise, we state a definition. As usual, we consider a discrete time dynamical system given by a map f on a phase space Ω, equipped with a metric d, and with an ergodic invariant measure μ. A real valued function K on Ω^n is said to be *componentwise Lipschitz* if, for any $1 \leq j \leq n$, the constant $L_j(K)$ defined by

$$L_j(K) = \tag{7.9}$$

$$\sup_{\substack{x_1,\dots,x_n,y \\ x_j \neq y}} \frac{\left|K\left(x_1,\dots,x_{j-1},x_j,x_{j+1},\dots,x_n\right) - K\left(x_1,\dots,x_{j-1},y,x_{j+1},\dots,x_n\right)\right|}{d(x_j,y)}$$

is finite. In other words, we assume the function to be Lipschitz in each component with a uniformly bounded Lipschitz constant.

Exercise 7.6. *Consider on Ω^n the distance*

$$d_{\ell^1}\left((x_1,\dots,x_n),(y_1,\dots,y_n)\right) = \sum_{j=1}^{n} d(x_j,y_j) .$$

Show that K is Lipschitz on Ω^n with respect to the distance d_{ℓ^1} if and only if it is componentwise Lipschitz. Relate the sequence of numbers $L_j(K)$ and the d_{ℓ^1} Lipschitz constant.

Definition 7.7. *We say that the measure μ satisfies the* exponential inequality *for the map f if there are two constants $C_1 > 0$ and $C_2 > 0$ such that for any integer $n > 0$, and for any componentwise Lipschitz function K of n variables we have*

$$\int \mathrm{e}^{K\left(x, f(x), \ldots, f^{n-1}(x)\right) - \mathbb{E}(K)} \mathrm{d}\mu(x) \le C_1 \mathrm{e}^{C_2 \sum_{j=1}^{n} L_j(K)^2} , \tag{7.10}$$

where

$$\mathbb{E}(K) = \int K\big(x, f(x), \ldots, f^{n-1}(x)\big) \mathrm{d}\mu(x) .$$

Remark 7.8. Rescaling the variables in the statement of (7.1), we see the relation between it and the more general equation (7.10).

Remark 7.9.

i) Comparing with the definition (7.7) of the pressure, one sees that the left-hand side is the same kind of integral as in (7.10) with $K = S_n(g)$. The idea of this definition is to generalize the ergodic sums S_n used in the large deviation results. A special case of K would be

$$K\big(x, f(x), \ldots, f^{n-1}(x)\big) = S_n(g)(x) = \sum_{j=0}^{n-1} g(f^j(x)) ,$$

and then

$$\mathbb{E}(K) = n \int g(x) \mathrm{d}\mu(x) .$$

In this case, one has

$$L_j(K) = \sup_{x_j \neq y} \frac{|g(f^{j-1}(x_j)) - g(f^{j-1}(y))|}{|f^{j-1}(x_j) - f^{j-1}(y)|} = \sup_{x \neq y} \frac{|g(x) - g(y)|}{|x - y|} \equiv L(g) ,$$

and therefore $\sum_{j=1}^{n} L_j(K)^2$ is equal to $nL(g)^2$. In other words, we require in Definition 7.7 an estimation for a larger class of functions, not only for ergodic sums.

ii) On the other hand, contrary to the case of the pressure, we require only an upper bound, not the existence of a precise limit.

iii) The estimates provide the right order of dependence in n. For example, if $X_1, \ldots, X_n$ are n random variables with zero average and exponential moment, one gets

$$\mathbb{E}\left(\mathrm{e}^{\sum_{j=1}^{n} X_j}\right) = \prod_{j=1}^{n} \mathbb{E}\left(\mathrm{e}^{X_j}\right) ,$$

which is indeed a product of n terms.

iv) It is also worth emphasizing that the estimate is valid for any n, not only as a limit.

v) One can use Hölder norms instead of Lipschitz functions, and in the case of discrete random variables which take only finitely many values, one uses the oscillation.
vi) One often exploits the exponential inequality by using a Chebyshev inequality (see examples below).
vii) An inequality which is weaker than the exponential inequality (7.10) is known as the *Devroye inequality*. It reads:

$$\mathrm{Var}(K) \equiv \int K\big(x, f(x), \dots, f^{n-1}(x)\big)^2 \,\mathrm{d}\mu(x) - \mathbb{E}\big(K\big)^2 \leq C \sum_{j=1}^{n} L_j(K)^2 \,.$$

In the case of dynamical systems, the exponential inequality has been established for mixing acim of piecewise expanding maps of the interval (see (Collet, Martínez, and Schmitt 2002; Doukhan and Louhichi 1999) and (Dedecker and Prieur 2005), see also (Rio 2000) for related processes). A similar estimate for the variance using weaker assumptions has been obtained in (Chazottes, Collet, and Schmitt 2005a), and some applications are discussed in (Chazottes, Collet, and Schmitt 2005b).

7.3.1 Concentration

The *concentration* phenomenon has been known for a long time (in some sense already by Gibbs and others) and was revived 10 years ago by Talagrand mostly in the context of independent random variables (see (Talagrand 1995), and (Ledoux 2001)).

Concentration is a basically well-known phenomenon in large dimension d. It was illustrated by Talagrand by saying that in large dimension (d large), sets of measure $1/2$ (here, a set S) are big in the sense that a small neighborhood B_ε of S has almost full measure. In a sense, it is similar to the statement that a full sphere of high dimension has all its volume within a "skin" of size ε near the surface.

There is a relation between the exponential estimate of Sect. 7.3 and concentration. This relation is somewhat analogous to the relation between pressure and large deviation function.

As an example, we consider a subshift of finite type (see Example 2.20) on the finite alphabet $\mathcal{A}$. We consider on the phase space a metric d_ζ (see (5.14)) denoted below by d. For a fixed integer $p > 0$, let S be a measurable subset of Ω^p for which

$$\alpha(S) \equiv \mu\left(\left\{x \,\middle|\, \big(x, f(x), \dots, f^{p-1}(x)\big) \in S\right\}\right) > 0 \,.$$

For a given $\varepsilon > 0$, denote by B_ε the neighborhood of S, given by

$$B_\varepsilon = \left\{ (x_1, \dots, x_p) \,\middle|\, \exists (y_1, \dots, y_p) \in S \text{ such that } \frac{1}{p}\sum_{j=1}^{p} d(x_j, y_j) \leq \varepsilon \right\} \,.$$

Let K_S be the function defined on Ω^p by

$$K_S(x_1, \dots, x_p) = \inf_{(y_1,\dots,y_p)\in S} \frac{1}{p}\sum_{j=1}^{p} d(x_j, y_j) \,.$$

This function measures the average distance to S of the different components of the vector $(x_1, \ldots, x_p)$. Note that with this definition, we have $B_\varepsilon = \{K_S \le \varepsilon\}$. It is easy to verify that K_S is a componentwise Lipschitz function with all the Lipschitz constants L_j equal to $1/p$.

Exercise 7.10. *Prove this assertion.*

It is useful to introduce the notation

$$K_{S,p}(x) = K_S\big(x, f(x), \ldots, f^{p-1}(x)\big) .$$

Assume that the measure μ satisfies the exponential inequality (7.10). Then we get for any real number β,

$$\int \mathrm{e}^{\beta K_{S,p}(x)} \mathrm{d}\mu(x) \le C_1 \mathrm{e}^{\beta \mathbb{E}(K_{S,p})} \mathrm{e}^{\beta^2 C_2/p} .$$

By Chebyshev's inequality, we immediately derive

$$\mu\big(\{x \mid K_{S,p}(x) > \varepsilon\}\big) \le C_1 \mathrm{e}^{-\beta\varepsilon} \mathrm{e}^{\beta \mathbb{E}(K_{S,p})} \mathrm{e}^{\beta^2 C_2/p} . \tag{7.11}$$

It remains to estimate $\mathbb{E}(K_{S,p})$. For the particular case at hand this can be done easily. Observe that $-\beta K_S$ is also componentwise Lipschitz, and we can apply the exponential inequality to this function. Equivalently we can change β into $-\beta$. We get

$$\int \mathrm{e}^{-\beta K_{S,p}(x)} \mathrm{d}\mu(x) \le C_1 \mathrm{e}^{-\beta \mathbb{E}(K_{S,p})} \mathrm{e}^{\beta^2 C_2/p} .$$

We now observe that if $\big(x, f(x), \ldots, f^{p-1}(x)\big)$ belongs to S, then

$$K_{S,p}(x) = 0 .$$

Therefore, we find

$$\alpha(S) \le \int \mathrm{e}^{-\beta K_{S,p}(x)} \mathrm{d}\mu(x) \le C_1 \mathrm{e}^{-\beta \mathbb{E}(K_{S,p})} \mathrm{e}^{\beta^2 C_2/p} ,$$

which implies (with $\alpha = \alpha(S)$),

$$\mathrm{e}^{\beta \mathbb{E}(K_{S,p})} \le C_1 \alpha^{-1} \mathrm{e}^{\beta^2 C_2/p} .$$

Combining with (7.11), we get

$$\mu\big(\{x \mid K_{S,p}(x) > \varepsilon\}\big) \le C_1^2 \alpha^{-1} \mathrm{e}^{2\beta^2 C_2/p} \mathrm{e}^{-\beta\varepsilon} .$$

Since this is true for any real β, we can take the optimal value

$$\beta = \frac{\varepsilon p}{4C_2}$$

and obtain the bound

$$\mu(B_\varepsilon^{\mathrm{c}}) = \mu\big(\{x \mid K_{S,p}(x) > \varepsilon\}\big) \le C_1^2 \alpha(S)^{-1} \mathrm{e}^{-\varepsilon^2 p/(8C_2)} .$$

We now see the concentration phenomena. Even if $\alpha = \alpha(S)$ is small (for example, equal to $1/2$), for any fixed (small) ε, if p is large enough ($p \gg \varepsilon^{-2} \log(\alpha^{-1})$), the set $B_\varepsilon^{\mathrm{c}}$ has a very small measure.

7.4 The Formalism of Statistical Mechanics

The ideas of 1-dimensional statistical mechanics have played an important role in the study of statistical properties of dynamical systems. Sinai, Ruelle, and Bowen observed that for uniformly hyperbolic systems one can construct some invariant measures using a coding. We briefly sketch how the analogy with statistical mechanics emerges.

Recall the definition of a 1-dimensional Gibbs state: One considers a finite alphabet $\mathcal{A}$, and the shift $\mathcal{S}$ on the phase space $\Omega = \mathcal{A}^{\mathbb{Z}^+}$. A shift invariant measure μ is a *Gibbs state* if there exists a Hölder continuous potential φ such that for any cylinder set x_q^p in Ω:

$$\mu\big(C(x_q^p)\big) \approx \mathrm{e}^{-(p-q+1)P_\varphi}\ \mathrm{e}^{\sum_{j=q}^{p} \varphi(\mathcal{S}^j(\mathbf{y}))}\ , \tag{7.12}$$

where $\mathbf{y} \in x_q^p$ and P_φ is the pressure. We refer to (5.15) for the precise statement.

To simplify the discussion, we consider the special case of a regular map f of the circle and we assume that there is a Hölder continuous coding Φ from Ω to the circle (see Sect. 3.1) conjugating f and $\mathcal{S}$. More precisely, we assume that we are given a partition into intervals $I_1, \dots, I_{|\mathcal{A}|}$ which has the Markov property (for example, assume for simplicity that for each j, $f(I_j)$ is the whole circle). Recall that if μ is an invariant measure for f, $\mu \circ \Phi$ is invariant under the shift, and vice versa.

Let μ be an absolutely continuous invariant measure with a density ϱ, which is bounded and strictly positive, namely there is a constant $c > 1$ such that $c^{-1} \le \varrho \le c$. Consider now a cylinder set x_0^p with p large. Then $J = \Phi(x_0^p)$ is a small interval. Moreover, f^p restricted to J is injective, and $f^p(J)$ is an interval of size of order 1 (it is one of the fundamental intervals I_j). We have from the bounds on the density

$$c^{-1}\,|J| \le \mu(J) \le c\,|J|\ ,$$

where $|J|$ is the length of the interval J (a slight abuse of notation with the cardinality of $|\mathcal{A}|$). On the other hand, by the mean value theorem, we find

$$\big|f^p(J)\big| = \big|f^{p\prime}(\xi)\big|\ |J| \tag{7.13}$$

for some $\xi \in J$. This implies by the chain rule,

$$\mu(J) \approx \frac{1}{\big|f^{p\prime}(\xi)\big|} = \prod_{j=0}^{p-1} \frac{1}{\big|f'\big(f^j(\xi)\big)\big|} = \mathrm{e}^{-\sum_{j=0}^{p-1} \log\left|f'\left(f^j(\xi)\right)\right|}\ .$$

This expression is very similar to the formula (7.12) if we define the potential φ by

$$\varphi(\mathbf{x}) = -\log\big|f'\big(\Phi(\mathbf{x})\big)\big|\ . \tag{7.14}$$

It is this potential which provides the connection to interaction potentials in statistical mechanics. The $|\mathcal{A}|$ symbols which make up Ω are identified with $|\mathcal{A}|$ different spins.

(When $|\mathcal{A}| = 2$ these are called *Ising spins*, for larger $|\mathcal{A}|$ this is called the $|\mathcal{A}|$-state *Potts model*.)

We can now reverse the argument. Assume that the map f has a coding Φ. Then we can construct the potential φ on Ω, using formula (7.14). If we can prove that there exists a Gibbs state (from statistical mechanics) for this potential, then its image by Φ^{-1} will be a candidate for an absolutely continuous invariant measure for f. In particular, if Φ is Hölder continuous and $|f'|$ is Hölder and strictly positive, then φ is also Hölder.

Note that Gibbs states for other potentials will also give rise to invariant measures but which are not absolutely continuous.

One thing is still missing. In the definition of Gibbs state (7.12), one can take any point $\mathbf{y}$ in the cylinder set x_q^p, whereas in formula (7.13), a particular point ξ of the interval J emerges. Assume now that the map f is expanding, namely there is a constant $\alpha > 1$ such that on each of the basic intervals I_j we have

$$|f'| > \alpha \, ;$$

this is what we called a piecewise expanding map. Let now $y = \Phi(\mathbf{y}) \in J$. Since for any $0 \le j < p$, f^j is injective, we have, again by the mean value theorem,

$$\left|f^j(\xi) - f^j(y)\right| \le \alpha^{-(p-j)} \left|f^p(\xi) - f^p(y)\right| \le 2\pi\alpha^{-(p-j)} \, .$$

From the chain rule we deduce

$$\frac{f^{p\prime}(\xi)}{f^{p\prime}(y)} = \prod_{j=0}^{p-1} \frac{f'(f^j(\xi))}{f'(f^j(y))} \, .$$

On the other hand, since f was assumed regular, there is a constant M such that for any z in a fundamental interval I_j, $\left|f''(z)\right| \le M$. Therefore, if z_1 and z_2 belong the same fundamental interval, we have

$$\left|\log\left|f'(z_1)\right| - \log\left|f'(z_2)\right|\right| \le \int_{z_1}^{z_2} \left|\frac{f''(z)}{f'(z)}\right| \mathrm{d}z \le \frac{M}{\alpha}\left|z_1 - z_2\right| ,$$

which leads to

$$\mathrm{e}^{-M|z_1 - z_2|/\alpha} \le \left|\frac{f'(z_1)}{f'(z_2)}\right| \le \mathrm{e}^{M|z_1 - z_2|/\alpha} \, .$$

We conclude immediately that

$$\mathrm{e}^{-2\pi M/(\alpha-1)} \le \left|\frac{f^{p\prime}(\xi)}{f^{p\prime}(y)}\right| \le \mathrm{e}^{2\pi M/(\alpha-1)} \, . \tag{7.15}$$

Such an estimate is often called a *bounded distortion* argument. In other words, up to a controlled constant, we can take in formula (7.13) any point inside the interval J, as in the definition of a Gibbs state.

Exercise 7.11. *Show that if f is a piecewise expanding map of the circle with a coding Φ as above, and μ is a Gibbs state for the potential $\varphi = -\log\left|f' \circ \Phi\right|$, then $\mu \circ \Phi^{-1}$ is an absolutely continuous invariant measure for f with a bounded density which is also strictly positive.*

As we have seen above, the Hölder condition on the potential comes from a Hölder condition on the derivative of the map. This Hölder condition is a translation of the condition of *short range interaction* in Statistical Mechanics, which in 1-dimensional systems ensures the uniqueness of the Gibbs state (and some other nice properties like decay of correlations). In other words, these conditions ensure the absence of phase transitions.

For short-range interactions, the existence, uniqueness, and properties of the Gibbs state are often derived using a Perron–Frobenius operator somewhat similar to the operator (5.10). For potentials φ defined on $\mathcal{A}^{\mathbb{Z}^+}$, the Perron–Frobenius operator P is defined on functions ψ on $\mathcal{A}^{\mathbb{Z}^+}$ by

$$P\psi\left(x_0^\infty\right) = \sum_{\sigma \in A} \mathrm{e}^{\varphi\left(\sigma,\, x_0,\, x_1,\, \ldots\right)} \psi\left(\sigma,\, x_0,\, x_1,\, \ldots\right) .$$

This is very similar to formula (5.10) if it is written as

$$Pg(x) = \sum_{y,\, f(y)=x} \mathrm{e}^{-\log\left|f'(y)\right|} g(y)$$

and (not surprisingly) the potential $-\log\left|f'\right|$ appears. One can indeed show that the two operators are conjugated. We refer to (Bowen 1975) and (Ruelle 2004) for details.

Examples of expanding C^1 maps (whose derivatives are not Hölder) have been constructed which have several absolutely continuous invariant measures intermingled in a nontrivial way. They are the equivalent of low-temperature systems with several pure phases. We refer to (Góra and Schmitt 1989) and (Quas 1996) for the details of the construction.

A similar approach holds in higher dimension. As we have seen in Sect. 6.2, SRB measures are absolutely continuous along the unstable directions. When we have a coding, the potential to consider is now $-\log J^{\mathrm{u}}$, where J^{u} is the Jacobian of the map in the unstable direction. The technique which extends the preceding results to higher dimension is based on the so-called volume lemma. We refer to (Bowen 1975) for the details. We note, however, that one still gets a subshift of finite type, namely in a sense a *1-dimensional* system of statistical mechanics. The reason is that the space dimension of the Ising–Potts model really corresponds to the time direction of the dynamical system (which is 1-dimensional, independently of the dimension of phase space).

For a complete presentation of the subject, see (Ruelle 2004), and for a presentation for a physics audience, see (Baladi, Eckmann, and Ruelle 1989). Note also that the thermodynamic formalism can be used to establish large deviation results.

7.5 Multifractal Measures

Even if the functions $x \mapsto \log \mu(B_r(x))/\log r$ converge almost surely when r tends to zero, this quantity can show for nonzero r interesting (and even measurable) fluctuations as a function of x (recall that $B_r(x)$ is the ball centered in x of radius r).

The study of these fluctuations is called the *multifractal analysis*. More precisely, for any $\alpha \geq 0$, let

$$\mathcal{E}_\alpha^+ = \left\{ x \,\middle|\, \limsup_{r \to 0} \frac{\log \mu(B_r(x))}{\log r} = \alpha \right\} ,$$

and

$$\mathcal{E}_\alpha^- = \left\{ x \,\middle|\, \liminf_{r \to 0} \frac{\log \mu(B_r(x))}{\log r} = \alpha \right\} .$$

One would like to characterize the "size" of these sets. This can be done, for example, using another measure (or family of measures), or using a dimension. The so-called *multifractal spectrum* is usually defined as the function $\alpha \mapsto d_{\mathrm{H}}(\mathcal{E}_\alpha^+)$ (or $\alpha \mapsto d_{\mathrm{H}}(\mathcal{E}_\alpha^-)$), where d_{H} is the Hausdorff dimension introduced earlier in (6.4).

Under suitable hypotheses, one can relate these quantities to some large deviation functions. We refer to (Collet, Lebowitz, and Porzio 1987; Eckmann and Procaccia 1986; Rand 1989; Vul, Sinaĭ, and Khanin 1984; Takens and Verbitski 1998; Theiler 1990) for more results and references. Most of the remainder of this section is devoted to the study of a simple example with which we illustrate the techniques.

Example 7.12. **The multifractal measures for the dissipative baker's map**
Consider the dissipative baker's map (2.6). As we have seen in Exercise 3.26, the attractor is the product of a segment by a Cantor set $\mathcal{K}$. This Cantor set can be constructed as follows. Consider the two contractions f_1 and f_2 of the interval given by

$$f_1(y) = \frac{y}{4} \quad \text{and} \quad f_2(y) = \frac{2+y}{3} .$$

The Cantor set is obtained by applying all the infinite compositions of f_1 and f_2 to the interval. This is often called an *iterated function system*; see e.g. (Mauldin 1995). Namely, one defines for each n a set $\mathcal{K}_n$ consisting of 2^n disjoint intervals, and constructed recursively as follows. First $\mathcal{K}_0 = [0,1]$, and then for any $n \geq 1$,

$$\mathcal{K}_{n+1} = \{f_1(I) \,|\, I \in \mathcal{K}_n\} \cup \{f_2(I) \,|\, I \in \mathcal{K}_n\} . \tag{7.16}$$

The transverse measure μ of the SRB measure—see (6.2)—satisfies $\mu(I) = 2^{-n}$ for any $I \in \mathcal{K}_n$. And $\mathcal{K}$ is obtained as the intersection of all the $\mathcal{K}_n$.

Exercise 7.13. *Show that $\mathcal{K}_n$ is composed of 2^n disjoint intervals and that this construction uniquely defines the measure μ.*

Exercise 7.14. *Consider the map f of the unit interval defined by*

$$f(x) = \begin{cases} 4x\,, & \textit{for } 0 \le x \le 1/4 \\ 12(x-1/4)/5\,, & \textit{for } 1/4 < x < 2/3 \\ 3(x-2/3)\,, & \textit{for } 2/3 \le x \le 1 \end{cases} .$$

Show that the Cantor set $\mathcal{K}$ defined above is invariant under f. Show that f_1 and f_2 are two inverse branches of f. Show that μ is an ergodic invariant measure for this map (one can use a coding).

We will need a large deviation estimate for the sizes of the intervals in each $\mathcal{K}_n$. In the present situation, this is easy since the maps f_1 and f_2 have constant slope, whose inverses are relatively prime. Indeed, we have for any integer $0 \le p \le n$,

$$\mathrm{card}\Big\{I \in \mathcal{K}_n \,\Big|\, |I| = 4^{-p}\,3^{p-n}\Big\} = \binom{n}{p}\,,$$

where $\mathrm{card}(S)$ is the cardinality of the set S, and $|I|$ is the length of the interval I. For later purposes it is convenient to use the Stirling approximation. Let φ be the function

$$\varphi(s) = \frac{s+\log_2 3}{\log_2(4/3)}\log_2\left(-\frac{s+\log_2 3}{\log_2(4/3)}\right) - \frac{s+\log_2 4}{\log_2(4/3)}\log_2\left(\frac{s+\log_2 4}{\log_2(4/3)}\right)\,,$$

then for $s \in [-\log_2 4, -\log_2 3]$,

$$\mathrm{card}\Big\{I \in \mathcal{K}_n \,\Big|\, |I| \approx 2^{ns}\Big\} \approx 2^{n\,\varphi(s)}\,. \tag{7.17}$$

Exercise 7.15. *Consider the sequence of functions*

$$Z_n(\beta) = \sum_{I\in\mathcal{K}_n} |I|^\beta\,.$$

Compute $Z_n(\beta)$, and derive that

$$F(\beta) = \lim_{n\to\infty}\frac{1}{n}\log_2 Z_n(\beta) = \log_2\left(4^{-\beta}+3^{-\beta}\right)\,.$$

Show that φ is the Legendre transform of F. Prove relation (7.17) using steepest descent. Show that $\varphi(s)$ is concave, and that its maximum is attained at $s = -\log_2\sqrt{12}$.

We can now state the following theorem.

Theorem 7.16. *For any $\alpha \in [1/\log_2 4, 1/\log_2 3]$ we have for the dissipative baker's map (2.6),*

$$d_{\mathrm{H}}\left(\mathcal{E}_\alpha^-\right) = d_{\mathrm{H}}\left(\mathcal{E}_\alpha^+\right) = \alpha\,\varphi(-1/\alpha)\,.$$

Proof (sketch). We give only a proof for $\alpha \leq \alpha_c = 1/\log_2 \sqrt{12}$. We refer the reader to the literature for the other range of α (for example (Collet, Lebowitz, and Porzio 1987; Rand 1989; Takens and Verbitski 1998)).

Let

$$\widetilde{\mathcal{E}}_\alpha = \left\{ x \,\middle|\, \alpha \text{ is an accumulation point of } \frac{\log \mu\big(B_r(x)\big)}{\log r} \text{ when } r \to 0 \right\} .$$

Note that $\mathcal{E}_\alpha^\pm \subset \widetilde{\mathcal{E}}_\alpha$. To construct the upper bound for $d_{\mathrm{H}}\big(\widetilde{\mathcal{E}}_\alpha\big)$, we use a particular covering by balls. Indeed, let $x \in \widetilde{\mathcal{E}}_\alpha$. Then by definition, there exists an infinite sequence $\{r_j\}$ tending to zero such that

$$\lim_{j\to\infty} \frac{\log \mu\big(B_{r_j}(x)\big)}{\log r_j} = \alpha .$$

Since $x \in \mathcal{K}$, we can find for any j an integer n_j such that there is an interval $I \in \mathcal{K}_{n_j}$ satisfying

$$\mathcal{K} \cap B_{r_j}(x) \subset I$$

and an interval $J \in \mathcal{K}_{n_j+2}$ such that

$$J \subset B_{r_j}(x) .$$

Exercise 7.17. *Prove this statement, for example, by drawing a local picture of $\mathcal{K}_n$ around x.*

In particular, we have

$$2^{-n_j-2} \leq \mu\big(B_{r_j}(x)\big) \leq 2^{-n_j} .$$

From the condition $x \in \widetilde{\mathcal{E}}_\alpha$, we derive

$$2^{-n_j} \approx \mu\big(B_{r_j}(x)\big) \approx r_j^\alpha ,$$

and therefore

$$r_j \approx 2^{-n_j/\alpha} ,$$

and finally, because of the choice of I, we find

$$|I| \lessapprox 2^{-n_j/\alpha} .$$

Therefore, for any fixed N we have

$$\widetilde{\mathcal{E}}_\alpha \subset \bigcup_{n=N}^{\infty} \bigcup_{\substack{I\in\mathcal{K}_n \\ |I|\leq 2^{-n/\alpha}}} I .$$

If we assume $\alpha \leq \alpha_{\text{crit}} \equiv 1/\log_2 \sqrt{12}$, we have $1/\alpha \geq \log_2 \sqrt{12}$ and therefore $s = -1/\alpha \leq -\log_2 \sqrt{12}$.

If we use the above intervals I as a covering for $\widetilde{\mathcal{E}}_\alpha$, we get from the large deviation estimate (7.17) and the definition (6.3) that for any $d > \alpha\,\varphi(-1/\alpha)$

$$\mathcal{H}_d\big(2^{-N/\alpha}, \widetilde{\mathcal{E}}_\alpha\big) \leq \sum_{n=N}^{\infty} 2^{-n\,d/\alpha} 2^{n\,\varphi(-1/\alpha)} ,$$

which implies

$$\lim_{N\to\infty} \mathcal{H}_d\big(2^{-N/\alpha}, \widetilde{\mathcal{E}}_\alpha\big) = 0 ,$$

and therefore

$$d_{\mathrm{H}}\big(\widetilde{\mathcal{E}}_\alpha\big) \leq \alpha\,\varphi(-1/\alpha) . \tag{7.18}$$

This is the desired upper bound.

We next prove a lower bound on $d_{\mathrm{H}}\left(\mathcal{E}_\alpha^- \cap \mathcal{E}_\alpha^+\right)$. For this purpose we use Corollary 6.23 and we start by constructing a measure with suitable properties. This could be done using a coding and the measure would appear as a Bernoulli measure; here, we construct it directly. We construct actually a one-parameter family of measures $\mu^{(p)}$, $p \in (0,1)$ (and we define as usual $q = 1 - p$). The construction is recursive, using the sequence of intervals in $\mathcal{K}_n$. The measure $\mu^{(p)}$ will be a probability measure and we therefore set $\mu^{(p)}\big([0,1]\big) = 1$. Assume now that $\mu^{(p)}(I)$ has already been defined for every $I \in \mathcal{K}_n$. As we have seen in the recursive construction (7.16), the intervals in $\mathcal{K}_{n+1}$ are obtained from those of $\mathcal{K}_n$ by applying f_1 or f_2. We now define $\mu^{(p)}$ on the intervals of $\mathcal{K}_{n+1}$ by

$$\mu^{(p)}(J) = \begin{cases} p\,\mu^{(p)}(I) , & \text{if } J = f_1(I) \\ q\,\mu^{(p)}(I) , & \text{if } J = f_2(I) \end{cases} .$$

The reader can check that this recursive construction defines a Borel measure on $[0,1]$ with support $\mathcal{K}$. In particular, $\mu = \mu^{(1/2)}$. Consider now the map f of Exercise 7.14. It is easy to verify that its Lyapunov exponent for the measure $\mu^{(p)}$ is equal to

$$\lambda_p = p \log 4 + q \log 3 .$$

Exercise 7.18. *Prove this statement.*

We define a set $A \subset \mathcal{K}$ by

$$A = \left\{ x \,\middle|\, \liminf_{r\to 0} \frac{\log \mu^{(p)}\big(B_r(x)\big)}{\log r} \geq \frac{-p\log p - q\log q}{\lambda_p} \right\} .$$

It follows from the ergodic theorem (Theorem 5.38) and the Shannon–McMillan–Breiman Theorem (Theorem 6.2) that $\mu^{(p)}(A) = 1$.

Exercise 7.19. *Prove this statement using the construction of the sets I and J above and observing that f^{n_j-2} is monotone and continuous on I and satisfies $\big|f^{n_j-2}(I)\big| = \mathcal{O}(1)$, which implies $|I| \approx 1/\big|f^{n_j\prime}(x)\big|$ for any $x \in I$.*

We now fix an $\alpha \in [1/\log_2 4, 1/\log_2 3]$, and we choose p such that

$$p = -\frac{\alpha + \log_2 3}{\log_2(4/3)} ,$$

or equivalently

$$\alpha = \frac{-\lambda_p}{\log 2} .$$

This p will be in $[0, 1]$. By construction, for this p one has $A \subset \mathcal{E}_\alpha^- \cap \mathcal{E}_\alpha^+$, and therefore

$$\mu^{(p)}\left(\mathcal{E}_\alpha^- \cap \mathcal{E}_\alpha^+\right) = 1 .$$

Therefore, with this choice of p, we get, using the corollary of Frostman's Lemma (Lemma 6.23),

$$d_{\mathrm{H}}\left(\mathcal{E}_\alpha^- \cap \mathcal{E}_\alpha^+\right) \geq \frac{-p\log p - q\log q}{\lambda_p} = \alpha\,\varphi(-1/\alpha) .$$

Since this quantity is equal to the upper bound (7.18), the result follows. □

Exercise 7.20. *Write a complete proof of the above result by putting all the necessary ε and δ.*

The maximal value for the Hausdorff dimension of $\mathcal{E}_\alpha^\pm$ is the dimension of the invariant set. On the other hand, if α is equal to the Lyapunov exponent (divided

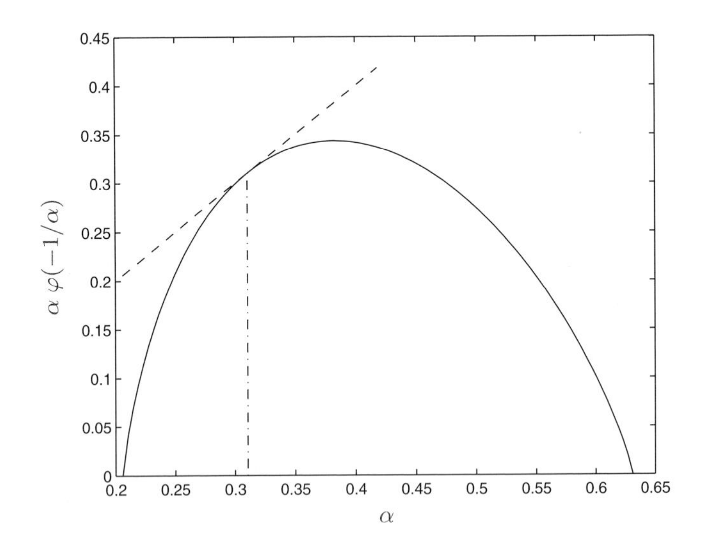

Fig. 7.4. Hausdorff dimension of the sets $\mathcal{E}_\alpha^\pm$ as a function of α for the maps (7.19). The contact point is at $\alpha = 1/\sqrt{\log_2(87)}$. This is also the inverse of the Lyapunov exponent for the choice of $p = 1/2$, namely, $87 = 3 \cdot 29$

by $\log 2$), then the $\mathcal{E}_\alpha^\pm$ have full measure (see Exercise 7.19). This implies that the curves $\alpha \to \mathcal{E}_\alpha^\pm$ should be below the first diagonal and tangent to it in one point. This is illustrated in Fig. 7.4, which shows the graph of $d_{\mathrm{H}}(\mathcal{E}_\alpha^-) = d_{\mathrm{H}}(\mathcal{E}_\alpha^+)$ when α varies. This figure was produced with the two maps

$$f_1(y) = \frac{y}{29} \quad \text{and} \quad f_2(y) = \frac{2+y}{3}\,, \tag{7.19}$$

so that one can see more clearly the curves than with $f_1(y) = y/4$. One can construct for this example a map f in analogy to Exercise 7.14 and a Cantor set $\mathcal{K}$.

Theorem 7.16 has been generalized in various directions, in particular to maps which have not piecewise constant slope. In the analog of Exercise 7.19 one uses a distortion estimate (of the kind of (7.15)) for the variations of the differential of large iterates on small intervals. We refer to the previously mentioned references for details.

8

Other Probabilistic Results

A number of questions of probabilistic origin have been investigated in the class of stochastic processes generated by dynamical systems. We describe some of the results below.

8.1 Entrance and Recurrence Times

One of the oldest questions in ergodic theory is the problem of entrance time in a (small) set. For a (measurable) subset A of the phase space Ω, one defines the (first) *entrance time* $\tau_A(x)$ of the trajectory of x in the set A by

$$\tau_A(x) = \inf \big\{ n > 0 \,\big|\, f^n(x) \in A \big\} .$$

For x in A this number is called the *return time*. Note that the function τ_A is a measurable function on the phase space with values in the integers.

Remark 8.1. Note that from the above definition, $\tau_A(x) \geq 1$ even if $x \in A$. One can modify the definition by imposing $\tau_A(x) = 0$ if $x \in A$. Our definition makes some results below simpler.

Historically, this quantity appeared in Boltzmann's ideas about ergodic theory, when he asked about the time it would take for all the molecules in a room to concentrate in only half of the available volume. His idea was that events of small probability occur only on a very large time scale.

One of the first few general results around these questions is the Poincaré recurrence theorem which dates from the same period of heated discussions about the foundations of statistical mechanics (see also Fig. 2.5).

Theorem 8.2. *Let f be a transformation on a phase space Ω, and let μ be an invariant probability measure for f. Then for each measurable set A of positive measure, there is a finite integer $n \geq 1$ such that $\mu\big(A \cap f^n(A)\big) > 0$.*

Proof. The proof is by contradiction. Assume that there exists a measurable set A with $\mu(A) > 0$ and such that for all n we have $\mu\big(A \cap f^n(A)\big) = 0$.

Exercise 8.3. *Show that for any set A and any integers m and n one finds $A = f^n\big(f^{-n}(A)\big)$, $A \subset f^{-n}\big(f^n(A)\big)$ and $f^{-n-m}(A) = f^{-m}\big(f^{-n}(A)\big)$ (use formula (5.3)).*

By the invariance of the measure we have for any integer n,

$$\mu\big(f^{-n}(A) \cap A\big) \le \mu\big(f^{-n}(A \cap f^n(A))\big) = \mu\big(A \cap f^n(A)\big) = 0 \ .$$

Therefore, since for any integer m, we have

$$f^{-m}\big(A \cap f^{-n}(A)\big)\big) = f^{-m}(A) \cap f^{-n-m}(A) \ ,$$

we conclude that for any integers m and n

$$\mu\big(f^{-m}(A) \cap f^{-n-m}(A)\big) = 0 \ .$$

Let

$$B_p = \cup_{n=0}^{p} f^{-n}(A) \ ,$$

since the sets $f^{-n}(A)$ have an intersection of measure zero and the same measure $\mu(A)$, we find

$$\mu\big(B_p\big) = p\, \mu(A) \ .$$

This is only possible as long as $p\, \mu(A) \le 1$ since A is of positive measure. Hence, we have a contradiction with our assumptions on A, and the theorem is proved. □

Remark 8.4. Note that it follows from the proof of the theorem that $A \cap f^n(A) \neq \emptyset$ (and in fact $\mu\big(A \cap f^n(A)\big) > 0$) for at least one n such that $1 \le n \le 1/\mu(A)$.

Exercise 8.5. *Show that for almost any initial condition in a set of positive measure A, the orbit returns infinitely often to A.*

The next important result follows from the ergodic theorem, and says that the entrance time is a well-defined function.

Theorem 8.6. *If A is such that $\mu(A) > 0$, and μ is ergodic, then τ_A is almost surely finite.*

Proof. The proof is by contradiction. Let B be a measurable set of positive measure where $\tau_A = \infty$. We have for almost every x in B

$$\lim_{n\to\infty} \frac{1}{n} \sum_{j=1}^{n} \chi_A\big(f^n(x)\big) = 0 \ .$$

On the other hand by ergodicity, this quantity should be almost surely equal to $\mu(A) > 0$, hence a contradiction. □

This result says in particular that if an event A has a nonzero probability, it will occur with probability 1, at some time. The ergodic theorem tells us that if one considers a large time interval $[0, T]$, and a (measurable) set A of positive measure, then the number of times a typical trajectory has visited A (the event A has occurred) is of the order $\mu(A)T$. In other words, $1/\mu(A)$ is a time scale associated to the set A, and one may wonder if this time scale is related to the entrance time. A general theorem in this direction due to Mark Kac.

Exercise 8.7. *Show that almost surely*

$$\tau_A(x) = 1 + \sum_{j=2}^{\infty} \prod_{m=1}^{j-1} \chi_{A^c}(f^m(x)) \,.$$

Theorem 8.8. *Let f be a transformation on a phase space Ω, and μ be an ergodic invariant probability measure for f. Then for each measurable set A of positive measure,*

$$\int_A \tau_A \mathrm{d}\mu = 1.$$

Proof. We give a proof in a particular case, and refer to the literature (for example (Krengel 1985)) for the general case. Using Exercise 8.7 it immediately follows that almost surely

$$\chi_{A^c}(f(x))\tau_A(f(x)) = \tau_A(x) - 1 = \chi_A(x)\tau_A(x) + \chi_{A^c}(x)\tau_A(x) - 1 \,.$$

If we now assume that the function τ_A is integrable, the result follows by integrating both sides of this equality and using the invariance of the measure. □

A general lower bound on the expectation of τ_A is given in the next theorem.

Theorem 8.9. *For any measurable set A with $\mu(A) > 0$ we have*

$$\mathbb{E}(\tau_A) \geq \frac{1}{2\mu(A)} - 3/2.$$

Note that the bound is not useful for $\mu(A) > 1/5$ since by definition $\tau_A \geq 1$.

Proof. Recall that

$$\mathbb{E}(\tau_A) = \sum_{p=1}^{\infty} \mathbb{P}\left(\tau_A \geq p\right) .$$

We now bound each term in the sum from below. Noting that

$$\{x \mid \tau_A(x) \geq p\} = \left\{ x \,\middle|\, \prod_{j=1}^{p-1} \chi_{A^c}(f^j(x)) = 1 \right\} ,$$

we have

$$\mathbb{P}\left(\tau_A \geq p\right) = \int \mathrm{d}\mu(x) \prod_{j=1}^{p-1} \chi_{A^c}\left(f^j(x)\right) = \int \mathrm{d}\mu(x) \prod_{j=1}^{p-1} \left(1 - \chi_A\left(f^j(x)\right)\right) .$$

It is left to the reader to verify that

$$\prod_{j=1}^{p-1} \left(1 - \chi_A\left(f^j(x)\right)\right) \geq 1 - \sum_{j=1}^{p-1} \chi_A\left(f^j(x)\right) .$$

Therefore, for any $p \geq 1$ we have

$$\mathbb{P}\left(\tau_A \geq p\right) \geq 1 - (p-1)\mu(A) .$$

Note that this bound is not very useful if $p > 1 + 1/\mu(A)$. Let q be a positive integer to be chosen optimally later. We have

$$\mathbb{E}(\tau_A) \geq \sum_{p=1}^{q} \mathbb{P}\left(\tau_A \geq p\right) \geq \sum_{p=1}^{q} \left(1 - (p-1)\mu(A)\right) = q - \frac{q(q-1)\mu(A)}{2} .$$

Since this estimate is true for any integer q, we can look for the largest right-hand side. We take $q = [1/\mu(A)]$, which is near the optimum and get the estimate using $1/\mu(A) - 1 \leq q \leq 1/\mu(A)$. □

In general, one cannot say much more about the random variable τ_A without making some hypothesis on the dynamical system or on the set A. We now give an estimate on the tail of the distribution of τ_A in the case of dynamical systems with decay of correlations like in (5.34).

Theorem 8.10. *Assume the dynamical system defined by the map f and equipped with the ergodic invariant measure μ satisfies the estimate (5.34) with $C_{\mathcal{B}_1,\mathcal{B}_2}(n)$ decaying exponentially fast. Namely, there are two constants $C > 0$ and $0 < \varrho < 1$ such that for any integer n*

$$C_{\mathcal{B}_1,\mathcal{B}_2}(n) \leq C\varrho^n .$$

Let A be a measurable set such that $0 < \mu(A) < 1$, and such that the characteristic function of A belongs to $\mathcal{B}_1$. Assume also that $\mathcal{B}_2 = \mathrm{L}^1(\mathrm{d}\mu)$. Then

$$\mathbb{P}\left(\tau_A > n\right) \leq \mathrm{e}^{\gamma_A \, n\, \mu(A)/\log \mu(A)} ,$$

where $\gamma_A > 0$ depends only on the $\mathcal{B}_1$ norm of the characteristic function of A.

Note that $\log \mu(A)$ is negative and therefore the upper bound decays exponentially fast with n. This estimate says that τ_A is unlikely to be much larger than $-(\log \mu(A))/\mu(A)$. Note also that in this estimate there is no restriction on the set A except that its characteristic function belongs to $\mathcal{B}_1$. All these assumptions are satisfied in the case of piecewise expanding maps of the interval when A is an interval and μ the Lebesgue measure (see Theorem 5.57). In this case $\mathcal{B}_1$ is the set of functions of bounded variation, and the norm of the characteristic function of an interval is independent of the interval.

Proof. From the definition (see also the previous proof) we deduce

$$\mathbb{P}\left(\tau_A > n\right) = \int \prod_{j=1}^{n} \chi_{A^c} \circ f^j(x)\, \mathrm{d}\mu(x)\,.$$

Let k be an integer to be fixed later, and let $m = [n/k]$. We have obviously

$$\mathbb{P}\left(\tau_A > n\right) \leq \int \prod_{l=1}^{m} \chi_{A^c} \circ f^{lk}(x)\, \mathrm{d}\mu(x)\,.$$

Using (5.34) with our hypothesis (and the invariance of the measure) we get

$$\mathbb{P}\left(\tau_A > n\right) \leq \int \left(1 - \chi_A\right) \prod_{l=1}^{m-1} \chi_{A^c} \circ f^{lk}(x)\, \mathrm{d}\mu(x)$$

$$\leq \left(1 - \mu(A) + C\varrho^k \|\chi_A\|_{\mathcal{B}_1}\right) \int \prod_{l=1}^{m-1} \chi_{A^c} \circ f^{lk}(x)\, \mathrm{d}\mu(x)\,.$$

We now choose k as the smallest integer for which

$$C\varrho^k \|\chi_A\|_{\mathcal{B}_1} \leq \frac{1}{2}\mu(A)\,.$$

Iterating the estimate we obtain

$$\mathbb{P}\left(\tau_A > n\right) \leq \left(1 - \mu(A)/2\right)^m \leq \mathrm{e}^{-m\,\mu(A)/2}\,,$$

and the result follows. □

Theorem 8.10 deals with an upper bound. In the case of small sets A a lot of work has been devoted to the study of asymptotic laws, that is, upper and lower bounds. One can expect to see emerging something similar to the exponential law by analogy with the standard Poisson limit theorem (see (Feller 1957, 1966)) provided there is a fast enough decorrelation.

We mention here only the case of piecewise expanding maps of the interval. See Fig. 8.1 for an example.

Theorem 8.11. *Let f be a piecewise expanding map of the interval, and μ a mixing acim with density h. There is a set B of full measure such that if $\{A_n\}$ is a sequence of intervals of length tending to zero and accumulating to a point $b \in B$, then the sequence of random variables $\mu(A_n)\tau_{A_n}$ converges in law to an exponential random variable of parameter 1. In other words, for any fixed number $s > 0$ we have*

$$\lim_{n\to\infty} \mathbb{P}\left(\tau_{A_n} > s/\mu(A_n)\right) = \mathrm{e}^{-s}\,.$$

We sketch below a proof of this theorem modulo a technical point for which we refer the reader to the literature. This proof is based on an idea of Kolmogorov for proving the central limit theorem in the i.i.d case (see (Borovkov 2004)).

We recall first a simple version of the Poisson limit theorem.

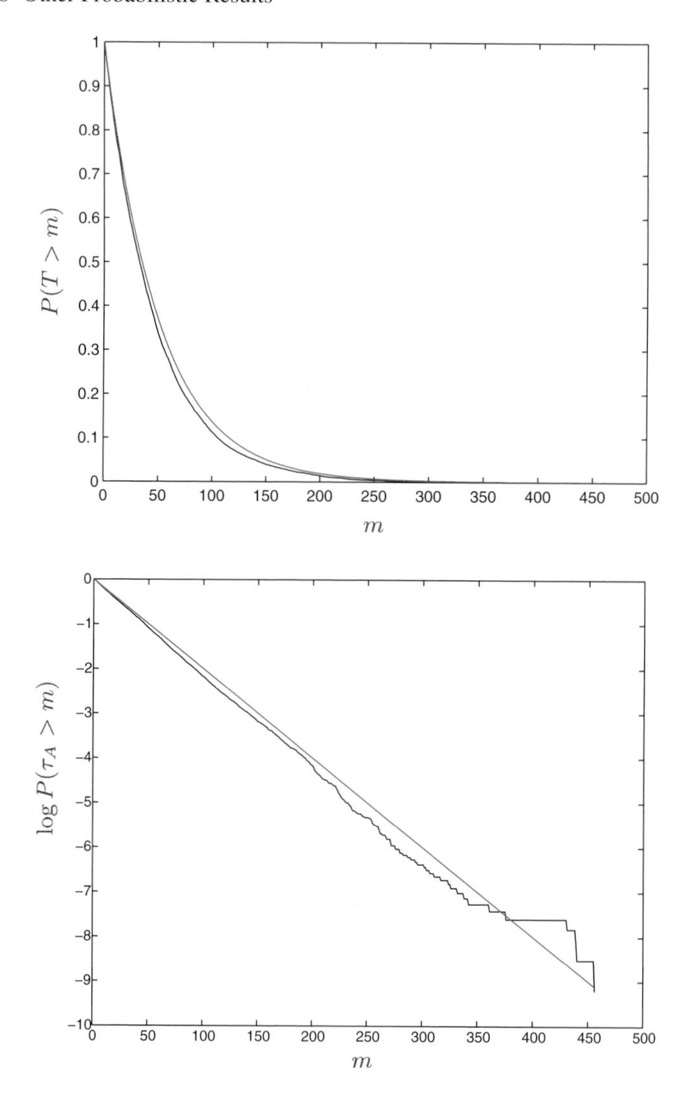

Fig. 8.1. An illustration of Theorem 8.11 for the map $f : x \mapsto 3x \pmod 1$. We take a small set $A = [\sqrt{1/2} - 0.01, \sqrt{1/2} + 0.01]$. For 10,000 randomly chosen initial points, we compute the number of orbits that have not landed in A before m iterations. The vertical axis is the probability that the entrance time τ_A exceeds m. The top panel shows the result on a linear scale, and the bottom panel shows the result on a logarithmic scale. We show both the experimental and theoretical curves

Exercise 8.12. *Assume a sequence $\{X_n\}$ of i.i.d random variables takes values 0 and 1 with probability $1-\varepsilon$ and ε respectively. Let N be the smallest integer such that $X_n = 1$. Show that for any fixed $s > 0$*

$$\lim_{\varepsilon \to 0} \mathbb{P}\left(N > \left[s\varepsilon^{-1}\right]\right) = \mathrm{e}^{-s} .$$

Proof (of Theorem 8.11). To alleviate the notation, we denote by ε the positive number $\mu(A)$. Let s be a positive number and let $n = [s/\varepsilon]$. Obviously,

$$\mathbb{P}\left(\tau_A > n\right) = (1-\varepsilon)^n + \sum_{q=0}^{n-1} \left((1-\varepsilon)^{n-q-1}\mathbb{P}\left(\tau_A > q+1\right) - (1-\varepsilon)^{n-q}\mathbb{P}\left(\tau_A > q\right)\right) .$$

If ε tends to zero, the first term converges to e^{-s}, and we need only to prove that the second term (the sum) converges to zero. Since for $q \geq 1$,

$$\mathbb{P}\left(\tau_A > q\right) = \mathbb{E}\left(\prod_{j=1}^{q} \chi_{A^c} \circ f^j\right) ,$$

we have from the invariance of the measure

$$\begin{aligned}
&(1-\varepsilon)^{n-q-1}\mathbb{P}\left(\tau_A > q+1\right) - (1-\varepsilon)^{n-q}\mathbb{P}\left(\tau_A > q\right) \\
&= (1-\varepsilon)^{n-q-1}\left(\mathbb{E}\left(\prod_{j=1}^{q+1} \chi_{A^c} \circ f^j\right) - (1-\varepsilon)\mathbb{E}\left(\left(\chi_A + \chi_{A^c}\right)\prod_{j=1}^{q} \chi_{A^c} \circ f^j\right)\right) \\
&= (1-\varepsilon)^{n-q-1}\left(\varepsilon\, \mathbb{E}\left(\prod_{j=1}^{q+1} \chi_{A^c} \circ f^j\right) - (1-\varepsilon)\mathbb{E}\left(\chi_A \prod_{j=1}^{q} \chi_{A^c} \circ f^j\right)\right) \\
&= (1-\varepsilon)^{n-q-1}\left(\varepsilon\, \mathbb{E}\left(\prod_{j=1}^{q} \chi_{A^c} \circ f^j\right) - \mathbb{E}\left(\chi_A \prod_{j=1}^{q} \chi_{A^c} \circ f^j\right)\right) \\
&\quad + (1-\varepsilon)^{n-q-1}\mathcal{O}(1)\varepsilon^2 ,
\end{aligned} \tag{8.1}$$

where the last term comes from the estimates

$$0 \leq \mathbb{E}\left(\chi_A \prod_{j=1}^{q+1} \chi_{A^c} \circ f^j\right) \leq \mathbb{E}\left(\chi_A\right) = \varepsilon$$

and

$$\mathbb{E}\left(\prod_{j=1}^{q} \chi_{A^c} \circ f^j\right) - \varepsilon \leq \mathbb{E}\left(\prod_{j=1}^{q+1} \chi_{A^c} \circ f^j\right) \leq \mathbb{E}\left(\prod_{j=1}^{q} \chi_{A^c} \circ f^j\right) .$$

Note that the right-hand side of (8.1) is of modulus less than $2\varepsilon + \varepsilon^2$ (drop the product of the characteristic functions and use $\varepsilon > 0$). Let k_A be an integer such that

$A \cap f^j(A) = \emptyset$ for $j = 1, \ldots, k_A$ (it is enough to assume that the intersection has measure zero). Therefore for $q > k_A$ we have

$$\mathbb{E}\left(\chi_A \prod_{j=1}^{q} \chi_{A^c} \circ f^j\right) = \mathbb{E}\left(\chi_A \prod_{j=k_A+1}^{q} \chi_{A^c} \circ f^j\right).$$

For the other term, we use *Bonferoni's inequality*, namely

$$1 - \sum_{j=1}^{k_A} \chi_A \circ f^j \leq \prod_{j=1}^{k_A} \chi_{A^c} \circ f^j \leq 1 - \sum_{j=1}^{k_A} \chi_A \circ f^j + \sum_{1 \leq r \neq s \leq k_A} \chi_A \circ f^r \chi_A \circ f^s . \quad (8.2)$$

Exercise 8.13. *Prove these inequalities.*

This implies immediately for $q > k_A$,

$$\left| \mathbb{E}\left(\prod_{j=1}^{q} \chi_{A^c} \circ f^j\right) - \mathbb{E}\left(\prod_{j=k_A+1}^{q} \chi_{A^c} \circ f^j\right) \right| \leq \varepsilon k_A^2 .$$

We can now use the decay of correlations in the form of equation (5.34) for $\chi_A \in \mathcal{B}_1$ and $\prod_{j=k_A+1}^{q+1} \chi_{A^c} \circ f^j \in \mathcal{B}_2$. For example, for a mixing acim of a piecewise expanding map of the interval we can take for A an interval whose characteristic function belongs to the space of functions of bounded variation ($\mathcal{B}_1$) and for $\mathcal{B}_2$ it is enough to take L^∞. We get

$$\left| \varepsilon \mathbb{E}\left(\prod_{j=k_A+1}^{q} \chi_{A^c} \circ f^j\right) - \mathbb{E}\left(\chi_A \prod_{j=k_A+1}^{q} \chi_{A^c} \circ f^j\right) \right| \leq C_{\mathcal{B}_1, \mathrm{L}^\infty}(k_A) \left\| \chi_A \right\|_{\mathcal{B}_1} .$$

Combining all the above estimates we get for $n - 1 \geq q > k_A$,

$$\left| (1-\varepsilon)^{n-q-1} \mathbb{P}\left(\tau_A > q+1\right) - (1-\varepsilon)^{n-q} \mathbb{P}\left(\tau_A > q\right) \right|$$

$$\leq (1-\varepsilon)^{n-q-1} \left(\varepsilon^2 (1 + k_A^2) + C_{\mathcal{B}_1, \mathrm{L}^\infty}(k_A) \|\chi_A\|_{\mathcal{B}_1}\right) .$$

On the other hand, as was already observed, we have immediately from (8.1),

$$\left| (1-\varepsilon)^{n-q-1} \mathbb{P}\left(\tau_A > q+1\right) - (1-\varepsilon)^{n-q} \mathbb{P}\left(\tau_A > q\right) \right| \leq 2\varepsilon + \varepsilon^2 .$$

This bound, which is not as good as the previous one, will be used for small q ($q \leq k_A$) when the sharper estimate does not hold. Summing over q (from 1 to k_A using the rough bound, and from k_A to n using the more elaborated one) we get for $n > k_A$

$$\left| \sum_{q=0}^{n-1} \left((1-\varepsilon)^{n-q-1} \mathbb{P}\left(\tau_A > q+1\right) - (1-\varepsilon)^{n-q} \mathbb{P}\left(\tau_A > q\right) \right) \right|$$

$$\le \mathcal{O}(1)\Big(\varepsilon k_A + \varepsilon k_A^2 + \varepsilon^{-1} C_{\mathcal{B}_1,\mathrm{L}^\infty}(k_A) \|\chi_A\|_{\mathcal{B}_1}\Big) \, .$$

The theorem follows for any sequence $\{A_n\}$ of sets for which the right-hand side of the above estimate tends to zero. For a mixing acim of a piecewise expanding map of the interval, $\|\chi_A\|_{\mathbf{BV}} \le 3$, and $C_{\mathcal{B}_1,\mathrm{L}^\infty}(k_A)$ decays exponentially fast in k_A (recall $\mathcal{B}_1 = \mathbf{BV}$ and see estimate (5.35)). It is therefore enough to ensure that $k_A = \mathcal{O}(1)(-\log \mu(A)) = \mathcal{O}(1)\log \varepsilon^{-1}$. We refer to (Collet and Galves 1995) and Exercise 8.14 for the details. □

Exercise 8.14. *Consider the full shift $\mathcal{S}$ on two symbols $\{0,1\}$ and the product measure $(1/2,1/2)$ (or the map $2x \pmod 1$ with the Lebesgue measure). Let $\mathcal{C}_{n,k}$ ($1 \le k \le n-1$) be the set of cylinders C of length n (starting at position 1) such that $C \cap \mathcal{S}^k C \neq \emptyset$ (recall that $\mathcal{S}$ is the shift). For $C \in \mathcal{C}_{n,k}$ show that the last $n-k$ symbols are determined by the k first ones. Show that $\mu(\mathcal{C}_{n,k}) \le 2^{k-n}$ (use that $|\mathcal{C}_{n,k}| \le 2^k$). For a cylinder set C of length n, define for $1 \le q \le n-1$ the function*

$$f_{C,\,q} = \chi_C \prod_{j=q}^{n-1} \left(1 - \chi_C \circ \mathcal{S}^j\right) .$$

Let

$$R_q^n = \left\{\mathbf{x} \,\middle|\, \mathcal{S}^j(\mathbf{x}) \in C(\mathbf{x}) \text{ for some } q \le j \le n-1\right\} ,$$

where $C(\mathbf{x})$ is the cylinder of length n containing $\mathbf{x}$. Show that $f_{C,q}$ is the characteristic function of $C \cap \left(R_q^n\right)^{\mathrm{c}}$. Derive that

$$\mu\left(\left(R_q^n\right)^{\mathrm{c}}\right) = \int \sum_C f_{C,q} \, \mathrm{d}\mu \, .$$

Show that (see the Bonferoni inequality (8.2))

$$f_{C,\,q} \ge \chi_C - \sum_{j=q}^{n-1} \chi_C \cdot \chi_C \circ \mathcal{S}^j \, .$$

Conclude that if $C \in \mathcal{C}_{n,k}$ for some $q \le k \le n-1$, then

$$\int f_{C,\,q} \, \mathrm{d}\mu \ge \mu(C) - n 2^{-n-q}$$

(use that $\chi_C \chi_C \circ \mathcal{S}^j$ is the characteristic function of a cylinder set of length $n+j$). Using the previous estimate of $\mu(C)$ for $C \in \mathcal{C}_{n,k}$ with $k < n/2$, show that there is a constant $c > 0$ such that for any n

$$\mu\left(R_1^n\right) \le c\, n\, 2^{-n/2} \, .$$

For $k = 2m$, let K be a cylinder of length m which does not belong to $\mathcal{C}_{m,j}$ for any $1 \le j < m$. Show that the cylinder $K \cap \mathcal{S}^m K$ belongs to $\mathcal{C}_{k,m}$. Show that $\mu(\mathcal{C}_{k,k/2}) \ge 2^{-k/2}$ and compare with the previous bound.

Periodic orbits of small period prevent k_A from being large, for example, if A contains a fixed point, then $A \cap f(A) \neq \emptyset$. Those sets that recur too fast prevent in some sense enough loss of memory to ensure a limiting exponential law. Some results can nevertheless be obtained in these situations; we refer to (Hirata 1993) and (Coelho and Collet 1994) for some examples. By looking at the successive entrance times, one can prove under adequate hypotheses that the limiting process when the set becomes smaller and smaller is a Poisson point process (or a marked Poisson point process). We recall that a (homogeneous) Poisson point process on $\mathbb{R}^+$ is given by a random (strictly) increasing sequence of points $\{x_n\}$ with $x_1 = 0$ and such that the random variables $x_{n+1} - x_n$ are i.i.d with exponential distribution. See (Hirata 1993), and (Coelho and Collet 1994; Collet 1996; Collet, Galves, and Schmitt 1992; Collet and Galves 1993; Hirata, Saussol, and Vaienti 1999; Abadi 2001; Denker, Gordin, and Sharova 2004) for more details, results, and references. The approach of two typical trajectories, namely the entrance time into a neighborhood of the diagonal for the product system leads to a marked Poisson process at least for piecewise expanding maps of the interval. We refer to (Coelho and Collet 1994) for details and to (Kontoyiannis 1998) for related results.

An interesting case of a set A is the first cylinder of length n of a coded sequence. We recall that the return time is then related to the entropy of the system by the Theorem 6.7 of Ornstein–Weiss. We refer to (Ornstein and Weiss 1993; Collet, Galves, and Schmitt 1999a; Kontoyiannis 1998; Chazottes and Collet 2005) for more results on this case and the study of the fluctuations.

The possible asymptotic laws for entrance times are discussed in (Durand and Maass 2001) and (Lacroix 2002).

8.2 Number of Visits to a Set

Given a measurable set A of positive measure, it is natural to ask how many times a typical orbit visits A during a time interval $[0,T]$ (it is convenient to allow T to be any positive real number). This *number of visits* denoted by $N_A[0,T]$ is obviously given by

$$N_A[0,T](x) = \sum_{0 \le j \le T} \chi_A\big(f^j(x)\big)$$

for the orbit of the point x. Birkhoff's ergodic theorem 5.38 tells us that if one considers a large time interval $[0,T]$, and a (measurable) set A of positive measure, then the number of times a typical trajectory has visited A (the event A has occurred) is of the order of $\mu(A)T$ where μ is the ergodic measure. In other words, $1/\mu(A)$ is a time scale associated to the set A. For a fixed A, we have discussed at length the behavior of $N_A[0,T] - \mu(A)T$ for large T (see e.g. the discussion of the central limit theorem in Chap. 7).

Another kind of asymptotic statistics appears when $\mu(A)$ is taken smaller and smaller and simultaneously T is of order $1/\mu(A)$. In that case, on a "large" interval of time of scale $1/\mu(A)$ one expects to see only a few occurrences of the event A.

It turns out that for certain classes of dynamical systems, this can be quantified in a precise way. It follows for example that if the process of successive entrance times converges in law to a Poisson point process, the suitably scaled number of visits converges in law to a Poisson random variable, as discussed in Sect. 8.1.

8.3 Extremes

An interesting quantity to look at is the distance of a typical finite orbit to a given point or the minimal distance between two finite orbits. Consider a dynamical system with a metric d on the phase space Ω. Assume a point x_0 has been chosen once and for all. One can look at the successive distances of points on the orbit of x from the point x_0, namely $d(x_0, x), \ldots, d(x_0, f^n(x))$. This is a sequence of positive numbers and the *record* is the smallest one. This number depends, of course, on x_0, n and x, and we define it more precisely by

$$R_{n,x_0}(x) = \sup_{0 \le j \le n} \log 1/d\big(x_0, f^j(x)\big) .$$

The logarithm is a convenient way to transform a small number into a large one but other means can be used to do that (for example the inverse). The logarithm is more convenient to formulate the results. If we consider a measure μ on the phase space for the distribution of x (for example an invariant measure), then R_{n,x_0} becomes a random variable. The law of R_{n,x_0} is related to entrance times. Indeed, it follows at once from the definition that if $R_{n,x_0}(x) < s$, then $d\big(x_0, f^j(x)\big) > \mathrm{e}^{-s}$ for $0 \le j \le n$. In other words,

$$\Big\{x \,\Big|\, R_{n,x_0}(x) < s\Big\} = \Big\{x \,\Big|\, \tau_{B_{\mathrm{e}^{-s}}(x_0)}(x) > n\Big\} \bigcap B_{\mathrm{e}^{-s}}(x_0)^{\mathrm{c}} .$$

This allows one to pass from results on entrance times to records and vice versa. We now give a simple estimate on the record related to the *capacitary dimension.*

Definition 8.15. *Let μ be a finite measure on a metric space Ω with a metric d, and let x_0 be a point in Ω. The capacitary dimension d_{Cap} of μ at x_0 is defined by*

$$d_{\mathrm{Cap}}^{x_0}(\mu) = \sup\left\{\beta \ge 0 \,\middle|\, \int \frac{1}{d(x, x_0)^\beta}\, \mathrm{d}\mu(x) < \infty\right\} .$$

We refer to (Kahane 1985) and (Mattila 1995) for more results on this quantity. Note that it can be infinite (for example, if x_0 does not belong to the support of μ).

Exercise 8.16. *Let f be a diffeomorphism of a compact Riemannian manifold M. If μ is an invariant ergodic measure for f, show that $d_{\mathrm{Cap}}^{x}(\mu)$ is μ-almost surely constant (show that this is an invariant function of x).*

Theorem 8.17. *For a fixed x_0 and an invariant measure μ, we have*

$$\limsup_{n \to \infty} \frac{1}{\log n} \mathbb{E}\big(R_{n,x_0}\big) \le \frac{1}{d_{\mathrm{Cap}}^{x_0}(\mu)} .$$

This result is similar to the general estimate (8.9).

Proof. We first apply the technique of *Pisier's inequality*, namely, for a fixed number $\beta > 0$ we have

$$e^{\beta R_{n,x_0}(x)} \leq \sum_{j=0}^{n} e^{-\beta \log d\left(f^j(x),x_0\right)} = \sum_{j=0}^{n} \frac{1}{d\left(f^j(x),x_0\right)^{\beta}} .$$

Integrating over μ and using the invariance of this measure, we get

$$\mathbb{E}\left(e^{\beta R_{n,x_0}}\right) \leq n \int \frac{1}{d(x,x_0)^{\beta}} \, d\mu(x) .$$

This bound is interesting only if $\beta < d_{\mathrm{Cap}}^{x_0}(\mu)$ which is the condition for the right-hand side to be finite. Using *Jensen's inequality* (the logarithm of the integral is larger than or equal to the integral of the logarithm) we get

$$e^{\beta \mathbb{E}\left(R_{n,x_0}\right)} \leq n \int \frac{1}{d(x,x_0)^{\beta}} \, d\mu(x) ,$$

and taking logarithms we obtain

$$\mathbb{E}\left(R_{n,x_0}\right) \leq \frac{\log n}{\beta} + \frac{1}{\beta} \log\left(\int \frac{1}{d(x,x_0)^{\beta}} \, d\mu(x)\right) .$$

This implies for any $\beta < d_{\mathrm{Cap}}^{x_0}(\mu)$,

$$\limsup_{n\to\infty} \frac{1}{\log n} \mathbb{E}\left(R_{n,x_0}\right) \leq \frac{1}{\beta}$$

and the result follows. □

It is in general more difficult to obtain a lower bound.

We observe that if we know that the entrance time into a ball $B_r(x_0)$ of radius r around a point x_0 converges in law when $r \to 0$ (after suitable normalization) to an exponential distribution, then we can often say something on the asymptotic distribution of the record (suitably normalized). Namely, if there is a sequence $\{\Lambda_r\}$ such that

$$\lim_{r\to\infty} \mathbb{P}\left(\tau_{B_r(x_0)} > \Lambda_r s\right) = e^{-s}$$

then we get

$$\mathbb{P}\left(\tau_{B_r(x_0)} > \Lambda_r s\right) = \mathbb{P}\left(-\log\left(d(f^j(x),x_0)\right) < -\log r \;\middle|\; 1 \leq j \leq \Lambda_r s\right)$$

$$= \mathbb{P}\left(R_{[\Lambda_r s],x_0} < -\log r\right) \longrightarrow_{r\to\infty} e^{-s} , \tag{8.3}$$

where we assumed that the measure μ has no atom in x_0. Assume now, as happens frequently, that

$$\Lambda_r \approx r^{-a} .$$

Note that if we are in a regular situation,

$$\mu\big(B_r(x_0)\big) \approx r^{d_{\mathrm{H}}(\mu)}$$

and

$$\Lambda_r \approx \frac{1}{\mu\big(B_r(x_0)\big)} \approx r^{-d_{\mathrm{H}}(\mu)} .$$

It follows from (8.3) by defining $n = [\Lambda_r s]$, that

$$\lim_{n\to\infty} \mathbb{P}\left(R_{n,x_0} < \frac{1}{a}\log n + v\right) = \mathrm{e}^{-\mathrm{e}^{-av}}$$

which is *Gumble's law* (see (Galambos 1987)). We refer to (Collet 2001) for more details.

8.4 Quasi-Invariant Measures

Another notion related to entrance times is that of dynamical systems with holes. Let A be a measurable subset of the phase space Ω and imagine that the trajectory under the map f is killed when it enters A for the first time (the system leaks out through A). Various interesting questions arise for the trajectories that have survived up to a time n. In particular, the set of initial conditions that never enter A (in other words, which survive forever) is given by

$$\mathscr{R} = \bigcap_{j=0}^{\infty} f^{-j}\left(A^{\mathrm{c}}\right) .$$

It often occurs that this set is small, for example of measure zero for some interesting measure. A simple example is given by the one-parameter family of maps of the interval $[0, s]$ (for $s > 1$).

$$f_s(x) = \begin{cases} s(1-|1-2x|) \,, & \text{for } 0 \le x \le 1 \\ x \,, & \text{for } 1 < x \le s \end{cases} . \tag{8.4}$$

The image of the interval $(1/(2s), 1-1/(2s))$ falls outside the interval $[0, 1]$; see Fig. 8.2.

Exercise 8.18. *Show that for the above maps f_s and $A = (1, s]$, the set $\mathscr{R}$ is a Cantor set of Lebesgue measure zero.*

It may sometimes occur that the set $\mathscr{R}$ is empty. From now on we always assume that this is not the case.

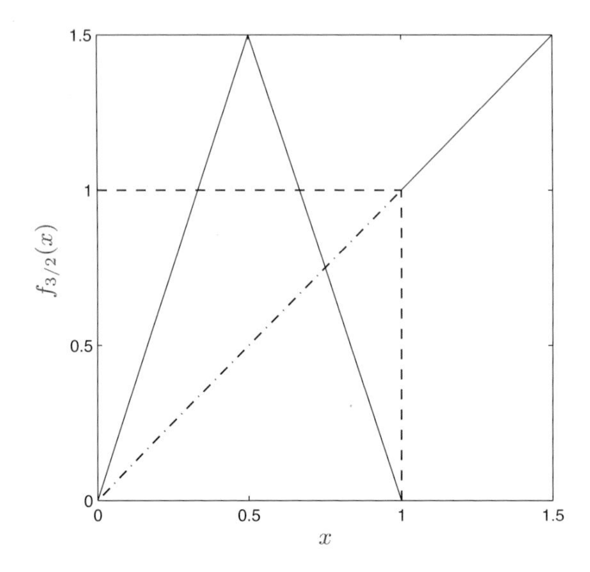

Fig. 8.2. The graph of the map $x \mapsto f_s(x)$ of formula (8.4) with $s = 3/2$

Assume that we are given a measure μ on the phase space Ω. We denote by Ω_n the set of initial conditions that have survived up to time n (in other words whose trajectory has not yet met A), namely

$$\Omega_n = \bigcap_{j=0}^{n} f^{-j}\left(A^c\right) .$$

This is obviously the subset of A^c for which $\tau_A > n$. As mentioned above, frequently, $\mu(\Omega_n)$ tends to zero when n tends to infinity.

One can look at the restriction of μ to the initial conditions in Ω_n, namely, define a sequence of probability measures $\{\mu_n\}$ by

$$\mu_n(B) = \frac{\mu\left(B \cap \Omega_n\right)}{\mu\left(\Omega_n\right)} ,$$

which is just the conditional measure on the set of initial conditions which have not escaped up to time n (here and in the sequel we assume that for any n, $\mu(\Omega_n > 0)$). If the limiting measure exists, it is supported by the set $\mathscr{R}$. Such measures have been studied extensively in particular in the case of complex dynamics (see (Brolin 1965)), and more generally in the case of repellers. We refer to (Collet, Martínez, and Schmitt 1994; 1997), (Chernov, Markarian, and Troubetzkoy 1998), and references therein.

Exercise 8.19. *Consider the map (8.4) and for μ the Lebesgue measure. Show that the sequence of measures $\{\mu_n\}$ converges (weakly) to the Cantor measure.*

Another interesting notion was introduced by Yaglom and is often called the *Yaglom limit*. It is connected with the notion of quasi-invariant measure (or quasi-invariant distribution). Consider the initial conditions that survived up to time n, namely which belong to Ω_n. The nth iterate of many of these initial conditions are points that are on the verge of escaping. Can one say something about their distribution? This leads one to consider a sequence $\{\mu_n\}$ of probability measures given by

$$\mu_n(B) = \frac{\mu\left(f^{-n}(B) \cap \Omega_n\right)}{\mu\left(\Omega_n\right)} \ .$$

Again, one can ask if there is a limit for this sequence (the so-called Yaglom limit). It turns out that these measures satisfy an interesting relation.

Lemma 8.20. *For any measurable set B and any integer n we have*

$$\mu_n\left(f^{-1}(A^{\mathrm{c}} \cap B)\right) = \frac{\mu(\Omega_{n+1})}{\mu(\Omega_n)} \mu_{n+1}(B) \ .$$

Exercise 8.21. *Give a proof of this lemma.*

Iterating this relation, one gets for any integers n and p,

$$\mu_n\left(\Omega_p \cap f^{-p}(B)\right) = \frac{\mu(\Omega_{n+p})}{\mu(\Omega_n)} \mu_{n+p}(B) \ .$$

In particular, if the ratio $\mu(\Omega_{n+1})/\mu(\Omega_n)$ converges to a number $\varrho > 0$, and the sequence $\{\mu_n\}$ converges in an adequate sense to a probability measure $\tilde{\mu}$, this measure satisfies the relation

$$\tilde{\mu}\left(f^{-1}(A^{\mathrm{c}} \cap B)\right) = \varrho\, \tilde{\mu}(B) \ . \tag{8.5}$$

Definition 8.22. *By analogy with (5.5), a measure $\tilde{\mu}$ satisfying the relation (8.5) for all B is called a* quasi-invariant measure *(or* quasi-invariant distribution*).*

Note that by taking $B = A^{\mathrm{c}}$ we get

$$\varrho = \frac{\tilde{\mu}\left(f^{-1}(A^{\mathrm{c}})\right)}{\tilde{\mu}(A^{\mathrm{c}})} \ .$$

Iterating (8.5) one gets

$$\tilde{\mu}\left(\Omega_p \cap f^{-p}(B)\right) = \varrho^p \tilde{\mu}(B) \ .$$

Taking $B = \Omega$ and using that μ is a probability measure we get

$$\tilde{\mu}(\Omega_p) = \varrho^p \ .$$

In other words, since

$$\tilde{\mu}(\Omega_p) = \tilde{\mu}(\tau_A > p)$$

we see that in the measure $\tilde{\mu}$, the entrance time is exactly exponentially distributed.

In the other direction, one can show the following theorem.

Theorem 8.23. *Let μ_0 be a probability measure on a compact set Ω_0. Assume that A is open or closed and $\mu_0(f^{-n}(\partial A)) = 0$ for any $n \in \mathbb{Z}$ (∂A denotes the boundary of A). Then, if moreover*

$$\limsup_{n\to\infty} \frac{\log \mu_0(\Omega_n)}{n} < 0 ,$$

there exists a quasi-invariant measure.

We refer to (Collet, Martínez, and Maume-Deschamps 2000; 2004a) for a proof and references.

Note that the quasi-invariant measure whose existence is ascertained by the theorem may not be unique and need not be simply related to the measure μ_0. We refer to (Collet, Martínez, and Maume-Deschamps 2000; 2004a) and (Chernov, Markarian, and Troubetzkoy 1998) for details and references.

Example 8.24. We come back to the case of the map f_s of (8.4), with the set $A = (1, s]$. Let μ be the measure on $[0, s]$, which is the Lebesgue measure on the interval $[0, 1]$ and vanishes on $(1, s]$. To check that μ is a quasi-invariant measure, it is enough to take $B \subset A^c = [0, 1]$ and by definition (8.5) to verify that

$$\frac{\mu(f^{-1}(B))}{\mu(B)}$$

does not depend on B. Since μ is a Borel measure, it is enough to verify this assumption for finite unions of intervals. Moreover, one readily checks that if this relation holds for any interval (contained in $[0, 1]$), it will also hold for any finite union of intervals.

If $B = [a, b]$ $(0 \le a < b \le 1)$, we conclude at once that

$$f^{-1}(B) = [a/(2s), b/(2s)] \cup [1 - b/(2s), 1 - a/(2s)] .$$

Therefore,

$$\mu(f^{-1}(B)) = \frac{b-a}{s}$$

and

$$\frac{\mu(f^{-1}(B))}{\mu(B)} = \frac{1}{s} ,$$

which does not depend on B.

8.5 Stochastic Perturbations

As we already mentioned several times, in concrete experiments (and to some extent also in numerical simulations) one obtains a unique record of a (finite piece of) trajectory of a dynamical system (or its image through an observable), often corrupted by noise. The effect of the noise can be modeled in many ways and we will deal here only with the simplest case. We assume that the phase space is contained in $\mathbb{R}^d$ and that there is a sequence of i.i.d random variables $\xi_0, \xi_1, \ldots$, such that instead of observing the orbit of an initial condition x, one observes sequences $\{x_n\}$ of points in the phase space given by

$$x_{n+1} = f(x_n) + \varepsilon\, \xi_n \ , \tag{8.6}$$

where ε is a fixed parameter. The process $\{x_n\}$ is called a *stochastic perturbation* of the dynamical system f. In most cases, the noise is small and it is convenient to normalize its amplitude by this parameter ε so that the ξ_n are of order 1. For example, one can impose that the variance of the ξ_n is equal to 1. In experimental contexts, this noise reflects the fact that extra effects neglected in the description of the experiment constantly perturb the system. A good example is temperature fluctuations. Note also that formula (8.6) describes a nonautonomous equation since the map also depends on time through the sequence $\{\xi_n\}$.

We henceforth assume that the average of ξ_n is equal to 0. This is a natural assumption; for example, the experimental noises are often modeled by white noises of zero average. If the average m of ξ_n is not zero, one can subtract it from the noise and add it to the transformation, namely replace ξ_n by $\xi_n - m$ and f by the map $x \mapsto f(x) + \varepsilon\, m$. In formula (8.6) this produces the same sequence $\{x_n\}$.

In the presence of noise there are two basic questions.

i) Can one recover to some extent the true trajectory out of the noisy one?
ii) Can one recover the statistical properties?

Concerning the first question, we discuss later denoising algorithms (see Remark 9.6). We also recall that the shadowing lemma allows one to construct a true orbit in the vicinity of a noisy one (see Theorem 4.63). The more ambitious complete reconstruction of a trajectory out of a noisy perturbation (of infinite length) was discussed in several papers by Lalley and Nobel (see (Lalley and Nobel 2000; Lalley 1999; Lalley and Nobel 2003)). We refer to the literature for the results.

Concerning the second question, namely recovering the statistical properties of the dynamical system out of noisy data, we first observe that due to the hypothesis of independence of the ξ_n in (8.6), the sequence $\{x_n\}$ defined in formula (8.6) is a Markov chain. Assume moreover that the random variables ξ_n have a density ϱ with respect to the Lebesgue measure. Then the transition probability is given by

$$p_\varepsilon\big(x_{n+1} \,\big|\, x_n\big) = \frac{1}{\varepsilon}\,\varrho\left(\frac{x_{n+1} - f(x_n)}{\varepsilon}\right) . \tag{8.7}$$

Exercise 8.25. *Prove this formula.*

Formally, when ε tends to zero, the right-hand side converges to $\delta\big(x_{n+1} - f(x_n)\big)$ (where δ is the Dirac distribution) and we recover the original dynamical system. This limiting procedure is however quite singular and requires some care.

It follows at once from (8.7) that in the case where the noise ξ_n has a density, any invariant measure for the Markov chain (x_n) is absolutely continuous (there may be several invariant measures if the chain is not recurrent).

It is natural to ask what are the accumulation points of the invariant measures of the chain, when the amplitude ε of the noise tends to zero. One should of course as usual avoid the phenomenon of escape of mass to infinity, although several papers have been devoted to this interesting question in the context of quasi-stationary measures; see (Collet, Martínez, and Schmitt 1999b; Collet and Martínez 1999; Ramanan and Zeitouni 1999). One way to avoid escape is by imposing a compact phase space (in that case some correction to the noise has to be made near the boundary).

Proposition 8.26. *Assume that the phase space is compact and that the map f of the dynamical system is continuous and the noise is bounded. Then any (weak) accumulation point of invariant measures of the Markov chain when the amplitude ε of the noise tends to zero is an invariant measure of f.*

Proof. We will cheat somewhat by ignoring the effects of the boundary. This can be fixed by somewhat lengthy and rather uninteresting technical details. Let $\{\varepsilon_n\}$ be a sequence tending to zero, and assume that we are given an associated sequence $\{\mu_{\varepsilon_n}\}$ of invariant measures (μ_{ε_n} is an invariant measure of the chain p_{ε_n}) that converges weakly to a measure μ. To show that μ is invariant, it is enough to prove (see Exercise 5.17) that for any continuous function g we have

$$\int g \circ f \, \mathrm{d}\mu = \int g \, \mathrm{d}\mu \, .$$

From the invariance of μ_{ε_n}, we find for any fixed continuous function g

$$\mathbb{E}\left(\int g\big(f(x) + \varepsilon_n \xi\big) \, \mathrm{d}\mu_{\varepsilon_n}(x) \right) = \int g(x) \, \mathrm{d}\mu_{\varepsilon_n}(x) \, , \tag{8.8}$$

where the expectation is over ξ. Since the phase space is compact, g is uniformly continuous. Therefore, since the noise is bounded and ε_n tends to zero, we can find for any $\eta > 0$ an integer N such that for any $n > N$ we have

$$\big| g\big(f(x) + \varepsilon_n \xi\big) - g\big(f(x)\big) \big| < \eta \, .$$

This implies for any $n > N$,

$$\left| \int g\big(f(x)\big) \, \mathrm{d}\mu_{\varepsilon_n} - \int g(x) \, \mathrm{d}\mu_{\varepsilon_n}(x) \right| < \eta \, ,$$

and the result follows by taking the limit $n \to \infty$ in equation (8.8). □

This leads naturally to the following definition.

Definition 8.27. *The invariant measure μ of a dynamical system is* stochastically stable *(with respect to the noise $\{\xi_n\}$) if it is an accumulation point of invariant measures of the stochastic processes (8.6) when the amplitude ε of the noise tends to zero.*

Even if the random variables ξ_n have a density (and we have seen that in this case the invariant measures of the chain are absolutely continuous), the accumulation points may not be absolutely continuous. This is necessarily the case if the map f has only attracting sets of Lebesgue measure zero. What comes closest to an absolutely continuous invariant measure is an SRB measure (see Definition 6.14) whose conditional measures on the local unstable manifolds are absolutely continuous. The stochastic stability of SRB measures has been indeed proven for Axiom A systems (see e.g. (Kifer 1988; 1997)) and in some nonuniformly hyperbolic cases (see (Young 1986) and references therein).

Theorem 8.28. *Let* (Ω, f) *be an Axiom A dynamical system. Let* Λ *be an attractor of* f *with a basin of attraction* U. *Let* $\{x_n\}$ *be a stochastic perturbation as in (8.6) with a bounded noise with continuous density. Then for small enough* ε *there is a unique invariant measure* μ_ε *for the process (8.6) supported in* U, *and this measure converges weakly to the SRB measure when* ε *tends to zero.*

We do not prove this theorem, and instead prove the stochastic stability of acim for piecewise expanding maps of the circle. The theory is analogous to the case of piecewise expanding maps of the interval, and in particular Theorem 5.57 applies. Considering the circle instead of the interval avoids unnecessary complications (which do not carry any interesting concepts). The problem is that on the interval if we use formula (8.6) for the definition of the stochastic perturbation, even if ξ_n is bounded, if the point $f(x_n)$ is too near to the boundary, the point $x_{n+1} = f(x_n)+\varepsilon_n$ could be outside of the interval. This is an interesting leaking problem somewhat related to the questions of Sect. 8.4 but we do not want to add this supplementary effect here.

Theorem 8.29. *Let* f *be a piecewise expanding map of the circle with a unique mixing absolutely continuous invariant measure with density* h. *Assume that the noise* ξ_n *has a density* ϱ *which is continuous and with compact support. Then for any* $\varepsilon > 0$ *small enough, the stochastic process*

$$\vartheta_{n+1} = f(\vartheta_n) + \varepsilon\xi_n \pmod{2\pi} \tag{8.9}$$

has a unique invariant measure μ_ε *which is absolutely continuous with density* h_ε *(which is a continuous function). When* ε *tends to zero,* μ_ε *converges weakly to* $h\,\mathrm{d}x$ *and more precisely* h_ε *converges to* h *in the Banach space* $\mathbf{BV}$ *of functions of bounded variation.*

Proof. Since the support of ϱ is bounded, there is a number $\varepsilon_0 > 0$ such that for any $\varepsilon \in [0, \varepsilon_0)$ we have almost surely and for any integer n the bound $\varepsilon|\xi_n| < \pi$. In particular, since for such an ε the function $\varrho(x/\varepsilon)$ vanishes outside the interval $(-\pi, \pi)$, we can consider this function as a function defined on the circle. We only consider below amplitudes ε of the noise belonging to the interval $\varepsilon \in [0, \varepsilon_0)$. Let R_ε be the operator defined on integrable functions of the circle by

$$R_\varepsilon g(\vartheta) = \frac{1}{\varepsilon}\int \varrho\left(\frac{\vartheta-\varphi}{\varepsilon}\right) g(\varphi)\mathrm{d}\varphi\,.$$

Using the *Young inequality* for g and $\varrho_\varepsilon(\varphi) = \varepsilon^{-1}\varrho(\varphi/\varepsilon)$, namely

$$\|g \star \varrho_\varepsilon\|_r \leq \|g\|_q \|\varrho_\varepsilon\|_p ,$$

with $p, q, r \in [1, \infty]$ and $1 + 1/r = 1/p + 1/q$, it follows at once that this operator is bounded in $\mathrm{L}^1(\mathrm{d}\vartheta)$ and $\mathrm{L}^\infty(\mathrm{d}\vartheta)$ with norm 1. It is also easy to verify that it is bounded in the Banach space $\mathbf{BV}$ also with norm 1.

Exercise 8.30. *Prove this last statement using the equivalent formula*

$$R_\varepsilon g(\vartheta) = \frac{1}{\varepsilon}\int \varrho(\psi/\varepsilon) g(\vartheta - \psi)\mathrm{d}\psi .$$

It is also easy to verify that R_ε tends strongly to the identity in the Banach spaces $\mathcal{B} = \mathrm{L}^1(\mathrm{d}\vartheta)$ and $\mathbf{BV}$. Namely for any $g \in \mathcal{B}$ we have

$$\lim_{\varepsilon \to 0} \left\|R_\varepsilon(g) - g\right\|_{\mathcal{B}} = 0 .$$

Exercise 8.31. *Prove this assertion. Hint: observe that*

$$R_\varepsilon(g)(\vartheta) - g(\vartheta) = \frac{1}{\varepsilon}\int \varrho(\psi/\varepsilon)\big(g(\vartheta - \psi) - g(\vartheta)\big)\mathrm{d}\psi ,$$

and approximate g by piecewise continuous functions.

As we explained earlier, the transition kernel of the stochastic process (8.9) is given by (8.7), and therefore any invariant measure $\mathrm{d}\mu_\varepsilon$ has a density h_ε given by

$$h_\varepsilon(\vartheta) = \frac{1}{\varepsilon}\int \varrho\left(\frac{\vartheta - f(\varphi)}{\varepsilon}\right) \mathrm{d}\mu_\varepsilon(\varphi) .$$

In particular, this function h_ε should satisfy the equation

$$h_\varepsilon(\vartheta) = \frac{1}{\varepsilon}\int \varrho\left(\frac{\vartheta - f(\varphi)}{\varepsilon}\right) h_\varepsilon(\varphi)\, \mathrm{d}\varphi .$$

This relation implies immediately that h_ε should be continuous and with support equal to the whole circle.

Exercise 8.32. *Prove these statements (use that ϱ is continuous).*

Performing a change of variables as in (5.8), one obtains immediately that h_ε should satisfy the equation

$$h_\varepsilon = R_\varepsilon P h_\varepsilon ,$$

where P is the Perron–Frobenius operator (5.9). Had R_ε converged in norm to the identity, we could have used perturbation theory to finish the proof. This is however not the case.

Exercise 8.33. *Prove that in the two Banach spaces $\mathcal{B}$ above, for any $\varepsilon > 0$ small enough,*

$$\left\|R_\varepsilon - I\right\|_{\mathcal{B}} \geq 1 .$$

To control the perturbation in a weaker sense, we recall that from Theorem 5.57 that one can write $P = P_0 + Q$, where P_0 is a rank one operator and Q is an operator with spectral radius $\sigma < 1$ (in $\mathbf{BV}$). Therefore,

$$R_\varepsilon P = R_\varepsilon P_0 + R_\varepsilon Q ,$$

and here again, $R_\varepsilon P_0$ is an operator of rank 1. We now analyze the resolvent of this operator. We first observe that the rank one operator $R_\varepsilon P_0$ has eigenvalue 1. Indeed, we have (see Theorem 5.57),

$$(P_0 g)(x) = h(x) \int g(y)\,\mathrm{d}y ,$$

and therefore, since R_ε preserves the integral, we find

$$\int (R_\varepsilon P_0 g)(x)\,\mathrm{d}x = \int (P_0 g)(x)\,\mathrm{d}x = \int g(y)\,\mathrm{d}y .$$

In other words, the integration with respect to the Lebesgue measure is a left eigenvector (eigenform) of eigenvalue one for the operator $R_\varepsilon P_0$. This is not surprising and reflects the fact that the kernel of R_ε is a transition probability. By a similar argument one also checks that for any integrable function g,

$$\int (R_\varepsilon Q g)(x)\,\mathrm{d}x = 0 ,$$

which implies $P_0 R_\varepsilon Q = 0$. Since $R_\varepsilon P_0$ is an operator of rank one with eigenvalue one, it is a projection.

Exercise 8.34. *Prove directly that $\left(R_\varepsilon P_0\right)^2 = R_\varepsilon P_0$.*

Therefore, for any complex number z with $z \neq 0, 1$ we have

$$\left(R_\varepsilon P_0 - z\right)^{-1} = \frac{R_\varepsilon P_0}{1-z} - \frac{I - R_\varepsilon P_0}{z} .$$

This implies that, for $z \neq 0, 1$, we can write

$$R_\varepsilon P - z = \left(R_\varepsilon P_0 - z\right)\Big(I - \left(R_\varepsilon P_0 - z\right)^{-1} R_\varepsilon Q\Big)$$

$$= \left(R_\varepsilon P_0 - z\right)\left(I - \frac{R_\varepsilon P_0}{1-z} R_\varepsilon Q + \frac{I - R_\varepsilon P_0}{z} R_\varepsilon Q\right) .$$

As we saw above, $P_0 R_\varepsilon Q = 0$, and we get

$$R_\varepsilon P - z = \left(R_\varepsilon P_0 - z\right)\left(I - \frac{R_\varepsilon Q}{z}\right) .$$

Assume for the moment $\|Q\|_{\mathbf{BV}} < 1$ (recall that it is only the spectral radius which is known to be smaller than one). We can try to obtain the resolvent of $R_\varepsilon P$ by using a Neumann series for the inverse of the last factor in the above formula. In other words, we will investigate the convergence of the sum of operators

$$K_\varepsilon(z) = \sum_{j=0}^{\infty} \left(\frac{R_\varepsilon Q}{z}\right)^j .$$

This series converges to a bounded operator in $\mathbf{BV}$ for any complex number z satisfying $|z| > \upsilon(\varepsilon) = \|R_\varepsilon Q\|_{\mathbf{BV}}$. In particular, if $|z| > \upsilon(\varepsilon)$ and $z \neq 1$, the operator $R_\varepsilon P - z$ is invertible, and its inverse is given by

$$(R_\varepsilon P - z)^{-1} = K_\varepsilon(z) \left(\frac{R_\varepsilon P_0}{1-z} - \frac{I - R_\varepsilon P_0}{z}\right) .$$

Furthermore, if $\upsilon(\varepsilon) < 1$, we conclude that the operator $R_\varepsilon P$ has one as a simple eigenvalue, and this is the only point in the spectrum outside the disk $|z| \leq \upsilon(\varepsilon)$. We can therefore express the (rank 1) spectral projection P_ε on this eigenvalue 1 by the formula (see (Kato 1984))

$$P_\varepsilon = \frac{1}{2\pi \mathrm{i}} \int_{|z-1|<(1-\upsilon(\varepsilon))/2} (R_\varepsilon P - z)^{-1} \,\mathrm{d}z$$

$$= \frac{1}{2\pi \mathrm{i}} \int_{|z-1|<(1-\upsilon(\varepsilon))/2} K_\varepsilon(z) \left(\frac{R_\varepsilon P_0}{1-z} + \frac{I - R_\varepsilon P_0}{z}\right) \mathrm{d}z .$$

The function $P_\varepsilon h$ is proportional to h_ε since P_ε is a rank 1 projection on this function. Moreover, since h is of integral 1 and the integration with respect to the Lebesgue measure is a left eigenvector of P_ε of eigenvalue 1, we have

$$\int (P_\varepsilon h)(x) \,\mathrm{d}x = \int h(x) \,\mathrm{d}x = 1 .$$

This implies immediately

$$P_\varepsilon h = h_\varepsilon .$$

Using the above formula for P_ε and the fact that $P_0 h = h$, we get from Cauchy's formula

$$h_\varepsilon = \frac{1}{2\pi \mathrm{i}} \int_{|z-1|<(1-\upsilon(\varepsilon))/2} K_\varepsilon(z) R_\varepsilon h \,\frac{\mathrm{d}z}{1-z} = K_\varepsilon(1) R_\varepsilon h .$$

We now observe that

$$h_\varepsilon = K_\varepsilon(1) R_\varepsilon h = R_\varepsilon h + \sum_{n=0}^{\infty} (R_\varepsilon Q)^n R_\varepsilon Q R_\varepsilon h .$$

Since R_ε converges strongly to the identity in $\mathbf{BV}$ and since $Qh = 0$, we have

$$\lim_{\varepsilon \searrow 0} Q R_\varepsilon h = 0 \,.$$

Since we assumed that $\|Q\|_{\mathbf{BV}} < 1$, and since we established that $\|R_\varepsilon\|_{\mathbf{BV}} = 1$, we conclude that

$$\lim_{\varepsilon \searrow 0} \sum_{n=0}^{\infty} (R_\varepsilon Q)^n R_\varepsilon Q R_\varepsilon h = 0 \,.$$

It then follows again from the strong convergence of R_ε to the identity in $\mathbf{BV}$ that

$$\lim_{\varepsilon \searrow 0} h_\varepsilon = h \,.$$

If $\|Q\|_{\mathbf{BV}} \geq 1$, since the spectral radius of Q is strictly smaller than 1, there is an equivalent norm where Q is a strict contraction (see (Kato 1984)). One equips $\mathbf{BV}$ with this new equivalent norm, and repeats the above argument. The details are left to the reader. □

We refer to (Young 1986; Kifer 1988; 1997) and (Blank and Keller 1997) for references and more results on this subject.

9

Experimental Aspects

Our aim in this chapter is mostly to describe measurement techniques which have a solid theoretical background. As usual, in Nature, one has no way of knowing whether the necessary mathematical assumptions are satisfied by a given system. But it turns out that it is very useful, in concrete examples, to act "as if" they were true. We do not enter into the many details which have been studied in the experimental-theoretical literature, such as optimal methods to work with not so abundant data, nonlinear fits to discover the evolution equations, and the like. There is a large literature on this subject; see e.g. (Grassberger, Schreiber, and Schaffrath 1991) and references therein.

Here, we explain some of the methods developed for the measurement of various interesting quantities for dynamical systems and more generally for stochastic processes. These methods can be applied to the output of numerical simulations or experimental data. They are usually targeted to a (large) unique realization of the process. More precisely, the system is observed through one, or very few, observables g (a real or vector valued function on the phase space) and for an initial condition x. One has a recording of the sequence

$$\{X_n\}_{0\le n\le N} = \{g(f^n(x))\}_{0\le n\le N} \ . \tag{9.1}$$

This is somewhat different from the usual statistical setting where one assumes to have a sequence of repeated independent observations. In the sequel we assume that an ergodic and invariant measure μ has been selected and that the initial condition x is typical for this measure. This is, of course, hard to verify in a numerical situation and even impossible in an experimental setting; this is why the Physical measure assumption (see Sect. 5.7) is relevant, because then the Lebesgue measure is the natural one from which to choose initial conditions. In any case, we are in the statistical situation of having the recording of the observation of a finite piece of one trajectory of a stochastic process. We assume for simplicity that the phase space is a bounded subset of $\mathbb{R}^d$ although the methods can be easily extended to the case of differentiable manifolds.

9.1 Correlation Functions and Power Spectrum

One of the easiest things to determine from the data is the (auto)correlation function. It follows immediately from the ergodic theorem and definition (5.24) that for μ-almost every x and for any square integrable observable g (with zero average) one obtains the correlation function as

$$C_{g,g}(k) = \lim_{n\to\infty} \frac{1}{n} \sum_{j=0}^{n-1} g\big(f^j(x)\big) g\big(f^{j+k}(x)\big) . \tag{9.2}$$

In Fig. 9.1 we show the correlation function computed with this formula for the map $x \mapsto 3x \pmod 1$ of the unit interval, and the observable $x - 1/2$. The figure shows two applications of formula (9.2), one with $n = 1024$ terms and the other one with $n = 32{,}768$ terms.

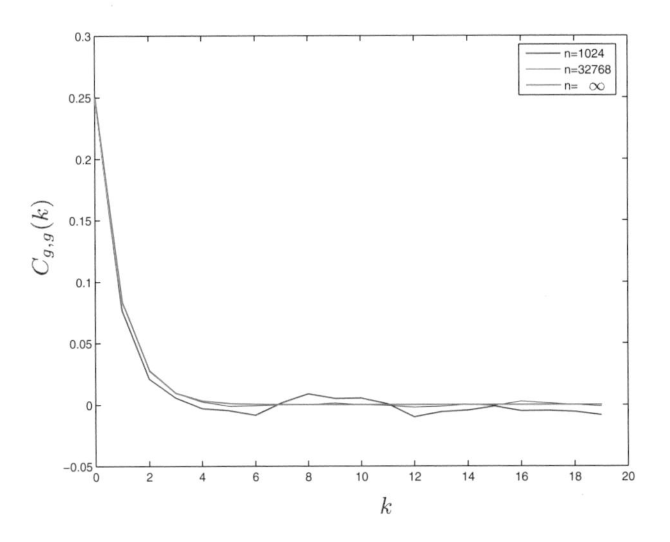

Fig. 9.1. Numerically determined correlation function for initial conditions chosen uniformly with respect to the Lebesgue measure, the map $3x \pmod 1$, and the observable $g(x) = x - 1/2$. The red line is for $n = 1024$, the blue for $n = 32{,}768$ in formula (9.2). We also show (green) the theoretical curve ($n = \infty$)

In practice, the measure μ is often unknown and therefore one does not know what to subtract from g to get a function with zero average. However, one can use again the ergodic theorem to estimate the average of g, namely

$$\int g \mathrm{d}\mu = \lim_{n\to\infty} \frac{1}{n} \sum_{j=0}^{n-1} g\big(f^j(x)\big) . \tag{9.3}$$

For particular classes of dynamical systems, the convergence rate can be estimated using various techniques; we refer to (Chazottes, Collet, and Schmitt 2005a) for some examples using the exponential estimate or the Devroye inequality.

Another quantity which is often used in practice is the *power spectrum* W_g (see (Kay and Marple 1981) or (Brockwell and Davis 1996)).

This is the Fourier transform of the correlation function, namely

$$W_g(u) = \sum_{k \geq 0} C_{g,g}(k) \mathrm{e}^{\mathrm{i}ku} .$$

If $\langle g \rangle$ denotes the average of g, using the ergodic theorem one is led to guess that the so-called *periodogram* $I_n(u, x)$ defined by

$$I_n(u, x) = \frac{1}{n} \left| \sum_{k=0}^{n-1} g(f^k(x)) - \langle g \rangle \mathrm{e}^{\mathrm{i}ku} \right|^2 \tag{9.4}$$

is a good approximation. If $\langle g \rangle$ is not known, it can be estimated as before using (9.3).

Unfortunately, although it is well-known that for any u one has

$$\lim_{n \to \infty} \int I_n(u, x) \mathrm{d}\mu(x) = W_g(u) ,$$

the periodogram does not converge μ-almost everywhere in x (it converges to a distribution valued process). This is illustrated in Fig. 9.2, which shows the periodogram

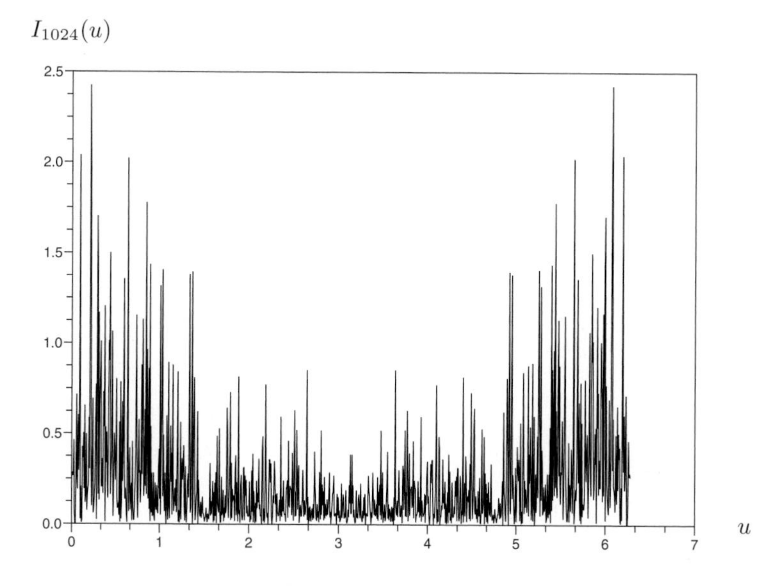

Fig. 9.2. Periodogram of the function $\chi_{[0,1/2]} - 1/2$, with $n = 1024$ in formula (9.4) for $u \in [0, 2\pi]$

(9.4) of the function $\chi_{[0,1/2]} - 1/2$ for the map $3x \pmod 1$ of the unit interval (and some random initial condition for the Lebesgue measure).

One can prove under adequate conditions that a much better behaved quantity to look at is the *integrated periodogram* J_n given by

$$J_n(u,x) = \int_0^u \frac{1}{n} \left| \sum_{k=0}^{n-1} \left(g(f^k(x))\right) - \langle g \rangle e^{iks} \right|^2 ds . \tag{9.5}$$

Namely, for any u, and for μ-almost every x one has

$$\lim_{n\to\infty} J_n(u,x) = \int_0^u W_g(s) ds .$$

Figure 9.3 shows the integrated periodogram of the function $\chi_{[0,1/2]} - 1/2$ for the map $3x \pmod 1$ of the unit interval (and random initial condition for the Lebesgue measure).

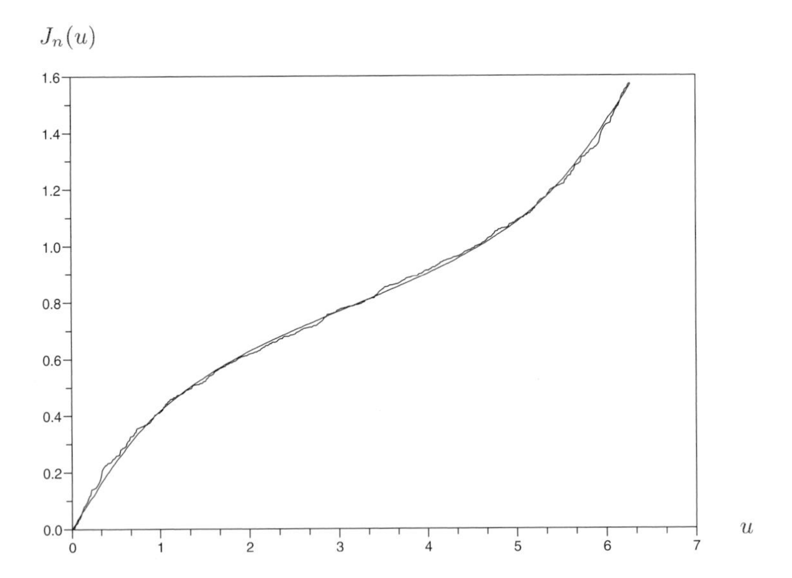

Fig. 9.3. Integrated periodogram of the function $\chi_{[0,1/2]} - 1/2$ for the map $3x \pmod 1$. The blue curve corresponds to $n = 32{,}768$ in formula (9.5) and the black curve to $n = 512$

We refer to (Chazottes, Collet, and Schmitt 2005a) for the case where the average of g has to be estimated from the data and for (uniform) estimates on the convergence rate under some assumptions on the dynamical system, using an exponential or a Devroye inequality. Other algorithms use local averages of the periodogram (see, for example, (Bardet, Doukhan, and Leon 2005)). An algorithm of Burg is based on the maximum entropy principle. We refer to (Lopes and Lopes 1998; 2002) for other results.

9.2 Resonances

In Sect. 5.6 we discussed the exponential decay of correlations. Here, we refine this discussion to the appearance of *resonances*, which are modulations of the rate of decay. These are not to be confused with the resonances which appear in the conjugation problems of Sect. 4.1.3. We want to illustrate two phenomena: The first is the modulation of the correlation function, while the second is the perhaps unexpected observation that the decay rate of the correlations is not necessarily equal to the inverse of the Lyapunov exponent.

Two examples are shown in Figs. 9.6 and in 9.7. The first allows for explicit calculations of all quantities, and the second is more like what happens in "typical" experiments. We begin with a (long) example.

Example 9.1. The map (which we call f) is drawn in Fig. 9.4 and is defined by

$$f(x) = \begin{cases} \lambda(x - x_c) + 1\,, & \text{if } x \le x_c \\ -\lambda(x - x_c) + 1\,, & \text{if } x \ge x_c \end{cases}\,,$$

where $x_c = \frac{2\lambda^2}{1+\lambda} + \frac{1-2\lambda^2}{\lambda^2}$. Note that it has a slope $\pm\lambda$, where $\lambda > 1$. Baladi obtained λ as follows: If we call $P_1, \dots, P_4$ the four pieces of the partition of $[0, 1]$ as shown on the bottom of Fig. 9.4, we see that $f(P_1) = P_2 \cup P_3$, $f(P_2) = f(P_3) = P_4$, and $f(P_4) = P_1 \cup P_2 \cup P_3$. Therefore, the transition matrix M (the Markov matrix) defined by $M_{i,j} = 1$ if $P_i \subset f(P_j)$ and zero otherwise is given by

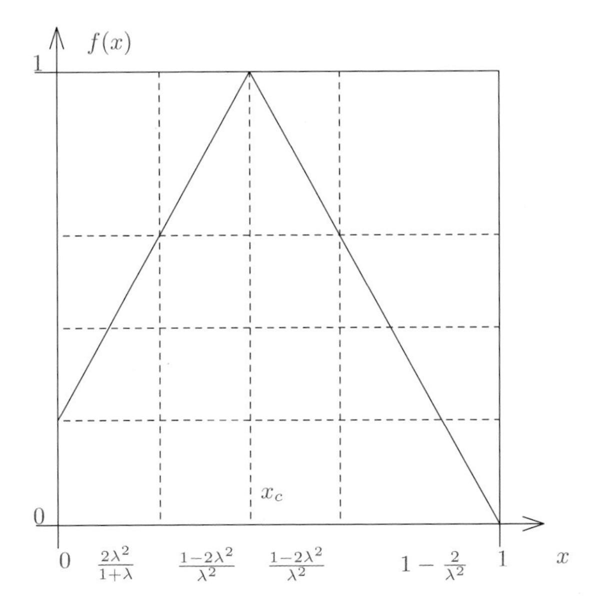

Fig. 9.4. The Baladi map. It is a Markov map

$$M = \begin{pmatrix} 0\,0\,0\,1 \\ 1\,0\,0\,1 \\ 1\,0\,0\,1 \\ 0\,1\,1\,0 \end{pmatrix} .$$

Its characteristic polynomial is

$$\lambda^4 - 2\lambda^2 - 2\lambda ,$$

and its eigenvalues are

$$\lambda \approx 1.76929 , \quad \lambda_{\mathrm{r},\pm} \approx -0.884846 \pm 0.58973\mathrm{i} , \text{ and } 0 .$$

The reader can check easily that the maximal eigenvalue is the right choice of λ. The correlation functions are given by

$$C_{F,G}(k) = \int \mathrm{d}x \, F(x) G\big(f^k(x)\big) h(x) ,$$

where the density h of the invariant measure (which is unique among the absolutely continuous invariant measures) is given by

$$h(x) = \begin{cases} \lambda^2/N \equiv \alpha , & \text{if } x < 2\lambda^2/(1+\lambda) \\ \lambda(1+\lambda)/N \equiv \beta , & \text{if } 2\lambda^2/(1+\lambda) < x < 2/\lambda^2 \\ 2(1+\lambda)/N \equiv \gamma , & \text{if } 2/\lambda^2 < x < 1 \end{cases} , \tag{9.6}$$

and $N = (2\lambda^3 - \lambda - 2)/\lambda^2$ is a normalization (Fig. 9.5). Changing variables to $y = f^{-1}(x)$, one gets

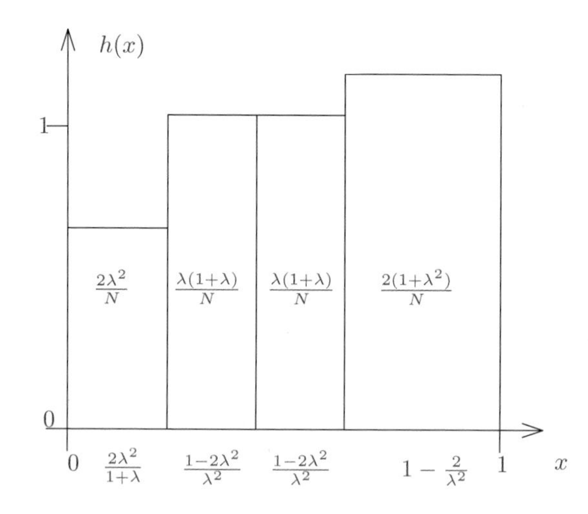

Fig. 9.5. The density h of the invariant measure. The normalization factor is $N = (2\lambda^3 - \lambda - 2)/\lambda^2$

$$C_{F,G}(k) = \int \mathrm{d}y \left(P^k(Fh)\right)(y)G(y) ,$$

where P is the Perron–Frobenius operator

$$(Pg)(y) = \sum_{x:f(x)=y} \frac{g(x)}{|f'(x)|} .$$

Note that since $|f'(x)| \equiv \lambda$ for our example, the Perron–Frobenius operator in this case equals $\lambda^{-1}M$ when acting on functions that are constant on the four pieces of the Markov partition. Therefore, on that space, its eigenvalues are given by

$$1 , \quad \frac{\lambda_{\mathrm{r},\pm}}{\lambda} \approx \frac{-0.884846 \pm \mathrm{i}\,0.58973}{1.76929} , \text{ and } 0 .$$

It follows that for generic observables *the correlation functions decay like*

$$|C_{F,G}(k)| \gtrsim C \left| \frac{\lambda_{\mathrm{r},\pm}}{\lambda} \right|^k . \tag{9.7}$$

This decay rate is *slower* than $C|1/\lambda|^k$ because $|\lambda_{\mathrm{r},\pm}| \approx 1.06320$. We illustrate these findings by numerical experiments in Fig. 9.6. The question is now whether

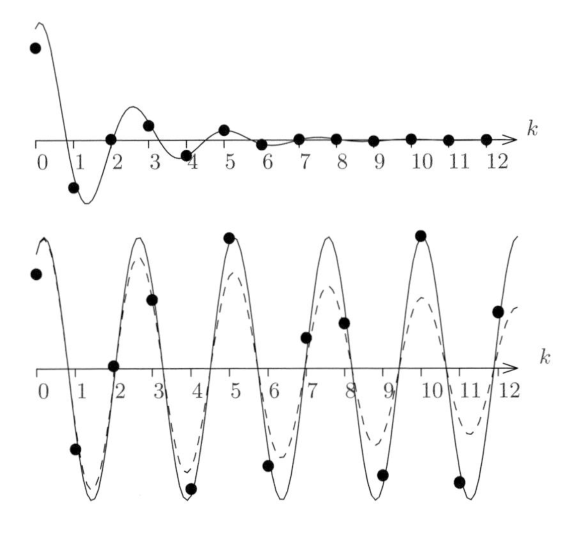

Fig. 9.6. A numerical study of the correlation function $C_{F,G}(k)$ (with $F(x) = G(x) = x - x_0$ where x_0 is the average $\int xh(x)\mathrm{d}x$ and h the invariant density of Fig. 9.5) for the 1-dimensional Baladi map, from $3 \cdot 10^7$ data points. The upper graph shows the computed values of $C_{F,G}(k)$ as dots and the theoretical curve $\mathrm{const.}\ \mathrm{Re}(\lambda_{\mathrm{r},+}/\lambda)^k$. The lower graph shows the same scaled vertically by $|\lambda_{\mathrm{r},+}/\lambda|^{-k}$, and as dashed lines, rescaled by λ^k. Since the data points fit the continuous curve, this shows that the decay rate is *not* given by the inverse of the expansion rate!

$C \neq 0$. The matrix M has an eigenvector $v_1 = (\alpha, \beta, \beta, \gamma)$ as defined in (9.6) corresponding to the eigenvalue λ, a 2-dimensional eigen-subspace corresponding to the eigenvalues $\lambda_{\mathrm{r},\pm}$ (spanned by some vectors v_2 and v_3), and a fourth eigendirection $v_4 = (0, 1, -1, 0)$ corresponding to the eigenvalue 0. These are also eigenspaces for $\lambda^{-1} P$. We see that *if the function* $F \cdot h$ *does not have any component in the subspace spanned by* v_2, v_3, *then* $C = 0$, *and the decay of* $C_{F,G}$ *is faster than described in (9.7)*. (Strictly speaking, we have shown this only for functions that are constant on the pieces of the partition. The proof of the general case is left to the reader.) In all other cases, $C \neq 0$ and (9.7) describes the relevant decay rate. We will therefore say that $\lambda_{\mathrm{r},\pm}/\lambda$ are **resonances** (see (Ruelle 1986; 1987; Baladi, Eckmann, and Ruelle 1989)) because they can be avoided by choosing observables (with zero average) in a subspace of codimension 2. See also (Collet and Eckmann 2004) for more details on relations between Lyapunov exponents, decay of correlations, and resonances. This ends Example 9.1.

One can see in the *power spectrum* (which is defined as the Fourier transform of the correlation function) peaks of periodic components and fat spectrum from noise of dynamical (chaotic) or experimental origin. For example, for the map $x \mapsto 1 - 1.7x^2$ of the interval $[-1, 1]$, Fig. 9.7 shows the correlation function and the power spectrum.

The *information correlation functions* are another useful tool for analyzing time series of dynamical systems. If U and V are two independent random variables, we have obviously

$$\mathbb{P}\left(U = a,\ V = b\right) = \mathbb{P}\left(U = a\right) \mathbb{P}\left(V = b\right) .$$

To test if the probability $\mathbb{P}\left(U = a,\ V = b\right)$ is a product, one uses the *relative entropy* between this measure and the product measure

$$D_{U,V}(a, b) = \sum_{a,b} \mathbb{P}\left(U = a\right) \mathbb{P}\left(V = b\right) \log \frac{\mathbb{P}\left(U = a,\ V = b\right)}{\mathbb{P}\left(U = a\right) \mathbb{P}\left(V = b\right)} .$$

One can also use the symmetrical expression

$$\sum_{a,b} \mathbb{P}\left(U = a,\ V = b\right) \log \frac{\mathbb{P}\left(U = a\right) \mathbb{P}\left(V = b\right)}{\mathbb{P}\left(U = a,\ V = b\right)} .$$

In practice, U is taken as the value of an observable at the initial time, and V its value at a later time $k : V = U \circ f^k$. The ergodic theorem is used to compute the probabilities (measure), it is simpler to use observables taking only finitely many values.

Example 9.2. For the map $x \mapsto 1.8x^2$ on the phase space $[-1, 1]$, we take, for example, the characteristic function of the interval $[0, 1] : U = \chi_{[0,1]}$. Then

$$\mathbb{P}\left(U = 1\right) = \lim_{N \to \infty} \frac{1}{N} \sum_{j=0}^{N-1} \chi_{[0,1]}\left(f^j(x)\right) ,$$

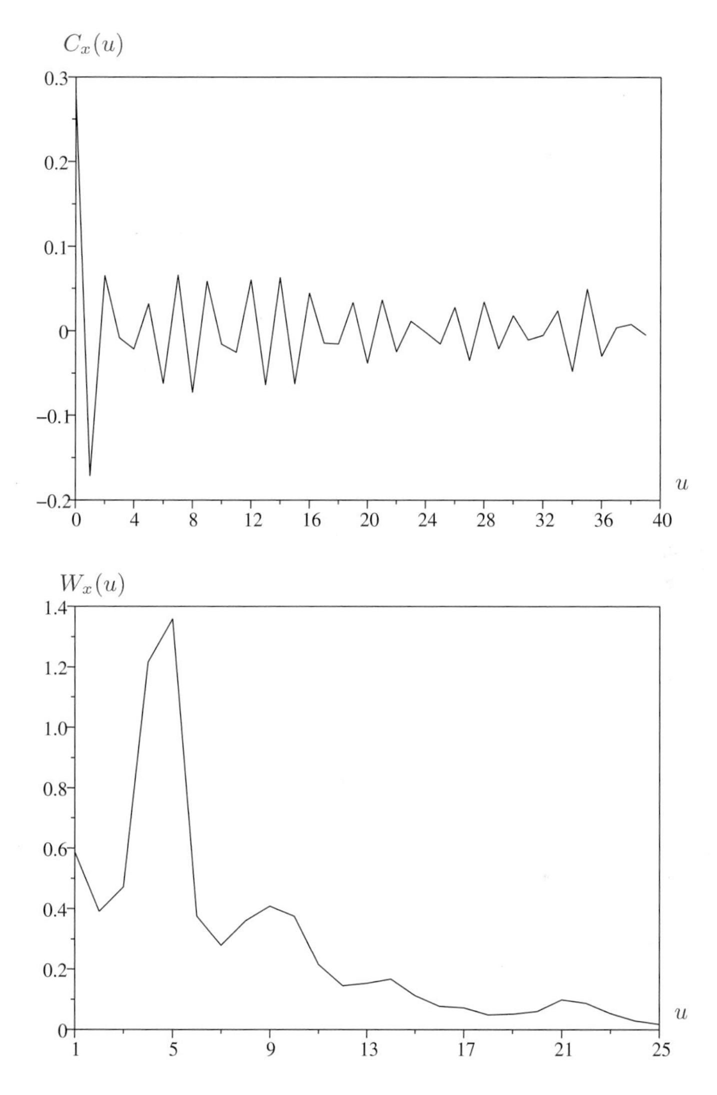

Fig. 9.7. Correlation function and power spectrum for the map $1 - 1.7x^2$. The observable is $g(x) = x$. The curves were obtained from 500 data points

and if $V = U \circ f^k$,

$$\mathbb{P}\,(U = 1, V = 1) = \lim_{N \to \infty} \frac{1}{N-k} \sum_{j=0}^{N-k-1} \chi_{[0,1]}\big(f^j(x)\big)\chi_{[0,1]}\big(f^{j+k}(x)\big)\ .$$

Figure 9.8 shows the information correlation $D_{U,V}(1,1)$ for the map $x \mapsto 1 - 1.8x^2$ of the interval $[-1,1]$, as a function of k (with $\mathbb{P}$ as defined above).

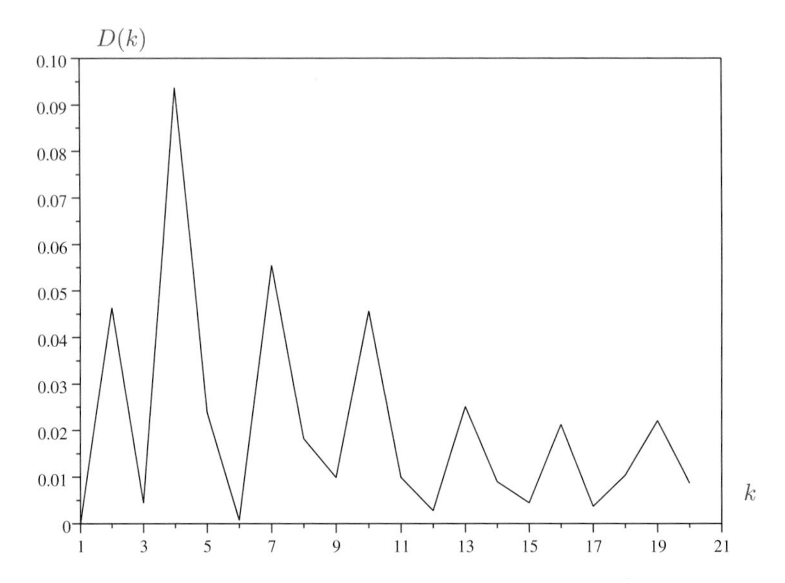

Fig. 9.8. Information correlation $D_{U,V}(1,1)$ for the map $1-1.8x^2$. The graph was obtained from 800 data points. See Example 9.2 for details

Exercise 9.3. *In the above example, show that all the quantities involved in the definition of D can be expressed in terms of $\mathbb{P}(U=1, V=1)$, $\mathbb{P}(U=1)$, and $\mathbb{P}(V=1)$. Express $\mathbb{P}(U=1, V=1)$ in terms of a correlation function.*

9.3 Lyapunov Exponents

Three main difficulties appear when one wants to measure Lyapunov exponents. Recall that from Oseledec's theorem 5.72, we need to compute, for the map f and an initial point x_0, the quantity $\left(M_n^{\mathrm{t}} M_n\right)^{1/2n}$, where

$$M_n = A_n \cdots A_1 \ ,$$

and where $A_j = \mathrm{D}_{x_{j-1}} f$, $x_j = f^j(x_0)$, and M_n^{t} denotes the transpose of M_n. We also remind the reader that in general the order matters in a product of matrices.

Remark 9.4.

i) When n is large, one needs to worry about the most efficient way to compute M_n, given that (the elements of) this matrix will in general grow exponentially fast with n (if there is a positive Lyapunov exponent).
ii) If the phase space is known but not the transformation, we would still like to estimate $A_j = \mathrm{D}_{x_{j-1}} f$.
iii) Reconstruct the information in the phase space if it is unknown (for example, in experimental contexts).

We will deal with the three questions one by one, but we first explain a simpler algorithm for the determination of the *maximal exponent* λ_0. The idea is that two typical nearby initial conditions x and y should separate exponentially fast with that rate (namely for typical x and y we should have $x - y \notin E_1(x)$, see (5.72)). Recall also that in practice we want to use the (finite) orbit of a point x, namely

$$\mathcal{O}_n(x) = \left\{x, f(x), \ldots, f^{n-1}(x)\right\} .$$

The algorithm consists in computing for a small fixed $\varepsilon > 0$ and several integers s the quantity

$$L_N(s, \varepsilon, x) = \frac{1}{N-s} \sum_{j=0}^{N-s} \log\left(\frac{1}{\left|\mathcal{U}_\varepsilon\left(f^j(x)\right)\right|} \sum_{y \in \mathcal{U}_\varepsilon\left(f^j(x)\right)} \left\|f^{j+s}(x) - f^s(y)\right\|\right) ,$$

where $\mathcal{U}_\varepsilon\left(f^j(x)\right)$ is the set of points y in the orbit $\mathcal{O}_n(x)$ at a distance less than ε of $f^j(x)$. The idea is that each term in the sum should be of order $\varepsilon e^{s\lambda_0}$. Hence, for s not too small, one expects to see a linear behavior as a function of s (because of the logarithm) with a slope giving a good approximation of λ_0. However, if s is too large, $\varepsilon e^{s\lambda_0}$ is larger than the size of the attractor, and we should observe a saturation. The summations in the formula serve to eliminate fluctuations, and it is advisable to compare the results for different values of ε.

Figure 9.9 shows the graph of $L_{2000}(s, \varepsilon, x)$ as a function of s for the Hénon map (see Example 2.31) and an initial condition chosen at random according to the Lebesgue measure.

We now discuss the general case where one would like to compute several (all) Lyapunov exponents. This is often numerically difficult, in particular if there are many exponents. Several methods have been proposed, and we will briefly explain one of them based on the *QR decomposition*. Recall that any real matrix M can be written as a product $M = QR$ with Q a real orthogonal matrix and R a real upper triangular matrix (see (Horn and Johnson 1994; Johnson, Palmer, and Sell 1987; Mera and Morán 2000)). This leads to a convenient recursive computation. Assume that we have the QR decomposition of the matrix M_n, namely $M_n = Q_n R_n$ (Q_n orthogonal, R_n upper triangular with positive diagonal elements). It follows from Theorem 5.72 that the logarithms of the diagonal elements of R_n (the eigenvalues) divided by n converge to the Lyapunov exponents.

The algorithm starts with Q_0 and R_0 the identity matrices. Then, $M_1 = A_1 Q_0$, which, by the QR decomposition can be written as $M_1 = Q_1 U_1$ with Q_1 orthogonal and U_1 upper triangular with positive diagonal. Define $R_1 = U_1 R_0$, which is again upper triangular. Now proceed by induction. Since $M_{n+1} = A_{n+1} M_n$, we have $M_{n+1} = A_{n+1} Q_n R_n$. Perform the QR decomposition of the matrix $A_{n+1} Q_n$, namely $A_{n+1} Q_n = Q_{n+1} U_n$ with Q_{n+1} orthogonal and U_n upper triangular. We then have $M_{n+1} = Q_{n+1} U_n R_n$ and we define the upper triangular matrix (with positive diagonal) $R_{n+1} = U_n R_n$. This leads to the QR decomposition of M_{n+1}. In practice it is therefore enough to keep at each step the matrix Q_n and to accumulate

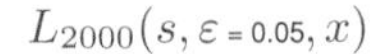

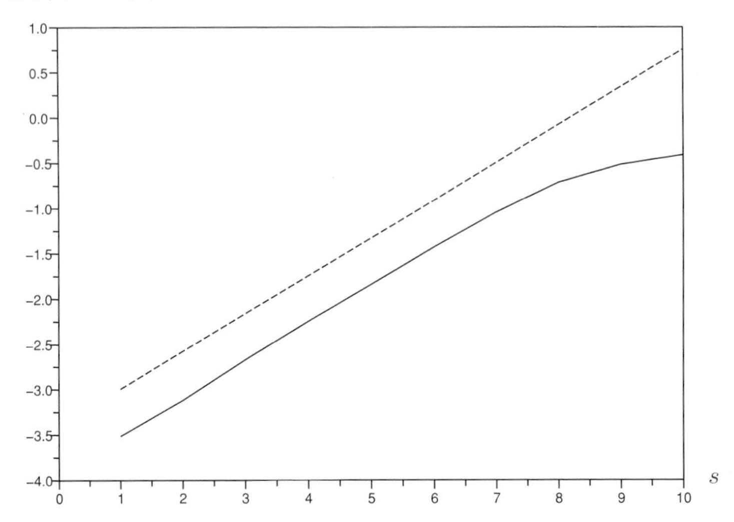

Fig. 9.9. The function $s \mapsto L_{2000}(s, \varepsilon = 0.05, x)$ for the Hénon map (full line) and a line of constant slope (dashed line)

the logarithms of the absolute values of the diagonal elements (eigenvalues) of the matrices U_n to find the Lyapunov exponents. Note that since one accumulates logarithms instead of multiplying numbers, numerical overflows (or underflows) are less likely. Since the matrices Q_n are orthogonal, they cannot overflow.

This method is illustrated in Fig. 9.10 for the Hénon map (see Example 2.31) with the parameters $a = 1.4$, $b = 0.3$. The figure shows how the result evolves when varying the number of iterations. The top curve corresponds to the positive exponent, the bottom curve to the negative one. In the case of the Hénon map, the determinant of the differential of the map is constant and equal to $-b$. Therefore, the sum of the exponents should be equal to $\log |b|$. This sum minus $\log |b|$ is the dashed curve in Fig. 9.10.

Remark 9.5. Some simple facts sometimes help for the computation or for checking the numerical accuracy.

i) If the map is invertible, the inverse map has the opposite exponents (for the same measure).
ii) If the volume element of the phase space is preserved by the dynamics (the Jacobian of the map is of modulus one, or the vector field is divergence free), the sum of the exponents is zero. This generalizes immediately to the case of constant Jacobian (respectively constant divergence) as in the Hénon map.
iii) For symplectic maps, if λ is a Lyapunov exponent, $-\lambda$ is also a Lyapunov exponent.

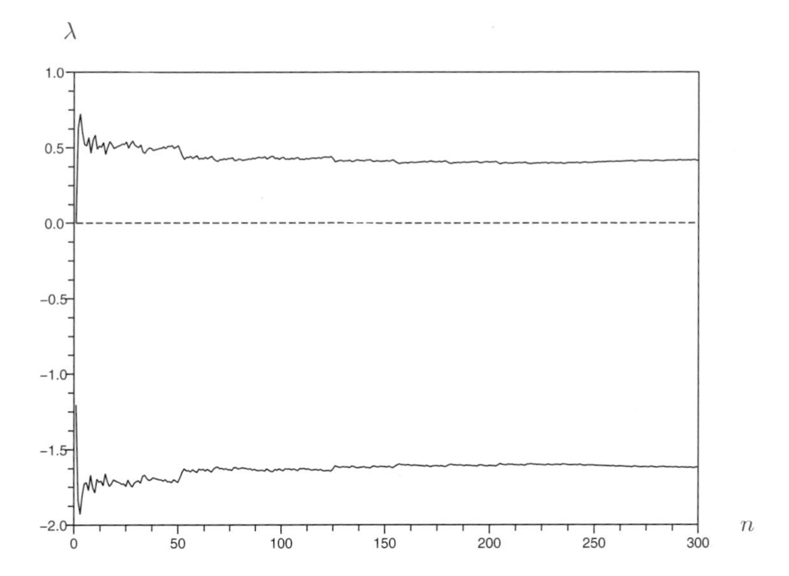

Fig. 9.10. The two Lyapunov exponents for the Hénon map (solid lines) and their sum minus $\log|b|$, as a function of the number of iterations used in the algorithm

iv) For continuous time dynamical systems one can make a discrete time sampling and use the above technique. There is also a direct QR method for continuous time systems.

Frequently, the differential $\mathrm{D}_x f$ of the map f is not known explicitly. In that case, one can try to use Taylor's formula. However, two difficulties appear. One we already mentioned is that only a single finite orbit $\mathcal{O}_n$ is available. The other one is that in practice the data are often corrupted by some noise (numerical or experimental). In other words, instead of having a sequence $x_1, \dots, x_n$ with $f(x_j) = x_{j+1}$, one observes (records) $y_1, \dots, y_n$ with

$$y_j = x_j + \varepsilon_j \ . \tag{9.8}$$

It is often assumed that the ε_j form a sequence of independent Gaussian random variables. This is a convenient assumption but not always verified.

The idea of one of the algorithms to estimate the differential is as follows, in the case without noise. Let $\mathcal{O}_n$ be an orbit of length n. Let x be a point where we want to estimate $\mathrm{D}_x f$ (x need not belong to $\mathcal{O}_n$). Let $\mathscr{U}_x$ be a small neighborhood of x. Assume that we have a point x_ℓ of $\mathcal{O}_n$ in $\mathscr{U}_x$, this means that x_ℓ is near x. By Taylor's formula we find

$$x_{\ell+1} = f(x_\ell) = f(x) + \mathrm{D}_x f(x_\ell - x) + \mathcal{O}(\|x_\ell - x\|^2) \ .$$

Neglecting the quadratic term, this can be written

$$x_{\ell+1} = c + A(x_\ell - x) , \tag{9.9}$$

where $c = f(x)$ is a vector in the phase space $\mathbb{R}^d$ and $A = \mathrm{D}_x f$ is a $d \times d$ matrix with real entries. If we know c (if x is in the orbit $\mathscr{O}_n$), we need to recover only A and therefore we need d^2 independent equations for the entries. These equations come from relations (9.9) for different ℓ if we have at least d points of the orbit $\mathscr{O}_n$ inside the neighborhood $\mathscr{U}_x$, since each relation (9.9) leads to d linear equations (we may need more points if some of these equations are not independent). If c is unknown (if x is not in the orbit $\mathscr{O}_n$), we can at the same time recover $c = f(x)$ and $A = \mathrm{D}_x f$ but then we need at least $d+1$ points of the orbit inside the neighborhood $\mathscr{U}_x$. In the presence of noise, we have using (9.8),

$$\begin{aligned} y_{\ell+1} &= \varepsilon_{\ell+1} + f(x_\ell) = \varepsilon_{\ell+1} + f(y_\ell - \varepsilon_\ell) \\ &= \varepsilon_{\ell+1} + f(x) + \mathrm{D}_x f(y_\ell - x) - \mathrm{D}_x f \varepsilon_\ell + \mathscr{O}(\|y_\ell - x\|^2) + \mathscr{O}(\|\varepsilon_\ell\|^2). \end{aligned}$$

Neglecting the quadratic terms, we have with the previous notations

$$y_{\ell+1} = c + A(y_\ell - x) + \varepsilon_{\ell+1} - \mathrm{D}_x f \varepsilon_\ell .$$

This is the classical problem of estimating an affine transformation in the presence of noise. Note that from our assumptions, $\varepsilon_{\ell+1}$ and $\mathrm{D}_x f \varepsilon_\ell$ are independent and Gaussian with zero average. To solve this estimation problem, one can, for example, use a least square algorithm. In the case where $f(x)$ is known this leads to

$$A = \mathrm{argmin}_B \sum_{y_j \in \mathscr{U}_x} \left\| y_{j+1} - f(x) - B(y_j - x) \right\|^2 .$$

Here $\mathrm{argmin}_z H(z)$ is the value of the argument z for which $H(z)$ attains its minimum. If, finally, even the map $f(x)$ is unknown, one can use the algorithm

$$(c, A) = \mathrm{argmin}_{(b,B)} \sum_{y_j \in \mathscr{U}_x} \left\| y_{j+1} - b - B(y_j - x) \right\|^2 ,$$

to estimate $f(x)$ and $\mathrm{D}_x f$ simultaneously.

Remark 9.6.

i) This and related methods have been proposed to perform *noise reduction* since we recover the deterministic quantity $f(x)$ out of the noisy data $y_1, \ldots, y_n$.
ii) This algorithm also performs a *prediction* of the future orbit of the initial condition x. Having predicted $f(x)$, one can repeat the algorithm at this new point. We refer to (Kantz and Schreiber 2004) for more on this method and references.

Remark 9.7. The choice of the size of the neighborhood $\mathscr{U}_x$ is important and relates to two competing constraints.

i) The neighborhood $\mathscr{U}_x$ should be small enough so that quadratic terms are indeed negligible. Some authors have proposed to take the quadratic corrections into account but the numerics become of course much heavier. If $\mathscr{U}_x$ is too large, the nonlinearities are not negligible and fake exponents may appear. These spurious exponents are often multiples of other ones. If one observes this phenomenon of having an exponent multiple of another one, it is advisable to diminish the size of the neighborhood and see if this relation persists.

ii) If the neighborhood $\mathscr{U}_x$ is too small, the statistics in the least squares fit may become poor. The neighborhood may even contain fewer than the d points needed to solve the problem in the deterministic case.

iii) In practice one should use neighborhoods of various sizes and compare the results.

iv) We refer to (Eckmann and Ruelle 1985a; Ruelle 1989a) for the theory and to (Eckmann, Kamphorst, Ruelle, and Ciliberto 1986; Geist, Parlitz, and Lauterborn 1990; Kantz and Schreiber 2004) and the software (Tisean 2000; Tstool 2003) for more details on the numerical implementation and discussion of the results.

9.4 Reconstruction

Contrary to the case of numerical simulations, when one deals with experimental data, the phase space is not known in a true physical experiment, for example in fluid dynamics. For partial differential equations, for example, the phase space is often infinite dimensional but one has a finite dimensional attractor. As explained above, the system is observed through the time evolution of a real (or vector valued) observable, namely one has a recording of a real (or vector valued) sequence (9.1). The initial condition x is of course unavailable and the observation may be corrupted by noise. What can be done in this apparently adverse context? A *reconstruction* method that is based on the shift has been developed by F. Takens. One first fixes an integer d for the dimension of the space where the attractor and the dynamics are to be reconstructed (we will discuss the best choice later). From the data (9.1) one constructs a sequence of d dimensional vectors $Z_0, Z_1, \ldots, Z_{N-d}$ by

$$Z_0 = \begin{pmatrix} X_{d-1} \\ \vdots \\ X_0 \end{pmatrix}, \; Z_1 = \begin{pmatrix} X_d \\ \vdots \\ X_1 \end{pmatrix}, \cdots, \; Z_{N-d} = \begin{pmatrix} X_N \\ \vdots \\ X_{N-d+1} \end{pmatrix}.$$

In other words, we define a map Φ from the phase space Ω to $\mathbb{R}^d$ by

$$\Phi(x) = \begin{pmatrix} g\big(f^{d-1}(x)\big) \\ \vdots \\ g(x) \end{pmatrix}. \tag{9.10}$$

If the observable g is regular, Φ is regular but in general not invertible. This map (semi) conjugates the time evolution f and the shift $\mathcal{S}$, namely $\mathcal{S} \circ \Phi = \Phi \circ f$. Indeed

$$Z_{n-1} = \begin{pmatrix} X_{n+d-2} \\ \vdots \\ X_{n-1} \end{pmatrix} = \begin{pmatrix} g\big(f^{n+d-2}(x)\big) \\ \vdots \\ g\big(f^{n-1}(x)\big) \end{pmatrix}$$

and therefore

$$\mathcal{S}Z_{n-1} = \mathcal{S}\Phi\big(f^{n-1}(x)\big) = \begin{pmatrix} X_{n+d-1} \\ \vdots \\ X_n \end{pmatrix} = \begin{pmatrix} g\big(f^{n+d-1}(x)\big) \\ \vdots \\ g\big(f^{n}(x)\big) \end{pmatrix}$$

$$= \begin{pmatrix} g\big(f^{n+d-2}(f(x))\big) \\ \vdots \\ g\big(f^{n-1}(f(x))\big) \end{pmatrix} = \Phi\big(f^{n-1}(f(x))\big) .$$

Remark 9.8. One cannot expect to get information outside the attractor. Indeed, after a transient time (which may be quite short), the orbit of the initial condition x is very near to the attractor.

We discuss next some of the well-known results about the reconstruction of manifolds. A simplified formulation is as follows. A natural question is: given a (regular) manifold H of dimension d, in a space of dimension n (n could be infinite), what is the minimal dimension m for which one can reconstruct accurately H in $\mathbb{R}^m$? More precisely, find a map Φ from H to $\mathbb{R}^m$ which is regular, injective and whose inverse (defined only on the image) is regular (and injective). This is called an *embedding*.

An important theorem of Whitney says that this can be done if $m \geq 2d+1$. Note that this bound does not depend on n (see, for example, (Hirsch 1994)).

Another important (and more difficult) theorem of Nash (see (Nash 1956) for details) says that a compact (bounded) Riemannian manifold (i.e., a hypersurface with a metric) can be embedded in $\mathbb{R}^m$ with its metric if $m \geq d(3d+11)/2$.

Figure 9.11 is an example of a curve, ($d = 1$) in $\mathbb{R}^3$ ($n = 3$). Many projections have a double point but some do not. For the latter there is an embedding in $\mathbb{R}^2$.

Mañé (Mañé 1981) proved a result for the projections of manifolds (which extends to fractal sets) which is analogous to Whitney's theorem. If K is a set of dimension D (not necessarily integer), almost all projections over subspaces of (integer) dimension $m > 2D + 1$ are injective (they separate the points). The difference with Whitney's theorem is that here we use projections, Whitney's theorem says there is an embedding. In the preceding example of a curve one gets $m = 4$.

Unfortunately, all the results like Whitney's theorem, Nash's theorem, or Mañé's theorem for Cantor sets are not very useful in the present context because we require a mapping of the special form (9.10). Such special mappings have sometimes some unpleasant behaviors that should be avoided. For example, if a and b are two fixed points in Ω (namely $f(a) = a$ and $f(b) = b$) such that $g(a) = g(b)$, then it follows immediately from the definition (9.10) that $\Phi(a) = \Phi(b)$; hence, Φ is not injective. Similar anomalies occur for periodic orbits and this led F. Takens to consider the generic case for g *and* f.

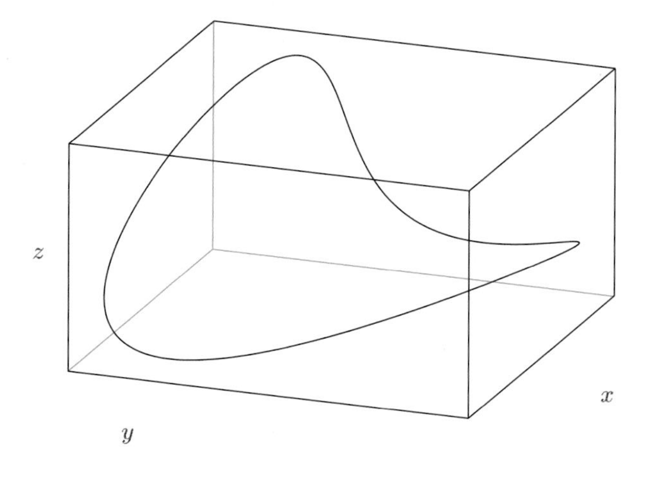

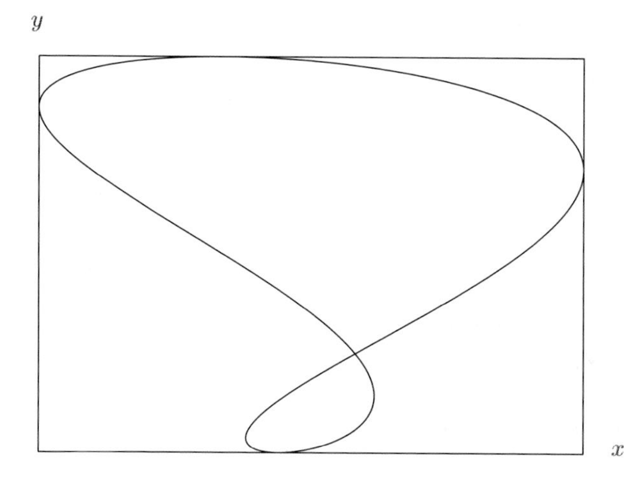

Fig. 9.11. Top: A curve in three dimensions. Bottom: Its projection onto the x-y plane has a double point

Remark 9.9. A similar reconstruction approach can be used in the case of continuous time, and similar difficulties occur. Assume that the flow φ_t has a periodic orbit C (cycle) which is stable and of (finite) period $\tau > 0$. Assume also that C is a C^0 submanifold. Define a discrete time evolution f by $f = \varphi_\tau$. Then every point of C is a fixed point of f. Since C is a closed curve and g is continuous, $g(C)$ is a segment and $\Phi(C)$ is a segment of the diagonal in $\mathbb{R}^d$. Hence, there are certainly at least two points of C which are identified through Φ even if g is not constant on C.

We now state the *Takens reconstruction theorem* (Takens 1981).

Theorem 9.10. *For a generic transformation f with an attractor $\mathscr{A}$ of (Hausdorff) dimension D, and for a generic real valued observable g, the map (9.10) is an embedding of $\mathscr{A}$ in $\mathbb{R}^d$ if $d > 2D$.*

The number d is called the *embedding dimension.*

Remark 9.11.

i) As we have seen earlier, attractors may have noninteger dimension.
ii) We have not specified the exact genericity conditions above, because there are several variants. In Takens' work generic is in the sense of Baire's second category (a countable intersection of open dense sets). The same result has been proven for other genericity conditions, for example of probabilistic nature (this has to be defined with some care because the spaces of maps and observables are in general infinite dimensional, see (Sauer, Yorke, and Casdagli 1991)). The genericity conditions avoid the problems we have mentioned above for the maps of the form (9.10). One can list a number of conditions for which the conclusions of Theorem 9.10 hold, but they are hard to check in specific examples. Theorem 9.10 states that "in general" things should work fine. This is why in experiments, the conditions of the theorem are tacitly assumed.
iii) Note that imposing symmetry properties (for example of the observable g) may break genericity.

In practice, one does not know a priori the dimension of the attractor, and therefore the dimension d of the reconstruction space to use. One tries several values for d and applies various criteria to select an adequate one. One of them is the false neighbors criterion. If d is too small, then by accident two widely separated points a and b of the phase space may be mapped to nearby points by Φ (recall the example of the projection of a 3-dimensional curve on a plane in Fig. 9.11, very often the projected curve self-intersects even if the original curve does not). However, this accidental effect should not persist for the iterates by the map f of the points a and b. At least some of the first few iterates should be reasonably separated from each other. In practice, one selects a number $\delta > 0$ much smaller than the size of the image of the attractor, and considers all the pairs of points at a distance less than δ. For each such pair, one looks at the image pair, and if for one pair one gets a large distance, then the reconstruction dimension is suspected to be too small.

Remark 9.12. Note that by continuity of Φ and f, if we consider two nearby initial conditions, their first few iterates should be close. Therefore, if there is a pair of points at distance less than δ, the images should not be far apart. If they are, we have the above problem of a too small embedding dimension.

In principle, there is no problem in using a large reconstruction dimension. However, the statistics of the different measurements will be poorer. When discussing the measurement of dimension, we will come across another criterion to select a reconstruction dimension. It is wise to use as many criteria as available (see also the Kaplan–Yorke bound in Theorem 6.27).

We refer to (Takens 1981; Sauer, Yorke, and Casdagli 1991; Kennel and Abarbanel 2002) and (Stark 2000) for proofs and references and to (Gibson, Farmer, Casdagli, and Eubank 1992; Kantz and Schreiber 2004) for more on the implementation of these techniques.

For partial differential equations, there is a reconstruction involving the (physical) space. Consider a nonlinear PDE in a bounded space domain $\mathcal{D}$. For example, the *Ginzburg–Landau equation*

$$\partial_t A = \Delta A + A - A|A|^2$$

with $A : \mathbb{R}^+ \times \mathcal{D} \to \mathbb{C}$ and some boundary condition on $\partial\mathcal{D}$. The phase space Ω is infinite dimensional. For example, the space of regular functions on $\mathcal{D}$ (with square summable Laplacian): $\int_{\mathcal{D}} \mathrm{d}x \big(|A|^2(t,x) + |\nabla A|^2(t,x)\big) < \infty$. Then a semiflow is well-defined (but going backward in time is mostly explosive). One can prove that there is a finite dimensional attracting set (in the infinite dimensional phase space). The dimension of this attracting set is proportional to the size of the domain $\mathcal{D}$ (provided there is some uniformity on the boundary condition).

Moreover, there exists an inertial manifold (see Sect. 3.3 for the definition). Recall that this is an invariant manifold (hypersurface) of finite (integer) dimension that attracts transversally all the orbits and contains the attractor. Points of the attractor can be distinguished by a (fine enough) discrete set in $\mathcal{D}$ (see (Jones and Titi 1993)). In other words, there is a number $\eta > 0$ such that if a set of points M in $\mathcal{D}$ does not leave any hole of size η, then if two functions A_1 and A_2 of the attractor are equal at each point of M, they are equal everywhere in $\mathcal{D}$. This result is known under certain conditions on the equation and for regular domains. It is still an open problem for the Navier–Stokes equation, even in dimension 2. The number η depends on the equation, and in particular of the size of the coefficients. The cardinality of M is proportional to the volume.

9.5 Measuring the Lyapunov Exponents

Here, we want to describe how the method of Sect. 9.3 is applied to an experimental context in which a reconstruction is the only way to get information. One chooses an embedding dimension d as in Sect. 9.4. One applies first the technique explained above to reconstruct the system. Note that because of the special form of the reconstruction, the tangent (reconstructed) map has a special form

$$\begin{pmatrix} 0 & 1 & 0 & \cdots & 0 \\ 0 & 0 & 1 & \cdots & 0 \\ \vdots & \vdots & \vdots & \ddots & \vdots \\ a_1 & a_2 & a_3 & \cdots & a_d \end{pmatrix}.$$

Remark 9.13.

i) We have seen that fake exponents may appear as multiples of true exponents if the nonlinearities are not negligible.

ii) In the reconstruction, fake exponents can also appear because of the numerical noise in the transverse directions of the reconstructed attractor.
iii) The reconstruction dimension is often larger than the dimension of the attractor, and the negative Lyapunov exponents are polluted by noise (fake zero exponents may appear).
iv) These fake exponents are in general unstable by time inversion (if the map is invertible one can look for the exponents of the inverse). It is also useful to explore changes in the various parameters of the algorithms.
v) It is in general difficult to determine several exponents, especially the negative ones.
vi) For the positive exponents, one can try to reduce the dimension to the number of positive exponents. For this purpose, one can use only some components of the reconstruction. For example, if one uses a reconstruction dimension $d = pq$ with p and q integers, one can consider the subspaces

$$\left(g(x), g(f^{q-1}(x)), \ldots, g(f^{pq-1}(x))\right) ,$$

when trying to measure the p largest exponents.

We refer to (Mera and Morán 2000) for conditions and proofs that show when the above methods converge to the correct result, and to (Abarbanel 1996; Kantz and Schreiber 2004; Schreiber 1999) for more results, methods, and references.

9.6 Measuring Dimensions

We now come to the question of measuring the dimension of an ergodic invariant measure μ. The reader should be aware that there are many different definitions around, and we will just concentrate on a few. In particular, if one considers a dynamical system with an invariant measure μ, there are two basic choices: The dimension of the *support* of the measure, and the dimension of measure itself. This second choice is the more natural one from the physicist's point of view, since it takes into account not only which points are visited by the orbit, but also how often a point (or its neighborhood) is visited. Therefore, we stick to the second choice here.

The idea is the following. Assume that for most points x in the support of the measure

$$\mu\big(B_r(x)\big) \approx r^{d_{\mathrm{H}}(\mu)} ,$$

then obviously

$$d_{\mathrm{H}}(\mu) \approx \frac{\log \mu\big(B_r(x)\big)}{\log r} .$$

Remark 9.14. In some cases the above assumption fails, see (Ledrappier and Misiurewicz 1985).

To increase the statistics, a better formula would be something like

$$d_{\mathrm{H}}(\mu) = \lim_{r \searrow 0} \int \frac{\log \mu\big(B_r(x)\big)}{\log r} \mathrm{d}\mu(x) . \tag{9.11}$$

Numerical simulations performed on several systems suggest that this quantity converges slowly to its limit, moreover it is not easy to compute from a numerical time series (one finite orbit). For these reasons another definition is often adopted, which was proposed in (Grassberger and Procaccia 1984). It is usually called the *Grassberger–Procaccia algorithm.*

Consider the function of $r > 0$

$$C_2(r) = \int \mu\big(B_r(x)\big) \, \mathrm{d}\mu(x) . \tag{9.12}$$

One defines the *correlation dimension* as

$$d_2(\mu) = \lim_{r \searrow 0} \frac{\log C_2(r)}{\log r}$$

if the limit exists. A first guess is that this quantity should be equal to $d_{\mathrm{H}}(\mu)$ since the integrands in (9.11) and (9.12) are similar. It turns out that $d_2(\mu) \leq d_{\mathrm{H}}(\mu)$ (because of Jensen's inequality—the logarithm of the integral is larger than or equal to the integral of the logarithm—and the negativity of $\log r$ for small r).

Exercise 9.15. *Consider the map* $x \mapsto 2x \pmod 1$ *and the ergodic invariant measure* μ_p *of Exercise 5.21. Show that for* $p \neq 1/2$, $d_2(\mu) \neq d_{\mathrm{H}}(\mu)$.

The quantity $C_2(r)$ is statistically rather stable and easy to compute from a numerical time series as we shall see now.

Assume that we are given a trajectory in the phase space $x_1, x_2, \ldots$ (an orbit). We will apply twice the ergodic theorem to obtain a (asymptotic) formula for the function $C_2(r)$.

By the ergodic theorem 5.38 we have almost surely

$$\int \mu\big(B_r(x)\big) \, \mathrm{d}\mu(x) = \lim_{n \to \infty} \frac{1}{n} \sum_{j=1}^{n} \mu\big(B_r(x_j)\big) .$$

We can now use again the ergodic theorem to determine $\mu\big(B_r(x_j)\big)$. Let θ denote the Heaviside function on $\mathbb{R}$ (the function equal to 1 on the positive real numbers and equal to 0 on the negative ones). We have

$$\mu\big(B_r(x_j)\big) = \lim_{m \to \infty} \frac{1}{m} \sum_{k=1}^{m} \theta\big(r - d(x_k, x_j)\big) ,$$

where d is the distance on the phase space. We can now put these two formulas together and obtain the following algorithm:

$$C_2(r) = \lim_{N \to \infty} \frac{1}{N^2} \sum_{p=1}^{N} \sum_{q=1}^{N} \theta\bigl(r - d(x_p, x_q)\bigr) . \tag{9.13}$$

We mention a rigorous result in the context of the above formal manipulation.

Theorem 9.16. *Formula (9.13) holds almost surely at any point where $C_2(r)$ is continuous.*

We refer to (Manning and Simon 1998) and (Serinko 1996) for a proof.

We now discuss some of the practical issues in using formula (9.13) to measure the dimension. To give a first idea, we show in Fig. 9.12 the function $C_2(r)$ for the Hénon map (see Example 2.31). For more extensive discussions, see (Grassberger, Schreiber, and Schaffrath 1991; Kantz and Schreiber 2004).

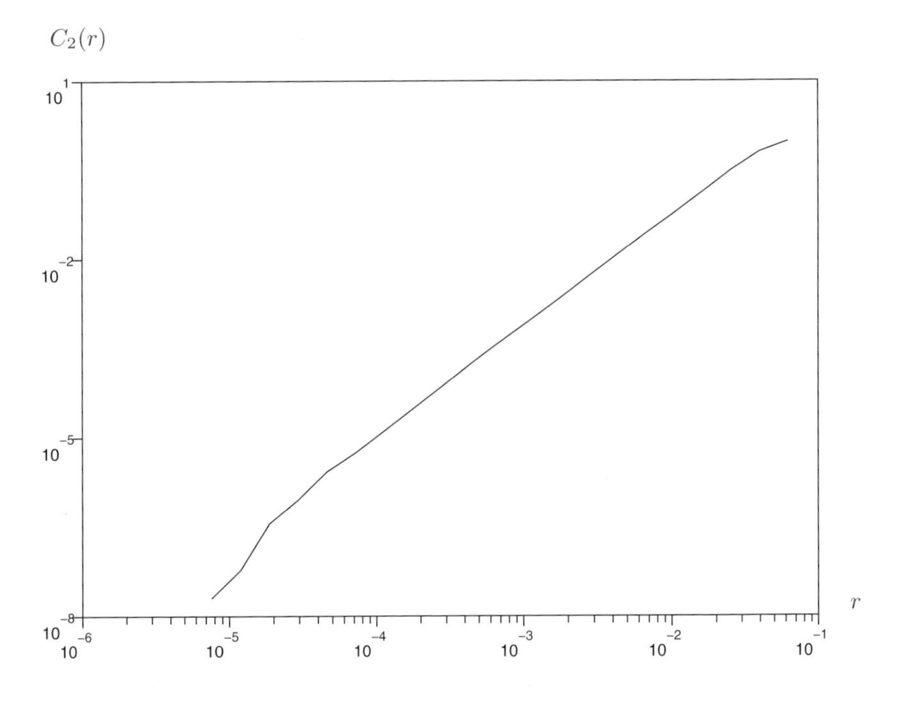

Fig. 9.12. The function $C_2(r)$ for the Hénon map. The dimension $d_2(\mu)$ is then obtained by fitting a linear curve to the log–log plot

Remark 9.17. Several practical difficulties in measuring the correlation dimension $d_2(\mu)$ can appear:

i) Strong local correlations may occur. For example, if one samples a flow $x_j = \varphi_{j\tau}(x_0)$. If τ is too small, by continuity there is a strong correlation between x_j and x_{j+1}. To avoid these problems, the summation is extended only on those p

and q with $|p - q| > K$, K large enough, of order $1/s$, where s is an estimate of the decorrelation time. This time scale can be obtained by looking at the signal (for example, the change of side in the Lorenz system), by looking at the decay of correlation function or by other means. For example, in (Fraser and Swinney 1986), the natural time is defined as the one that maximizes mutual information.

ii) Another difficulty is the presence of noise. A white noise corresponds to an attractor of infinite dimension. If the noise has amplitude ε, its influence will be noticeable only if r is of order ε or smaller. Above this value, one expects that the noise averages out. One can hope for a power law behavior of $C_2(r)$ only for a range of values of r which must be
   a) small enough for the (dynamical) fluctuations of $\log \mu\big(B_r(x)\big) / \log r$ to be small and
   b) large enough to be above the noise.

iii) One also uses other kernels than the Heaviside function, like, for example, a Gaussian.

For some "good" systems there is a theoretical bound on the fluctuations of C_2. In the general case one expects less good bounds.

For some systems where exponential or Devroye estimates are available, the standard deviation of $C_2(r, N)$ can be bounded by $N^{-1/2}r^{-1}$. If one wants to see the power law behavior above the dynamical fluctuations, one needs

$$r^{d_2(\mu)} \gg \frac{1}{N^{1/2}r} .$$

In other words,

$$r \gg \frac{1}{N^{1/(2d_2(\mu)+2)}} \qquad \text{or} \qquad N \gg \frac{1}{r^{2d_2(\mu)+2}} .$$

If one has an a priori idea on the value of $d_2(\mu)$, one can guess a value for N. For the Hénon map this gives $r \gg N^{-0.21}$, which leads to $N \sim$ 60,000 for $r > 0.1$. These bounds are often pessimistic (they are for the worst cases). But they may give an order of magnitude.

We refer to (Eckmann and Ruelle 1992; Procaccia 1988; Borovkova 1998; Borovkova, Burton, and Dehling 1999; Bosq, Guégan, and Léorat 1999; Cutler 1994; Keller 1997; Keller and Sporer 1996; Schouten, Takens, and van den Bleek 1994a;b; Takens 1996; 1998; Olofsen, Degoede, and Heijungs 1992) for more results and references.

From a practical point of view, one looks in the graph of $C_2(r)$ for a range of intermediate values of r where this function looks like a power of r. Logarithmic scales are convenient in a first approach. The method is commonly known as the *Grassberger–Procaccia algorithm* (Grassberger and Procaccia 1984).

With the exception of numerical simulations, one works with reconstructed attractors. It is useful to look at the results for different values of the reconstruction dimension. If the observed dimension (slope of the log–log plot of $C_2(r)$) is equal to the reconstruction dimension, this probably means that the dimension of the attractor

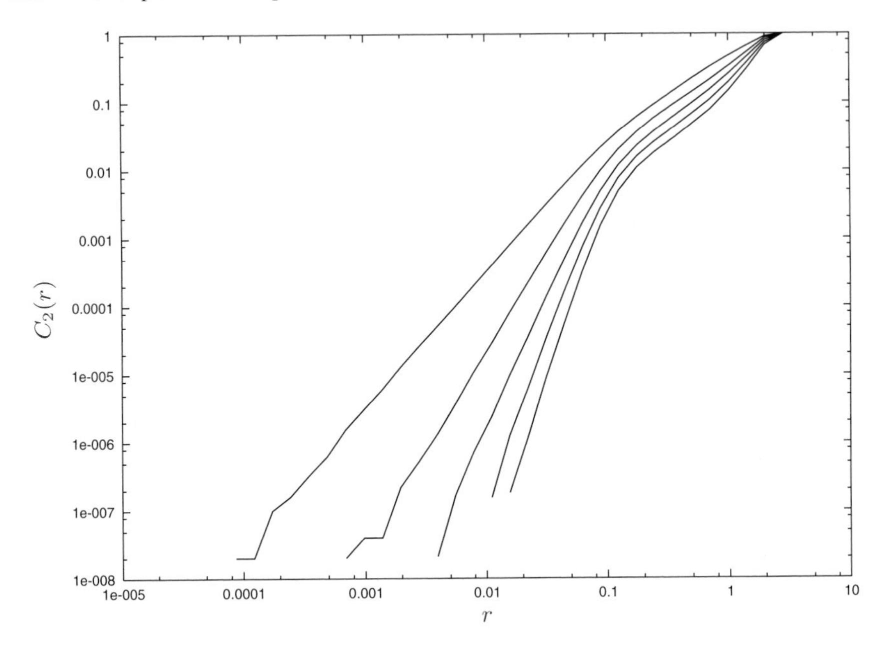

Fig. 9.13. The function $C_2(r)$ for the x component of the Hénon map and several reconstruction dimensions with noise of standard deviation 0.05

has not yet been reached. One should then increase the reconstruction dimension. It is useful to plot the measured dimension as a function of the reconstruction dimension and to look for a saturation when a reconstruction dimension is large enough to obtain an embedding. One should also verify the coherence of the different estimates of the reconstruction dimension such as false neighbors. The influence of the noise at small r is clearly visible in Fig. 9.13, which shows the function $C_2(r)$ for the x component of the Hénon map and several reconstruction dimensions with noise of standard deviation 0.05.

Estimating the slope of the function $C_2(r)$ is not an easy task, especially if the power law is valid only on a restricted range (limited by dynamical fluctuations at small—$\log r$ and by noise at large—$\log r$). Several statistical papers have dealt with this problem using linear interpolation, maximum likelihood, derivative estimates, and others. Here is one of the proposed formulas (with no claim that it is universally better than others):

$$d_2(\mu) = \frac{-N_s}{\sum_{i=1}^{N_s} \log(r_i) + N_\ell \log r_\ell} ,$$

where the power law range is $[r_\ell, r_u]$, and it contains N_s points $r_1, \ldots, r_{N_s}$, while the range $[0, r_\ell]$ contains N_ℓ points. Statistical studies of this estimator or of similar ones allow to construct confidence intervals, see (Ellner 1988; Olofsen, Degoede, and Heijungs 1992; Theiler and Lookman 1993).

9.7 Measuring Entropy

To measure the entropy, one uses the Brin–Katok formula (6.5). As in the case of the correlation dimension, to compute $\mu\big(V(x,\varepsilon,n)\big)$, we apply the ergodic theorem. For μ-almost all z,

$$\mu\big(V(x,\varepsilon,n)\big) = \lim_{N\to\infty} \frac{1}{N} \sum_{j=0}^{N-1} \chi_{V(x,\varepsilon,n)}\big(f^j(z)\big)\,,$$

$$= \lim_{N\to\infty} \frac{1}{N} \sum_{j=0}^{N-1} \prod_{\ell=0}^{n-1} \theta\big(\varepsilon - d\big(f^\ell(x), f^{\ell+j}(z)\big)\big)\,.$$

This formula is analogous to the formula used to compute the correlation dimension (9.13) and similar algorithms are used in practice. One then uses the Brin–Katok formula (6.5), namely

$$h_\mu(f) = \lim_{\delta\searrow 0} \liminf_{n\to\infty} -\frac{1}{n} \log \mu\big(V(x,f,\delta,n)\big)\,.$$

9.8 Estimating the Invariant Measure

As we have seen, one has rarely access to the phase space but only to an image of it. In particular, as was emphasized several times, we may have only a time series corresponding to the observation of the system through an observable g, namely a finite sequence of real numbers $y_1, \ldots, y_n$ satisfying

$$y_j = g\big(f^{j-1}(x)\big)$$

for an initial condition x on the phase space which in general is unknown. In particular, we have only an image μ_g of the invariant measure μ through g, namely

$$\mu_g = \mu \circ g^{-1}\,.$$

To determine μ_g, one can use the empirical distribution function

$$F_{n,g}(t) = \frac{1}{n} \sum_{j=1}^{n} \theta\big(t - y_j\big) = \frac{1}{n} \sum_{j=1}^{n} \theta\big(t - g\big(f^{j-1}(x)\big)\big)\,.$$

If the measure μ is ergodic, one concludes immediately from Birkhoff's ergodic theorem 5.38 that for μ-almost every x, this sequence of random functions converges to the *distribution function* of μ_g, $F_g(t) = \mu_g(-\infty, t]$. In the case of independent random variables, this is the Glivenco–Cantelli theorem (see, for example, (Borovkov 1998)). Again, in the case of independent random variables (under some minor hypotheses), the fluctuations are known. More precisely, a celebrated theorem of Kolmogorov establishes that the L^∞ norm of the difference between the empirical distribution function and the true one, multiplied by $n^{1/2}$ converges in law to a random variable whose distribution is known explicitly (see (Borovkov 1998)). This leads to the well-known Kolmogorov–Smirnov nonparametric test (see (Borovkov 1998)). One may wonder if similar results hold for dynamical systems. For piecewise expanding maps of the interval with an absolutely continuous invariant measure, and the observable $g(x) = x$ (which is enough in that case), this question was considered in (Collet, Martínez, and Schmitt 2002). There, it was proven that the process

$$\sqrt{n}\big(F_{n,g}(t) - F_g(t)\big)$$

converges to a *Gaussian bridge* (in general not a *Brownian bridge*, not even for the map $2x \pmod 1$)). By "bridge," one means a process that is tied down at both ends, not just at the beginning.

From this result one can derive powerful nonparametric tests. We refer to (Collet, Martínez, and Schmitt 2004b) for the details. In (Chazottes, Collet, and Schmitt 2005b) this question was also considered for some nonuniformly hyperbolic systems.

We derive here for the case of piecewise expanding maps of the interval a general estimate following from the exponential estimate (see also (Collet, Martínez, and Schmitt 2004b; Doukhan and Louhichi 1998; 2001) and (Prieur 2001b;c)).

Theorem 9.18. *Let f be a dynamical system on a phase space Ω, and let μ be an ergodic invariant measure satisfying the exponential estimate (7.10). Then there are two positive constants Γ_1 and Γ_2 such that for any real valued Lipschitz observable g for which $g^*\mu$ is absolutely continuous with bounded density and for which $\|g\|_{\mathrm{L}^\infty} \leq 1$ we have*

$$\mathbb{P}\left(\sup_t \big|F_{n,g}(t) - F_g(t)\big| > sn^{-1/4}\right) \leq \Gamma_1 \sqrt{n}\, s^{-2} \mathrm{e}^{-\Gamma_2 s^4} .$$

Remark 9.19. The assumption $\|g\|_{\mathrm{L}^\infty} \leq 1$ is only for convenience, one can rescale the result by replacing g by $g/\|g\|_{\mathrm{L}^\infty}$. Note that from this assumption, it is enough to consider $t \in [-1, 1]$ in the theorem.

Proof. The difficulty here is that the function θ entering the definition of the empirical distribution $F_{n,g}$ is not Lipschitz. To be able to apply nevertheless the exponential inequality, we will sandwich the function θ between two Lipschitz functions. For a positive number γ (small in the application), we define the function θ_γ by

$$\theta_\gamma(s) = \begin{cases} 0\,, & \text{if } s < 0 \\ s/\gamma\,, & \text{if } 0 \leq s \leq \gamma \\ 1, & \text{if } s \geq \gamma \end{cases} .$$

We now define the random function $F_{n,\gamma,g}$ by

$$F_{n,\gamma,g}(t) = \frac{1}{n} \sum_{j=1}^{n} \theta_\gamma\big(t - g \circ f^{j-1}\big) .$$

It is easy to verify that for any real s we have

$$\theta_\gamma(s) \le \theta(s) \le \theta_\gamma(s+\gamma)\,, \tag{9.14}$$

which immediately implies for any t,

$$F_{n,\gamma,g}(t) \le F_{n,g}(t) \le F_{n,\gamma,g}(t+\gamma)\,. \tag{9.15}$$

Note that $F_{n,\gamma,g}(t)$ is now a Lipschitz function in t with Lipschitz constant γ^{-1}. Therefore, if $\delta > 0$ is a (small) number to be optimally chosen later, we have

$$\left|\sup_t \Big(F_{n,\gamma,g}(t) - F_{\gamma,g}(t)\Big) - \sup_p \Big(F_{n,\gamma,g}(p\delta) - F_{\gamma,g}(p\delta)\Big)\right| \le 2\gamma^{-1}\delta\,, \tag{9.16}$$

where

$$F_{\gamma,g}(t) = \mathbb{E}\big(\theta_\gamma(t - g(\,\cdot\,))\big)\,.$$

This estimate is of course interesting only if $\gamma^{-1}\delta < 1$.

We now consider the function $K_{n,\gamma,g}$ of n variables $z_1, \ldots, z_n$ given by

$$K_{n,\gamma,g}(t) = \frac{1}{n}\sum_{j=1}^{n} \theta_\gamma\big(t - g(z_j)\big)\,.$$

This function is obviously componentwise Lipschitz, and it is easy to verify that for each variable its Lipschitz constant defined in (7.9) is bounded by

$$L_j(K_{n,\gamma,g}) \le \frac{\gamma^{-1}L_g}{n}\,,$$

where L_g is the Lipschitz constant of the observable g. Therefore, from Pisier's inequality, we get for any real number β

$$\mathbb{E}\left(\exp\Big(\beta \sup_{-\delta^{-1}(1+\gamma)\le p\le \delta^{-1}(1+\gamma)} \big(F_{n,\gamma,g}(p\delta) - F_{\gamma,g}(p\delta)\big)\Big)\right)$$

$$\le \sum_{-\delta^{-1}(1+\gamma)\le p\le \delta^{-1}(1+\gamma)} \mathbb{E}\left(\mathrm{e}^{\beta\left(F_{n,\gamma,g}(p\delta) - F_{\gamma,g}(p\delta)\right)}\right)\,,$$

where, as we have seen already several times, the expectation is taken over the measure

$$\mathrm{d}\mu(z_1)\prod_{j=1}^{n-1}\delta\big(z_{j+1} - f(z_j)\big)\,.$$

Using the exponential inequality (7.10), we get

$$\mathbb{E}\left(\exp\Big(\beta \sup_{-\delta^{-1}(1+\gamma)\le p\le \delta^{-1}(1+\gamma)} \big(F_{n,\gamma,g}(p\delta) - F_{\gamma,g}(p\delta)\big)\Big)\right)$$

$$\leq 2\delta^{-1}(1+\gamma)C_1 \mathrm{e}^{C_2\beta^2\gamma^{-2}L_g^2/n} .$$

If ε_1 is a (small) positive number, Chebyshev's inequality implies

$$\mathbb{P}\left(\sup_{-\delta^{-1}(1+\gamma)\leq p\leq\delta^{-1}(1+\gamma)} \left(F_{n,\gamma,g}(p\delta) - F_{\gamma,g}(p\delta)\right) > \varepsilon_1\right)$$

$$\leq 2\delta^{-1}(1+\gamma)C_1 \mathrm{e}^{-\beta\varepsilon_1}\mathrm{e}^{C_2\beta^2\gamma^{-2}L_g^2/n} ,$$

and this inequality holds for any real β. Using the estimate (9.16), this implies for any $\varepsilon_1 > 0$,

$$\mathbb{P}\left(\sup_t \left(F_{n,\gamma,g}(t) - F_{\gamma,g}(t)\right) > \varepsilon_1 + 2\gamma^{-1}\delta\right) \leq 2\delta^{-1}(1+\gamma)C_1 \mathrm{e}^{-\beta\varepsilon_1}\mathrm{e}^{C_2\beta^2\gamma^{-2}L_g^2/n} .$$

Define the function $\omega_g(\gamma)$ by

$$\omega_g(\gamma) = \sup_t \left(F_g(t) - F_{\gamma,g}(t)\right) .$$

We then have immediately from inequality (9.15)

$$\mathbb{P}\left(\sup_t \left(F_{n,\gamma,g}(t) - F_g(t)\right) > \varepsilon_1 + 2\gamma^{-1}\delta + \omega_g(\gamma)\right)$$

$$\leq 2\delta^{-1}C_1(1+\gamma)\mathrm{e}^{-\beta\varepsilon_1}\mathrm{e}^{C_2\beta^2\gamma^{-2}L_g^2/n} .$$

It follows immediately from (9.14) and the boundedness of the density of $g^*\mu$ that there is a constant $C_3 > 0$ such that for any $\gamma > 0$ we have

$$\omega_g(\gamma) \leq C_3\gamma ,$$

the optimal choice of δ and γ in the quantity

$$\varepsilon_1 + \gamma^{-1}\delta + \omega_g(\gamma)$$

is (up to constant factors)

$$\gamma = \delta^{1/2} = \varepsilon_1 .$$

The optimal choice of β is

$$\beta = \frac{n\varepsilon_1\gamma^2}{2C_2L_g^2} ,$$

and we get for some positive constants Γ_1 and Γ_2 independent of n and $s > 0$

$$\mathbb{P}\left(\sup_t \left|F_{n,g}(t) - F_g(t)\right| > sn^{-1/4}\right) \leq \Gamma_1\sqrt{n}\, s^{-2}\mathrm{e}^{-\Gamma_2 s^4} .$$

□

Although this estimate is asymptotically weaker than the convergence in law to a Gaussian bridge as in the Kolmogorov–Smirnov test, it allows however to obtain rather strong tests.

Results on kernel density estimates in the case of absolutely continuous invariant measures can be found in (Chazottes, Collet, and Schmitt 2005a; Bosq and Guégan 1995) and (Bosq 1995; Prieur 2001a).

References

Abadi, M. (2001). Exponential approximation for hitting times in mixing processes. *Math. Phys. Electron. J.*, 7:Paper 2, 19 pp. (electronic).

Abarbanel, H. D. I. (1996). *Analysis of observed chaotic data*. Institute for Nonlinear Science. Springer-Verlag, New York.

Afraimovich, V. and Hsu, S.-B. (2003). *Lectures on chaotic dynamical systems*, volume 28 of *AMS/IP Studies in Advanced Mathematics*. American Mathematical Society, Providence, RI.

Alligood, K. T., Sauer, T. D., and Yorke, J. A. (1997). *Chaos*. Textbooks in Mathematical Sciences. Springer-Verlag, New York. An introduction to dynamical systems.

Alves, J. F., Bonatti, C., and Viana, M. (2000). SRB measures for partially hyperbolic systems whose central direction is mostly expanding. *Invent. Math.*, 140:351–398.

Anosov, D. V. (1969). *Geodesic flows on closed Riemann manifolds with negative curvature.* Proceedings of the Steklov Institute of Mathematics, No. 90 (1967). Translated from the Russian by S. Feder. American Mathematical Society, Providence, RI.

Arnold, V. I. (1978). *Mathematical methods of classical mechanics*. Springer-Verlag, New York. Translated from the Russian by K. Vogtmann and A. Weinstein, Graduate Texts in Mathematics, 60.

Arnold, V. I. and Avez, A. (1967). *Problèmes ergodiques de la mécanique classique*. Monographies Internationales de Mathématiques Modernes, No. 9. Gauthier-Villars, Éditeur, Paris.

Avila, A. and Moreira, C. G. (2005). Statistical properties of unimodal maps: the quadratic family. *Ann. Math. (2)*, 161:831–881.

Baladi, V. (2000). *Positive transfer operators and decay of correlations*, volume 16 of *Advanced Series in Nonlinear Dynamics*. World Scientific Publishing Co. Inc., River Edge, NJ.

Baladi, V., Eckmann, J.-P., and Ruelle, D. (1989). Resonances for intermittent systems. *Nonlinearity*, 2:119–135.

Barbour, A. D., Gerrard, R. M., and Reinert, G. (2000). Iterates of expanding maps. *Probab. Theory Related Fields*, 116:151–180.

Bardet, J.-M., Doukhan, P., and Leon, J. (2005). A uniform central limit theorem for the periodogram and its applications to whittle parametric estimation for weakly dependent time series. URL `http://www.crest.fr/pageperso/doukhan/publications.htm`.

Barreira, L. (1997). Pesin theory. In: Hazewinkel, M., ed., *Encyclopaedia of mathematics. Supplement. Vol. I*, pages viii+587. Kluwer Academic Publishers, Dordrecht.

Barreira, L. and Pesin, Y. B. (2001). Lectures on Lyapunov exponents and smooth ergodic theory. In: *Smooth ergodic theory and its applications (Seattle, WA, 1999)*, volume 69 of *Proc. Sympos. Pure Math.*, pages 3–106. Amer. Math. Soc., Providence, RI. Appendix A by M. Brin and Appendix B by D. Dolgopyat, H. Hu and Pesin.

Barreira, L. and Pesin, Y. B. (2002). *Lyapunov exponents and smooth ergodic theory*, volume 23 of *University Lecture Series*. American Mathematical Society, Providence, RI.

Barreira, L. and Schmeling, J. (2000). Sets of "non-typical" points have full topological entropy and full Hausdorff dimension. *Israel J. Math.*, 116:29–70.

Bartlett, J. H. (1976). Stability of area-preserving mappings. In: *Transformations ponctuelles et leurs applications (Colloq. Internat. CNRS, No. 229, Toulouse, 1973)*, pages 155–175, 221–223. Éditions Centre Nat. Recherche Sci., Paris. Avec discussion.

Belickiĭ, G. R. (1978). Equivalence and normal forms of germs of smooth mappings. *Uspekhi Mat. Nauk (English translation: Russian Math. Surveys 33 (1978), 107–177)*, 33:95–155, 263.

Benedicks, M. and Carleson, L. (1991). The dynamics of the Hénon map. *Ann. Math. (2)*, 133:73–169.

Benedicks, M. and Young, L.-S. (1993). Sinai–Bowen–Ruelle measures for certain Hénon maps. *Invent. Math.*, 112:541–576.

Billingsley, P. (1999). *Convergence of probability measures*. Wiley Series in Probability and Statistics: Probability and Statistics. John Wiley & Sons Inc., New York, second edition. A Wiley-Interscience Publication.

Birkhoff, G. D. (1931). Proof of the ergodic theorem. *Proc. Nat. Acad. Sci. USA*, 17:656–660.

Bishop, E. (1967). *Foundations of constructive analysis*. McGraw-Hill Book Co., New York.

Blank, M. and Keller, G. (1997). Stochastic stability versus localization in one-dimensional chaotic dynamical systems. *Nonlinearity*, 10:81–107.

Bonatti, C. and Viana, M. (2000). SRB measures for partially hyperbolic systems whose central direction is mostly contracting. *Israel J. Math.*, 115:157–193.

Bonic, R. and Frampton, J. (1965). Differentiable functions on certain Banach spaces. *Bull. Amer. Math. Soc.*, 71:393–395.

Borovkov, A. A. (1998). *Mathematical statistics*. Gordon and Breach Science Publishers, Amsterdam. Translated from the Russian by A. Moullagaliev and revised by the author.

Borovkov, A. A. (2004). Kolmogorov and boundary value problems of probability theory. *Uspekhi Mat. Nauk*, 59:91–102.

Borovkova, S. (1998). *Estimation and prediction for nonlinear time series*. Ph.D. thesis, Groningen.

Borovkova, S., Burton, R., and Dehling, H. (1999). Consistency of the Takens estimator for the correlation dimension. *Ann. Appl. Probab.*, 9:376–390.

Bosq, D. (1995). Optimal asymptotic quadratic error of density estimators for strong mixing or chaotic data. *Statist. Probab. Lett.*, 22:339–347.

Bosq, D. and Guégan, D. (1995). Nonparametric estimation of the chaotic function and the invariant measure of a dynamical system. *Statist. Probab. Lett.*, 25:201–212.

Bosq, D., Guégan, D., and Léorat, G. (1999). Statistical estimation of the embedding dimension of a dynamical system. *Internat. J. Bifur. Chaos Appl. Sci. Engrg.*, 9:645–656.

Bowen, R. (1971). Entropy for group endomorphisms and homogeneous spaces. *Trans. Amer. Math. Soc.*, 153:401–414.

Bowen, R. (1975). *Equilibrium states and the ergodic theory of Anosov diffeomorphisms*. Springer-Verlag, Berlin. Lecture Notes in Mathematics, Vol. 470.

Bradley, R. C. (2005). Basic properties of strong mixing conditions. A survey and some open questions. *Probab. Surv.*, 2:107–144 (electronic). Update of, and a supplement to, the 1986 original.

Bressaud, X. and Liverani, C. (2002). Anosov diffeomorphisms and coupling. *Ergodic Theory Dynam. Systems*, 22:129–152.

Bressaud, X., Fernández, R., and Galves, A. (1999a). Speed of $\overline{d}$-convergence for Markov approximations of chains with complete connections. A coupling approach. *Stochastic Process. Appl.*, 83:127–138.

Bressaud, X., Fernández, R., and Galves, A. (1999b). Decay of correlations for non-Hölderian dynamics. A coupling approach. *Electron. J. Probab.*, 4:no. 3, 19 pp. (electronic).

Brin, M. and Katok, A. (1983). On local entropy. In: *Geometric dynamics (Rio de Janeiro, 1981)*, volume 1007 of *Lecture Notes in Math.*, pages 30–38. Springer, Berlin.

Brockwell, P. J. and Davis, R. A. (1996). *Introduction to time series and forecasting*. Springer Texts in Statistics. Springer-Verlag, New York. With 1 IBM-PC floppy disk (3.5 inch; HD).

Broise, A. (1996). Transformations dilatantes de l'intervalle et théorèmes limites. *Astérisque*, (238):1–109. Études spectrales d'opérateurs de transfert et applications.

Brolin, H. (1965). Invariant sets under iteration of rational functions. *Ark. Mat.*, 6:103–144 (1965).

Buzzi, J. (1997). Specification on the interval. *Trans. Amer. Math. Soc.*, 349:2737–2754.

Chazottes, J.-R. and Collet, P. (2005). Almost-sure central limit theorems and the Erdős-Rényi law for expanding maps of the interval. *Ergodic Theory Dynam. Systems*, 25:419–441.

Chazottes, J.-R., Collet, P., and Schmitt, B. (2005a). Devroye inequality for a class of non-uniformly hyperbolic dynamical systems. *Nonlinearity*, 18:2323–2340.

Chazottes, J.-R., Collet, P., and Schmitt, B. (2005b). Statistical consequences of the Devroye inequality for processes. Applications to a class of non-uniformly hyperbolic dynamical systems. *Nonlinearity*, 18:2341–2364.

Chernov, N., Markarian, R., and Troubetzkoy, S. (1998). Conditionally invariant measures for Anosov maps with small holes. *Ergodic Theory Dynam. Systems*, 18:1049–1073.

Coelho, Z. and Collet, P. (1994). Asymptotic limit law for the close approach of two trajectories in expanding maps of the circle. *Probab. Theory Related Fields*, 99:237–250.

Coelho, Z. and Parry, W. (1990). Central limit asymptotics for shifts of finite type. *Israel J. Math.*, 69:235–249.

Collet, P. (1996). Some ergodic properties of maps of the interval. In: *Dynamical systems (Temuco, 1991/1992)*, volume 52 of *Travaux en Cours*, pages 55–91. Hermann, Paris.

Collet, P. (2001). Statistics of closest return for some non-uniformly hyperbolic systems. *Ergodic Theory Dynam. Systems*, 21:401–420.

Collet, P. and Eckmann, J.-P. (1980). *Iterated maps on the interval as dynamical systems*, volume 1 of *Progress in Physics*. Birkhäuser Boston, Mass.

Collet, P. and Eckmann, J.-P. (1983). Positive Liapunov exponents and absolute continuity for maps of the interval. *Ergodic Theory Dynam. Systems*, 3:13–46.

Collet, P. and Eckmann, J.-P. (1997). Oscillations of observables in one-dimensional lattice systems. *Math. Phys. Electron. J.*, 3:Paper 3, 19 pp. (electronic).

Collet, P. and Eckmann, J.-P. (2004). Liapunov multipliers and decay of correlations in dynamical systems. *J. Statist. Phys.*, 115:217–254.

Collet, P. and Galves, A. (1993). Statistics of close visits to the indifferent fixed point of an interval map. *J. Statist. Phys.*, 72:459–478.

Collet, P. and Galves, A. (1995). Asymptotic distribution of entrance times for expanding maps of the interval. In: *Dynamical systems and applications*, volume 4 of *World Sci. Ser. Appl. Anal.*, pages 139–152. World Sci. Publishing, River Edge, NJ.

Collet, P. and Isola, S. (1991). On the essential spectrum of the transfer operator for expanding Markov maps. *Comm. Math. Phys.*, 139:551–557.

Collet, P. and Martínez, S. (1999). Diffusion coefficient in transient chaos. *Nonlinearity*, 12:445–450.

Collet, P., Lebowitz, J. L., and Porzio, A. (1987). The dimension spectrum of some dynamical systems. In: *Proceedings of the symposium on statistical mechanics of phase transitions—mathematical and physical aspects (Trebon, 1986)*, volume 47, pages 609–644.

Collet, P., Galves, A., and Schmitt, B. (1992). Unpredictability of the occurrence time of a long laminar period in a model of temporal intermittency. *Ann. Inst. H. Poincaré Phys. Théor.*, 57:319–331.

Collet, P., Martínez, S., and Schmitt, B. (1994). The Yorke–Pianigiani measure and the asymptotic law on the limit Cantor set of expanding systems. *Nonlinearity*, 7:1437–1443.

Collet, P., Martínez, S., and Schmitt, B. (1997). The Pianigiani–Yorke measure for topological Markov chains. *Israel J. Math.*, 97:61–70.

Collet, P., Galves, A., and Schmitt, B. (1999a). Repetition times for Gibbsian sources. *Nonlinearity*, 12:1225–1237.

Collet, P., Martínez, S., and Schmitt, B. (1999b). On the enhancement of diffusion by chaos, escape rates and stochastic instability. *Trans. Amer. Math. Soc.*, 351:2875–2897.

Collet, P., Martínez, S., and Maume-Deschamps, V. (2000). On the existence of conditionally invariant probability measures in dynamical systems. *Nonlinearity*, 13:1263–1274.

Collet, P., Martínez, S., and Schmitt, B. (2002). Exponential inequalities for dynamical measures of expanding maps of the interval. *Probab. Theory Related Fields*, 123:301–322.

Collet, P., Martínez, S., and Maume-Deschamps, V. (2004a). Corrigendum: "On the existence of conditionally invariant probability measures in dynamical systems" [Nonlinearity **13** (2000), no. 4, 1263–1274; mr1767958]. *Nonlinearity*, 17:1985–1987.

Collet, P., Martínez, S., and Schmitt, B. (2004b). Asymptotic distribution of tests for expanding maps of the interval. *Ergodic Theory Dynam. Systems*, 24:707–722.

Constantin, P. and Foias, C. (1988). *Navier-Stokes equations*. Chicago Lectures in Mathematics. University of Chicago Press, Chicago, IL.

Constantin, P., Foias, C., and Temam, R. (1988). On the dimension of the attractors in two-dimensional turbulence. *Phys. D*, 30:284–296.

Cornfeld, I. P., Fomin, S. V., and Sinaĭ, Y. G. (1982). *Ergodic theory*, volume 245 of *Grundlehren der Mathematischen Wissenschaften [Fundamental Principles of Mathematical Sciences]*. Springer-Verlag, New York. Translated from the Russian by A. B. Sosinskiĭ.

Crandall, M. G. and Rabinowitz, P. H. (1971). Bifurcation from simple eigenvalues. *J. Funct. Anal.*, 8:321–340.

Cutler, C. D. (1994). A theory of correlation dimension for stationary time series. *Philos. Trans. Roy. Soc. Lond. Ser. A*, 348:343–355.

de la Llave, R. (2001). A tutorial on KAM theory. In: *Smooth ergodic theory and its applications (Seattle, WA, 1999)*, volume 69 of *Proc. Sympos. Pure Math.*, pages 175–292. Amer. Math. Soc., Providence, RI.

de Melo, W. and van Strien, S. (1993). *One-dimensional dynamics*, volume 25 of *Ergebnisse der Mathematik und ihrer Grenzgebiete (3) [Results in Mathematics and Related Areas (3)]*. Springer-Verlag, Berlin.

Dedecker, J. and Prieur, C. (2005). New dependence coefficients. examples and applications to statistics. *Probab. Theory and Relat. Fields*, 132:203–236.

Dembo, A. and Zeitouni, O. (1997). Moderate deviations for iterates of expanding maps. In: *Statistics and control of stochastic processes (Moscow, 1995/1996)*, pages 1–11. World Sci. Publishing, River Edge, NJ.

Denker, M. (1989). The central limit theorem for dynamical systems. In: *Dynamical systems and ergodic theory (Warsaw, 1986)*, volume 23 of *Banach Center Publ.*, pages 33–62. PWN, Warsaw.

Denker, M., Grillenberger, C., and Sigmund, K. (1976). *Ergodic theory on compact spaces.* Springer-Verlag, Berlin. Lecture Notes in Mathematics, Vol. 527.

Denker, M., Gordin, M., and Sharova, A. (2004). A Poisson limit theorem for toral automorphisms. *Ill. J. Math.*, 48:1–20.

Devaney, R. L. (2003). *An introduction to chaotic dynamical systems.* Studies in Nonlinearity. Westview Press, Boulder, CO. Reprint of the second (1989) edition.

Dieudonné, J. (1968). *Éléments d'analyse. Tome I: Fondements de l'analyse moderne.* Traduit de l'anglais par D. Huet. Avant-propos de G. Julia. Nouvelle édition revue et corrigée. Cahiers Scientifiques, Fasc. XXVIII. Gauthier-Villars, Éditeur, Paris.

Dinaburg, E. I. (1971). A connection between various entropy characterizations of dynamical systems. *Izv. Akad. Nauk SSSR Ser. Mat.*, 35:324–366.

Doukhan, P. (1994). *Mixing*, volume 85 of *Lecture Notes in Statistics.* Springer-Verlag, New York. Properties and examples.

Doukhan, P. and Louhichi, S. (1998). Estimation de la densité d'une suite faiblement dépendante. *C. R. Acad. Sci. Paris Sér. I Math.*, 327:989–992.

Doukhan, P. and Louhichi, S. (1999). A new weak dependence condition and applications to moment inequalities. *Stochastic Process. Appl.*, 84:313–342.

Doukhan, P. and Louhichi, S. (2001). Functional estimation of a density under a new weak dependence condition. *Scand. J. Statist.*, 28:325–341.

Downarowicz, T. and Weiss, B. (2004). Entropy theorems along times when x visits a set. *Ill. J. Math.*, 48:59–69.

Durand, F. and Maass, A. (2001). Limit laws of entrance times for low-complexity Cantor minimal systems. *Nonlinearity*, 14:683–700.

Eckmann, J.-P. (1981). Roads to turbulence in dissipative dynamical systems. *Rev. Mod. Phys.*, 53:643–654.

Eckmann, J.-P. (1983). Savez-vous résoudre $z^3 = 1$? *La Recherche*, 14:260–262.

Eckmann, J.-P. and Procaccia, I. (1986). Fluctuations of dynamical scaling indices in nonlinear systems. *Phys. Rev.*, A34:659–661.

Eckmann, J.-P. and Ruelle, D. (1985a). Ergodic theory of chaos and strange attractors. *Rev. Mod. Phys.*, 57:617–656.

Eckmann, J.-P. and Ruelle, D. (1985b). Two-dimensional Poiseuille flow. *Physics Scripta*, T9:153–154.

Eckmann, J.-P. and Ruelle, D. (1992). Fundamental limitations for estimating dimensions and Lyapunov exponents in dynamical systems. *Phys. D*, 56:185–187.

Eckmann, J.-P., Kamphorst, S. O., Ruelle, D., and Ciliberto, S. (1986). Liapunov exponents from time series. *Phys. Rev. A (3)*, 34:4971–4979.

Ellis, R. S. (1985). *Entropy, large deviations, and statistical mechanics*, volume 271 of *Grundlehren der Mathematischen Wissenschaften [Fundamental Principles of Mathematical Sciences].* Springer-Verlag, New York.

Ellner, S. (1988). Estimating attractor dimensions from limited data: A new method, with error estimates. *Physics Letters A*, 133:128–133.

Falconer, K. J. (1986). *The geometry of fractal sets*, volume 85 of *Cambridge Tracts in Mathematics*. Cambridge University Press, Cambridge.

Feller, W. (1957, 1966). *An introduction to probability theory and its applications. Vols. I and II*. John Wiley and Sons, Inc., New York. 2nd ed.

Fogg, N. P. (2002). *Substitutions in dynamics, arithmetics and combinatorics*, volume 1794 of *Lecture Notes in Mathematics*. Springer-Verlag, Berlin. Edited by V. Berthé, S. Ferenczi, C. Mauduit and A. Siegel.

Foiaş, C. and Temam, R. (1979). Some analytic and geometric properties of the solutions of the evolution Navier-Stokes equations. *J. Math. Pures Appl. (9)*, 58:339–368.

Fraser, A. M. and Swinney, H. L. (1986). Independent coordinates for strange attractors from mutual information. *Phys. Rev.*, A33:1134–1140.

Galambos, J. (1987). *The asymptotic theory of extreme order statistics*. Robert E. Krieger Publishing Co. Inc., Melbourne, FL, second edition.

Gallavotti, G., Bonetto, F., and Gentile, G. (2004). *Aspects of Ergodic, Qualitative and Statistical Theory of Motion*. Springer-Verlag, New York.

Geist, K., Parlitz, U., and Lauterborn, W. (1990). Comparison of different methods for computing Lyapunov exponents. *Progr. Theoret. Phys.*, 83:875–893.

Gibson, J. F., Farmer, J. D., Casdagli, M., and Eubank, S. (1992). An analytic approach to practical state space reconstruction. *Phys. D*, 57:1–30.

Gikhman, I. I. and Skorokhod, A. V. (1996). *Introduction to the theory of random processes*. Dover Publications Inc., Mineola, NY. Translated from the 1965 Russian original, Reprint of the 1969 English translation, With a preface by Warren M. Hirsch.

Golubitsky, M. and Stewart, I. (2002). *The symmetry perspective*, volume 200 of *Progress in Mathematics*. Birkhäuser Verlag, Basel. From equilibrium to chaos in phase space and physical space.

Góra, P. and Schmitt, B. (1989). Un exemple de transformation dilatante et C^1 par morceaux de l'intervalle, sans probabilité absolument continue invariante. *Ergodic Theory Dynam. Systems*, 9:101–113.

Gordin, M. (1993). Homoclinic approach to the central limit theorem for dynamical systems. In: *Doeblin and modern probability (Blaubeuren, 1991)*, volume 149 of *Contemp. Math.*, pages 149–162. Amer. Math. Soc., Providence, RI.

Gordin, M. I. (1969). The central limit theorem for stationary processes. *Dokl. Akad. Nauk SSSR*, 188:739–741.

Gouëzel, S. (2005). Berry-Esseen theorem and local limit theorem for non uniformly expanding maps. *Ann. Inst. H. Poincaré Probab. Statist.*, 41:997–1024.

Grassberger, P. and Procaccia, I. (1984). Dimensions and entropies of strange attractors from a fluctuating dynamics approach. *Phys. D*, 13:34–54.

Grassberger, P., Schreiber, T., and Schaffrath, C. (1991). Nonlinear time sequence analysis. *Internat. J. Bifur. Chaos Appl. Sci. Engrg.*, 1:521–547.

Guckenheimer, J. and Holmes, P. (1990). *Nonlinear oscillations, dynamical systems, and bifurcations of vector fields*, volume 42 of *Applied Mathematical Sciences*. Springer-Verlag, New York. Revised and corrected reprint of the 1983 original.

Hardy, G. H. (1917). Weierstrass's non-differentiable function. *Trans. Amer. Math. Soc.*, 17:301–325.

Hénon, M. (1976). A two-dimensional mapping with a strange attractor. *Comm. Math. Phys.*, 50:69–77.

Hénon, M. and Heiles, C. (1964). The applicability of the third integral of motion: Some numerical experiments. *Astronom. J.*, 69:73–79.

Hirata, M. (1993). Poisson law for Axiom A diffeomorphisms. *Ergodic Theory Dynam. Systems*, 13:533–556.

Hirata, M., Saussol, B., and Vaienti, S. (1999). Statistics of return times: a general framework and new applications. *Comm. Math. Phys.*, 206:33–55.

Hirsch, M. W. (1994). *Differential topology*, volume 33 of *Graduate Texts in Mathematics*. Springer-Verlag, New York. Corrected reprint of the 1976 original.

Hirsch, M. W., Pugh, C. C., and Shub, M. (1977). *Invariant manifolds*. Springer-Verlag, Berlin. Lecture Notes in Mathematics, Vol. 583.

Hofbauer, F. and Keller, G. (1982). Ergodic properties of invariant measures for piecewise monotonic transformations. *Math. Z.*, 180:119–140.

Horn, R. A. and Johnson, C. R. (1994). *Topics in matrix analysis*. Cambridge University Press, Cambridge. Corrected reprint of the 1991 original.

Ivanov, V. V. (1996a). Geometric properties of monotone functions and the probabilities of random oscillations. *Sibirsk. Mat. Zh.*, 37:117–150, ii.

Ivanov, V. V. (1996b). Oscillations of averages in the ergodic theorem. *Dokl. Akad. Nauk*, 347:736–738.

Jaynes, E. T. (1982). On the rationale of maximum-entropy methods. *Proc. IEEE*, 70:939–952.

Johnson, R. A., Palmer, K. J., and Sell, G. R. (1987). Ergodic properties of linear dynamical systems. *SIAM J. Math. Anal.*, 18:1–33.

Jones, D. A. and Titi, E. S. (1993). Upper bounds on the number of determining modes, nodes, and volume elements for the Navier–Stokes equations. *Ind. Univ. Math. J.*, 42:875–887.

Kachurovskiĭ, A. G. (1996). Rates of convergence in ergodic theorems. *Uspekhi Mat. Nauk*, 51:73–124.

Kahane, J.-P. (1985). *Some random series of functions*, volume 5 of *Cambridge Studies in Advanced Mathematics*. Cambridge University Press, Cambridge, second edition.

Kalikov, S. (2000). *Outline of Ergodic Theory*. URL `http://www.math.umd.edu/~djr/kalikow.html`.

Kalikow, S. and Weiss, B. (1999). Fluctuations of ergodic averages. In: *Proceedings of the Conference on Probability, Ergodic Theory, and Analysis (Evanston, IL, 1997)*, volume 43, pages 480–488.

Kamae, T. (1982). A simple proof of the ergodic theorem using nonstandard analysis. *Israel J. Math.*, 42:284–290.

Kantz, H. and Schreiber, T. (2004). *Nonlinear time series analysis*. Cambridge University Press, Cambridge, second edition.

Kaplan, J. L. and Yorke, J. A. (1979). Chaotic behavior of multidimensional difference equations. In: *Functional differential equations and approximation of fixed points (Proc. Summer School and Conf., Univ. Bonn, Bonn, 1978)*, volume 730 of *Lecture Notes in Math.*, pages 204–227. Springer, Berlin.

Kato, T. (1984). *Perturbation Theory for Linear Operators*. Springer-Verlag, Berlin. Second corrected printing of the second edition.

Katok, A. (1980). Lyapunov exponents, entropy and periodic orbits for diffeomorphisms. *Inst. Hautes Études Sci. Publ. Math.*, (51):137–173.

Katok, A. and Hasselblatt, B. (1995). *Introduction to the modern theory of dynamical systems*, volume 54 of *Encyclopedia of mathematics and its applications*. Cambridge University Press, Cambridge. With a supplementary chapter by Katok and Leonardo Mendoza.

Katznelson, Y. and Weiss, B. (1982). A simple proof of some ergodic theorems. *Israel J. Math.*, 42:291–296.

Kay, S. and Marple, S. (1981). Spectrum analysis—a modern perspective. *Proceedings of the IEEE*, 69:1380–1419.

Keane, M. S. (1991). Ergodic theory and subshifts of finite type. In: *Ergodic theory, symbolic dynamics, and hyperbolic spaces (Trieste, 1989)*, Oxford Sci. Publ., pages 35–70. Oxford Univ. Press, New York.

Keane, M. S. and Petersen, K. (2006). Easy and nearly simultaneous proofs of the Ergodic Theorem and Maximal Ergodic Theorem. In: Denteneer, D., den Hollander, F., and Verbitskiy, E., eds., *Dynamics & Stochastics: Festschrift in Honor of M.S. Keane*, volume 48, pages 248–251. Institute for Mathematical Statistics, Lecture Notes–Monograph Series.

Keller, G. (1997). A new estimator for information dimension with standard errors and confidence intervals. *Stochastic Process. Appl.*, 71:187–206.

Keller, G. and Sporer, R. (1996). Remarks on the linear regression approach to dimension estimation. In: *Stochastic and spatial structures of dynamical systems (Amsterdam, 1995)*, Konink. Nederl. Akad. Wetensch. Verh. Afd. Natuurk. Eerste Reeks, 45, pages 17–27. North-Holland, Amsterdam.

Kennel, M. B. and Abarbanel, H. D. I. (2002). False neighbors and false strands: A reliable minimum embedding dimension algorithm. *Phys. Rev. E*, 66:026209(18).

Kifer, Y. (1988). *Random perturbations of dynamical systems*, volume 16 of *Progress in Probability and Statistics*. Birkhäuser Boston Inc., Boston, MA.

Kifer, Y. (1990). Large deviations in dynamical systems and stochastic processes. *Trans. Amer. Math. Soc.*, 321:505–524.

Kifer, Y. (1997). Computations in dynamical systems via random perturbations. *Discrete Contin. Dynam. Syst.*, 3:457–476.

Knuth, D. E. (1981). *The art of computer programming. Vol. 2*. Addison-Wesley Publishing Co., Reading, Mass., second edition. Seminumerical algorithms, Addison-Wesley Series in Computer Science and Information Processing.

Kontoyiannis, I. (1998). Asymptotic recurrence and waiting times for stationary processes. *J. Theoret. Probab.*, 11:795–811.

Krengel, U. (1985). *Ergodic theorems*, volume 6 of *de Gruyter Studies in Mathematics*. Walter de Gruyter & Co., Berlin. With a supplement by Antoine Brunel.

Lacroix, Y. (2002). Possible limit laws for entrance times of an ergodic aperiodic dynamical system. *Israel J. Math.*, 132:253–263.

Lalley, S. P. (1999). Beneath the noise, chaos. *Ann. Statist.*, 27:461–479.

Lalley, S. P. and Nobel, A. B. (2000). Denoising detereministic time series. URL `http://lanl.arxiv.org/abs/nlin.CD/0604052`.

Lalley, S. P. and Nobel, A. B. (2003). Indistinguishability of absolutely continuous and singular distributions. *Statist. Probab. Lett.*, 62:145–154.

Lanford, O. (1973). Entropy and equilibrium states in classical statistical mechanics. In: *Statistical mechanics and mathematical problems*, volume 20 of *Lecture Notes in Physics*, pages 1–113. Springer-Verlag, Berlin.

Lasota, A. and Yorke, J. (1973). Existence of invariant measures for piecewise monotonic transformations. *Trans. Amer. Math. Soc.*, 186:481–488.

L'Ecuyer, P. (1994). Uniform random number generation. *Ann. Oper. Res.*, 53:77–120. Simulation and modeling.

L'Ecuyer, P. (2004). Random number generation. In: *Handbook of computational statistics*, pages 35–70. Springer, Berlin.

Ledoux, M. (2001). *The concentration of measure phenomenon*, volume 89 of *Mathematical surveys and monographs*. American Mathematical Society, Providence, RI.

Ledrappier, F. and Misiurewicz, M. (1985). Dimension of invariant measures for maps with exponent zero. *Ergodic Theory Dynam. Systems*, 5:595–610.

Li, T. Y. and Yorke, J. A. (1975). Period three implies chaos. *Amer. Math. Monthly*, 82:985–992.

Liverani, C. (1995). Decay of correlations. *Ann. Math. (2)*, 142:239–301.

Lopes, A. and Lopes, S. (1998). Parametric estimation and spectral analysis of piecewise linear maps of the interval. *Adv. Appl. Probab.*, 30:757–776.

Lopes, A. O. and Lopes, S. R. C. (2002). Convergence in distribution of the periodogram of chaotic processes. *Stoch. Dyn.*, 2:609–624.

Lorenz, E. (1963). Deterministic nonperiodic flow. *J. Atmospheric Sci.*, 20:130–141.

Luzzatto, S. (2005). Stochastic-like behaviour in nonuniformly expanding maps. In: Hasselblatt, B. and Katok, A., eds., *Handbook of Dynamical Systems*, volume 1B. Elsevier.

Mañé, R. (1981). On the dimension of the compact invariant sets of certain nonlinear maps. In: *Dynamical systems and turbulence, Warwick 1980 (Coventry, 1979/1980)*, volume 898 of *Lecture Notes in Math.*, pages 230–242. Springer, Berlin.

Manneville, P. (1990). *Dissipative structures and weak turbulence*. Perspectives in Physics. Academic Press Inc., Boston, MA.

Manneville, P. (2004). *Instabilities, chaos and turbulence*. Imperial College Press, London. An introduction to nonlinear dynamics and complex systems.

Manning, A. and Simon, K. (1998). A short existence proof for correlation dimension. *J. Statist. Phys.*, 90:1047–1049.

Markus, L. and Meyer, K. R. (1974). *Generic Hamiltonian dynamical systems are neither integrable nor ergodic*. American Mathematical Society, Providence, RI. Memoirs of the American Mathematical Society, No. 144.

Marsaglia, G. (1992). The mathematics of random number generators. In: *The unreasonable effectiveness of number theory (Orono, ME, 1991)*, volume 46 of *Proc. Sympos. Appl. Math.*, pages 73–90. Amer. Math. Soc., Providence, RI.

Mattila, P. (1995). *Geometry of sets and measures in Euclidean spaces*, volume 44 of *Cambridge Studies in Advanced Mathematics*. Cambridge University Press, Cambridge. Fractals and rectifiability.

Mauldin, R. D. (1995). Infinite iterated function systems: theory and applications. In: *Fractal geometry and stochastics (Finsterbergen, 1994)*, volume 37 of *Progr. Probab.*, pages 91–110. Birkhäuser, Basel.

Melbourne, I. and Nicol, M. (2005). Almost sure invariance principle for nonuniformly hyperbolic systems. *Comm. Math. Phys.*, 260:131–146.

Mera, M. E. and Morán, M. (2000). Convergence of the Eckmann and Ruelle algorithm for the estimation of Liapunov exponents. *Ergodic Theory Dynam. Systems*, 20:531–546.

Merlevède, F., Peligrad, M., and Utev, S. (2006). Recent advances in invariance principles for stationary sequences. *Probability Surveys*, 3:1–36.

Milnor, J. (1985a). On the concept of attractor. *Comm. Math. Phys.*, 99:177–195.

Milnor, J. (1985b). Correction and remarks: "On the concept of attractor." *Comm. Math. Phys.*, 102:517–519.

Misiurewicz, M. (1981). Absolutely continuous measures for certain maps of an interval. *Inst. Hautes Études Sci. Publ. Math.*, (53):17–51.

Murray, J. D. (2002). *Mathematical biology. I*, volume 17 of *Interdisciplinary Applied Mathematics*. Springer-Verlag, New York, third edition. An introduction.

Murray, J. D. (2003). *Mathematical biology. II*, volume 18 of *Interdisciplinary Applied Mathematics*. Springer-Verlag, New York, third edition. Spatial models and biomedical applications.

Nash, J. (1956). The imbedding problem for Riemannian manifolds. *Ann. Math. (2)*, 63:20–63.

Nelson, E. (1969). *Topics in dynamics. I: Flows.* Mathematical Notes. Princeton University Press, Princeton, NJ.

Nitecki, Z. (1971). *Differentiable dynamics. An introduction to the orbit structure of diffeomorphisms.* The MIT Press, Cambridge, Mass.-London.

Novo, J., Titi, E. S., and Wynne, S. (2001). Efficient methods using high accuracy approximate inertial manifolds. *Numer. Math.*, 87:523–554.

Nowicki, T. (1998). Different types of nonuniform hyperbolicity for interval maps are equivalent. In: *European Congress of Mathematics, Vol. II (Budapest, 1996)*, volume 169 of *Progr. Math.*, pages 116–123. Birkhäuser, Basel.

Nowicki, T. and van Strien, S. (1991). Invariant measures exist under a summability condition for unimodal maps. *Invent. Math.*, 105:123–136.

O'Brien, G. L. (1983). Obtaining prescribed rates of convergence for the ergodic theorem. *Canad. J. Math.*, 35:1129–1146.

Olofsen, E., Degoede, J., and Heijungs, R. (1992). A maximum likelihood approach to correlation dimension and entropy estimation. *Bull. Math. Biol.*, 54:45–58.

Ornstein, D. S. (1974). *Ergodic theory, randomness, and dynamical systems.* Yale University Press, New Haven, Conn. James K. Whittemore Lectures in Mathematics given at Yale University, Yale Mathematical Monographs, No. 5.

Ornstein, D. S. and Weiss, B. (1993). Entropy and data compression schemes. *IEEE Trans. Inform. Theory*, 39:78–83.

Ott, E., Sauer, T., and Yorke, J. A. (1994). Coping with chaos. Part I. Background. In: *Coping with chaos*, Wiley Ser. Nonlinear Sci., pages 1–62. Wiley, New York.

Peinke, J., Parisi, J., Rössler, O. E., and Stoop, R. (1992). *Encounter with chaos.* Springer-Verlag, Berlin. Self-organized hierarchical complexity in semiconductor experiments.

Pesin, J. B. (1976). Families of invariant manifolds that correspond to nonzero characteristic exponents. *Izv. Akad. Nauk SSSR Ser. Mat.*, 40:1332–1379, 1440.

Petersen, K. (1983). *Ergodic theory*, volume 2 of *Cambridge studies in advanced mathematics.* Cambridge University Press, Cambridge.

Philipp, W. and Stout, W. (1975). Almost sure invariance principles for partial sums of weakly dependent random variables. *Mem. Amer. Math. Soc. 2*, (issue 2, 161):iv+140.

Plachky, D. and Steinebach, J. (1975). A theorem about probabilities of large deviations with an application to queuing theory. *Period. Math. Hungar.*, 6:343–345.

Prieur, C. (2001a). Density estimation for one-dimensional dynamical systems. *ESAIM Probab. Statist.*, 5:51–76 (electronic).

Prieur, C. (2001b). Estimation de la densité invariante de systèmes dynamiques en dimension 1. *C. R. Acad. Sci. Paris Sér. I Math.*, 332:761–764.

Prieur, C. (2001c). Density estimation for one-dimensional dynamical systems. *ESAIM Probab. Statist.*, 5:51–76 (electronic).

Procaccia, I. (1988). Is the weather complex or just complicated? *Nature*, 333:498.

Quas, A. N. (1996). Non-ergodicity for C^1 expanding maps and g-measures. *Ergodic Theory Dynam. Syst.*, 16:531–543.

Ramanan, K. and Zeitouni, O. (1999). The quasi-stationary distribution for small random perturbations of certain one-dimensional maps. *Stochastic Proc. Appl.*, 84:25–51.

Rand, D. A. (1989). The singularity spectrum $f(\alpha)$ for cookie-cutters. *Ergodic Theory Dynam. Syst.*, 9:527–541.

Reinhold, K. (2000). A smoother ergodic average. *Ill. J. Math.*, 44:843–859.

Rio, E. (2000). Inégalités de Hoeffding pour les fonctions lipschitziennes de suites dépendantes. *C. R. Acad. Sci. Paris Sér. I Math.*, 330:905–908.

Robinson, R. C. (2004). *An introduction to dynamical systems: Continuous and discrete.* Pearson Prentice Hall, Upper Saddle River, NJ.

Ruelle, D. (1977). Applications conservant une mesure absolument continue par rapport à dx sur $[0, 1]$. *Comm. Math. Phys.*, 55:47–51.

Ruelle, D. (1979). Ergodic theory of differentiable dynamical systems. *Inst. Hautes Études Sci. Publ. Math.*, (50):27–58.

Ruelle, D. (1984). Characteristic exponents for a viscous fluid subjected to time dependent forces. *Comm. Math. Phys.*, 93:285–300.

Ruelle, D. (1986). Resonances of chaotic dynamical systems. *Phys. Rev. Lett.*, 56:405–407.

Ruelle, D. (1987). Resonances for Axiom **A** flows. *J. Differential Geom.*, 25:99–116.

Ruelle, D. (1989a). *Chaotic evolution and strange attractors.* Lezioni Lincee. [Lincei Lectures]. Cambridge University Press, Cambridge. The statistical analysis of time series for deterministic nonlinear systems, Notes prepared and with a foreword by Stefano Isola.

Ruelle, D. (1989b). *Elements of differentiable dynamics and bifurcation theory.* Academic Press Inc., Boston, MA.

Ruelle, D. (2004). *Thermodynamic formalism.* Cambridge Mathematical Library. Cambridge University Press, Cambridge, second edition. The mathematical structures of equilibrium statistical mechanics.

Sauer, T., Yorke, J. A., and Casdagli, M. (1991). Embedology. *J. Statist. Phys.*, 65:579–616.

Schouten, J. C., Takens, F., and van den Bleek, C. M. (1994a). Estimation of the dimension of a noisy attractor. *Phys. Rev. E (3)*, 50:1851–1861.

Schouten, J. C., Takens, F., and van den Bleek, C. M. (1994b). Maximum-likelihood estimation of the entropy of an attractor. *Phys. Rev. E (3)*, 49:126–129.

Schreiber, T. (1999). Interdisciplinary application of nonlinear time series methods. *Phys. Rep.*, 308:64.

Serinko, R. J. (1996). Ergodic theorems arising in correlation dimension estimation. *J. Statist. Phys.*, 85:25–40.

Shannon, C. E. (1948). A mathematical theory of communication. *Bell System Tech. J.*, 27:379–423, 623–656.

Shannon, C. E. (1993). *Claude Elwood Shannon: Collected papers/ edited by N.J.A. Sloane, Aaron D. Wyner.* IEEE Press, New York. URL `http://cm.bell-labs.com/cm/ms/what/shannonday/paper.html`.

Sinai, Y. G. (1976). *Introduction to ergodic theory.* Princeton University Press, Princeton, N.J. Translated by V. Scheffer, Mathematical Notes, 18.

Smale, S. (1967). Differentiable dynamical systems. *Bull. Amer. Math. Soc.*, 73:747–817.

Stark, J. (2000). Observing complexity, seeing simplicity. *R. Soc. Lond. Philos. Trans. Ser. A Math. Phys. Eng. Sci.*, 358:41–61. Science into the next millennium: Young scientists give their visions of the future, Part II.

Świątek, G. (2001). Collet–Eckmann condition in one-dimensional dynamics. In: *Smooth ergodic theory and its applications (Seattle, WA, 1999)*, volume 69 of *Proc. Sympos. Pure Math.*, pages 489–498. Amer. Math. Soc., Providence, RI.

Takens, F. (1981). Detecting strange attractors in turbulence. In: *Dynamical systems and turbulence, Warwick 1980 (Coventry, 1979/1980)*, volume 898 of *Lecture Notes in Math.*, pages 366–381. Springer, Berlin.

Takens, F. (1996). Estimation of dimension and order of time series. In: *Nonlinear dynamical systems and chaos (Groningen, 1995)*, volume 19 of *Progr. Nonlinear Differential Equations Appl.*, pages 405–422. Birkhäuser, Basel.

Takens, F. (1998). The analysis of correlation integrals in terms of extremal value theory. *Bol. Soc. Brasil. Mat. (N.S.)*, 29:197–228.

Takens, F. and Verbitski, E. (1998). Generalized entropies: Rényi and correlation integral approach. *Nonlinearity*, 11:771–782.

Talagrand, M. (1995). Concentration of measure and isoperimetric inequalities in product spaces. *Inst. Hautes Études Sci. Publ. Math.*, (81):73–205.

Temam, R. (1997). *Infinite-dimensional dynamical systems in mechanics and physics*, volume 68 of *Applied Mathematical Sciences*. Springer-Verlag, New York, second edition.

Temam, R. (2001). *Navier–Stokes equations*. AMS Chelsea Publishing, Providence, RI. Theory and numerical analysis, Reprint of the 1984 edition.

Theiler, J. (1990). Estimating fractal dimension. *J. Opt. Soc. Amer. A*, 7:1055–1073.

Theiler, J. and Lookman, T. (1993). Statistical error in a chord estimator of correlation dimension: the 'rule of five'. *International Journal of Bifurcations and Chaos*, 3:2597–2600.

Tisean (2000). URL `http://www.mpipks-dresden.mpg.de/~tisean`.

Tstool (2003). URL `http://www.physik3.gwdg.de/tstool`.

Ulam, S. and von Neumann, J. (1967). On combination of stochastic and deterministic processes. *Bull. Amer. Math. Soc.*, 52:1120.

Vidal, C. and Lemarchand, H. (1997). *La réaction créatrice*, volume 36 of *Enseignement des Sciences*. Hermann, Paris.

Volterra, V. (1990). *Leçons sur la théorie mathématique de la lutte pour la vie*. Les Grands Classiques Gauthier-Villars. [Gauthier-Villars Great Classics]. Éditions Jacques Gabay, Sceaux. Reprint of the 1931 original.

von Neumann, J. (1932). Proof of the quasi-ergodic hypothesis. *Proc. Nat. Acad. Sci., USA*, 18:70–82.

Vul, E. B., Sinaĭ, Y. G., and Khanin, K. M. (1984). Feigenbaum universality and thermodynamic formalism. *Uspekhi Mat. Nauk*, 39:3–37.

Weierstrass, K. (1886). *Abhandlungen aus der Functionenlehre*. Julius Springer, Berlin.

Witkowski, N. (1995). La chasse à l'effet papillon. *Alliage*, (22):46–53.

Young, L. S. (1982). Dimension, entropy and Lyapunov exponents. *Ergodic Theory Dynam. Syst.*, 2:109–124.

Young, L.-S. (1986). Stochastic stability of hyperbolic attractors. *Ergodic Theory Dynam. Syst.*, 6:311–319.

Young, L.-S. (1990). Large deviations in dynamical systems. *Trans. Amer. Math. Soc.*, 318:525–543.

Young, L.-S. (1998). Statistical properties of dynamical systems with some hyperbolicity. *Ann. Math. (2)*, 147:585–650.

Young, L.-S. (1999). Recurrence times and rates of mixing. *Israel J. Math.*, 110:153–188.

Young, L.-S. (2002). What are SRB measures, and which dynamical systems have them? *J. Statist. Phys.*, 108:733–754. Dedicated to David Ruelle and Yasha Sinai on the occasion of their 65th birthdays.

Ziemian, K. (1985). Almost sure invariance principle for some maps of an interval. *Ergodic Theory Dynam. Syst.*, 5:625–640.

Index

(n, ε)-separated 34
(n, ε)-spanning 35
δ-shadow 65
ε-different before time n 34
ε-pseudo-orbit 65
A^c : complement of the set A 30
$\mathbb{E}$: expectation with respect to a measure 113
$\mathbb{N}$: positive integers 10
$\mathbb{P}$: probability 112
$\mathbb{Z}^+$: non-negative integers 10
$(\mathrm{mod} 1)$: modulo-1 operation 12
$B_r(x)$: ball of radius r around x 36
$[a, b)$: half-open interval 5
$\circ$: functional composition 10
θ : Heaviside function 207
$\dot{x}$: time derivative 1
$\prec$: absolute continuity 81
$\vee$: refinement of partition 32
iteration 10
$|A|$: cardinality of a finite set A 19

absolutely continuous 81
acim 81, 90
adapted norm 54
admissible sequences 77
affine map 46, 47
almost every 95
almost sure central limit theorem 147
almost sure invariance principle 147
alphabet 19
Anosov maps 62
atoms 27
attracting set 38
attractor 39
auto-correlation function 105
Axiom A 63

baker's map 20
baker's transformation 14
basin of attraction 38, 39
Belusov–Zhabotinsky reaction 23
Bernoulli measure 85, 89
Berry–Esseen inequality 144
Bonferoni inequality 170
bounded distortion 154
bounded variation 91
Bowen balls 128
box counting dimension 133
box dimension 133
bracket operation 72
Brownian bridge 212
Brownian motion 145
butterfly effect 69
BV 91

Cantor set 41
capacitary dimension 173
capacity 133
cardinality 31
Cat map 14
ceiling function 11
central limit theorem 142
Chebyshev inequality 148
coding 28
componentwise Lipschitz function 149
concentration 151

conjugation 11
correlation 105
correlation dimension 207
coupling 112
critical damping 6
cylinder subset 89

Devroye inequality 151
diffeomorphism 46
dimension of a measure 134
dissipative baker's map 21
distribution function 211
down-crossing 104
dynamical system 11

elliptic 9
embedding 202
embedding dimension 204
entrance time 163
ergodic 95
ergodic sum 105
ergodic theorem 93
expanding property 18
expansivity 69
expectation 113
exponential inequality 150

fiber 21
finer cover 32
fixed point 87
Frostman lemma 134
full shift 19
functional compositions 10

Gaussian bridge 212
generating partition 73, 124
geodesic flow 45
Gibbs state 92, 93, 153
Ginzburg–Landau equation 205
Grashof number 140
Grassberger–Procaccia algorithm 207, 209
Gumble's law 175

Hölder continuous with exponent 1 92
Hölder norms 57
Hénon map 22
Hartman–Grobman theorem 53
Hausdorff dimension 133
Heaviside function 207
homoclinic orbits 5
homoclinic point 69
horseshoe map 64
hyperbolic 9
hyperbolic fixed point 47
hyperbolic map 61
hyperbolic matrix 47
hyperbolic set 60

iid, independent identically distributed 89
incidence matrix 19
inertial manifold 43
information correlation functions 194
initial condition 10
integrated periodogram 190
invariance principle 147
invariant manifolds 49
invariant measure 80, 86
Ising spins 154
iterated function system 156
iteration 10
Ivanov's Theorem 104

Jensen inequality 174

Kaplan–Yorke formula 138
Kolmogorov–Sinai entropy 125
Koopman operator 92

large deviations 147
law of iterated logarithm 146
Lebesgue measure 88
Lebesgue point 84
Legendre transform of the pressure 148
linearization 46
linearized map 46
Lipschitz constant 92
Lipschitz continuous 92
Lipshitz norm 57
local stable manifold 49
local stable manifold at x 60
local stable manifold with exponent 60
local unstable manifold 49
logistic map 18
Lorenz system 23
Lotka–Volterra equation 24
Lyapunov dimension 138
Lyapunov exponents 121
Lyapunov metric 54

Markov partition 72
Markov property 18
maximal exponent 197
metric space 30
mixing 106
multifractal analysis 156
multifractal spectrum 156
mutually singular 82

noise reduction 200
nonwandering 63
number of visits 172

observable 10, 93
open cover 32
orbit 10
Oregonator 23
Oseledec's theorem 121

partial differential equation 24
partition 27
PDE 24
periodogram 189
Perron–Frobenius operator 90
phase portraits 5
phase space 10
Physical measure 115
piecewise expanding maps of the interval 17
Pisier's inequality 174
Poincaré section 11
potential 93
Potts model 154
power spectrum 194
prediction 200
preimage of order n 114
preimages 112
pressure 93
pressure function 148
probability measure 94
probability space 94

QR decomposition 197
quadratic family 18
quasi-invariant distribution 177
quasi-invariant measure 177

random number generators 18
reconstruction 201
record 173
rectangle 72
relative entropy 194
renewal process 92
resonance 59, 191
return time 163

semiflow 10
sensitive dependence on initial conditions 116
shadowing 65
shift 19, 28
short range interaction 155
skew-product 21
slaving 43
solenoid map 15
SRB 114
SRB measure 132
stable 9
stable fixed point 54
stable manifold 48, 51
stable manifold with exponent 131
standard map 21
states 10
stochastic perturbation 179
stochastic stability 180
strange attractors 39
strongly stable manifold 6
subadditive ergodic theorem 102
subadditivity property 33
subshift of finite type 19
supercritical 6
suspension flow 11
Swiss Chocolate Theorem 31

Takens reconstruction theorem 203
tangent bundle 45
tangent map 46
time evolution 10
topological entropy 33
trajectory 10
transition matrix 73
transverse measure 132
triangle inequality 30
turbulence 25

ultrametric 92
uniformly hyperbolic set 60
unilateral shift 19

unimodal map 18
unimodal maps 37
unstable fixed point 54
unstable manifold 48, 51
unstable manifold with exponent 131

vector field 6

wandering 63

Yaglom limit 177
Young inequality 182

Theoretical and Mathematical Physics

Concepts and Results in Chaotic Dynamics: A Short Course
By P. Collet and J.-P. Eckmann

The Theory of Quark and Gluon Interactions
4th Edition
By F. J. Ynduráin

Titles published before 2006 in *Texts and Monographs in Physics*

The Statistical Mechanics of Financial Markets
3rd Edition
By J. Voit

Magnetic Monopoles
By Y. Shnir

Coherent Dynamics of Complex Quantum Systems
By V. M. Akulin

Geometric Optics on Phase Space
By K. B. Wolf

General Relativity
By N. Straumann

Quantum Entropy and Its Use
By M. Ohya and D. Petz

Statistical Methods in Quantum Optics 1
By H. J. Carmichael

Operator Algebras and Quantum Statistical Mechanics 1
By O. Bratteli and D. W. Robinson

Operator Algebras and Quantum Statistical Mechanics 2
By O. Bratteli and D. W. Robinson

Aspects of Ergodic, Qualitative and Statistical Theory of Motion
By G. Gallavotti, F. Bonetto and G. Gentile

The Frenkel-Kontorova Model
Concepts, Methods, and Applications
By O. M. Braun and Y. S. Kivshar

The Atomic Nucleus as a Relativistic System
By L. N. Savushkin and H. Toki

The Geometric Phase in Quantum Systems
Foundations, Mathematical Concepts, and Applications in Molecular and Condensed Matter Physics
By A. Bohm, A. Mostafazadeh, H. Koizumi, Q. Niu and J. Zwanziger

Relativistic Quantum Mechanics
2nd Edition
By H. M. Pilkuhn

Physics of Neutrinos
and Applications to Astrophysics
By M. Fukugita and T. Yanagida

High-Energy Particle Diffraction
By E. Barone and V. Predazzi

Foundations of Fluid Dynamics
By G. Gallavotti

Many-Body Problems and Quantum Field Theory An Introduction
2nd Edition
By Ph. A. Martin, F. Rothen, S. Goldfarb and S. Leach

Statistical Physics of Fluids
Basic Concepts and Applications
By V. I. Kalikmanov

Statistical Mechanics A Short Treatise
By G. Gallavotti

Quantum Non-linear Sigma Models
From Quantum Field Theory to Supersymmetry, Conformal Field Theory, Black Holes and Strings
By S. V. Ketov

Perturbative Quantum Electrodynamics and Axiomatic Field Theory
By O. Steinmann

The Nuclear Many-Body Problem
By P. Ring and P. Schuck

Magnetism and Superconductivity
By L.-P. Lévy

Information Theory and Quantum Physics
Physical Foundations for Understanding the Conscious Process
By H. S. Green

Quantum Field Theory in Strongly Correlated Electronic Systems
By N. Nagaosa

Quantum Field Theory in Condensed Matter Physics
By N. Nagaosa

Conformal Invariance and Critical Phenomena
By M. Henkel

Statistical Mechanics of Lattice Systems
Volume 1: Closed-Form and Exact Solutions
2nd Edition
By D. A. Lavis and G. M. Bell

Statistical Mechanics of Lattice Systems
Volume 2: Exact, Series and Renormalization Group Methods
By D. A. Lavis and G. M. Bell

Fields, Symmetries, and Quarks
2nd Edition
By U. Mosel

Renormalization An Introduction
By M. Salmhofer

Multi-Hamiltonian Theory of Dynamical Systems
By M. Błaszak

Quantum Groups and Their Representations
By A. Klimyk and K. Schmüdgen

Quantum The Quantum Theory of Particles, Fields, and Cosmology
By E. Elbaz

Effective Lagrangians for the Standard Model
By A. Dobado, A. Gómez-Nicola, A. L. Maroto and J. R. Peláez

Scattering Theory of Classical and Quantum *N*-Particle Systems
By. J. Derezinski and C. Gérard

Quantum Relativity A Synthesis of the Ideas of Einstein and Heisenberg
By D. R. Finkelstein

The Mechanics and Thermodynamics of Continuous Media
By M. Šilhavý

Local Quantum Physics Fields, Particles, Algebras
2nd Edition
By R. Haag

Relativistic Quantum Mechanics and Introduction to Field Theory
By F. J. Ynduráin

Supersymmetric Methods in Quantum and Statistical Physics
By G. Junker

Path Integral Approach to Quantum Physics An Introduction
2nd printing
By G. Roepstorff

Finite Quantum Electrodynamics
The Causal Approach
2nd edition
By G. Scharf

From Electrostatics to Optics
A Concise Electrodynamics Course
By G. Scharf

Geometry of the Standard Model of Elementary Particles
By A. Derdzinski

Quantum Mechanics II
By A. Galindo and P. Pascual

Generalized Coherent States
and Their Applications
By A. Perelomov

The Elements of Mechanics
By G. Gallavotti

Essential Relativity Special, General, and Cosmological Revised
2nd edition
By W. Rindler